UMTS Signaling

UMTS Signaling

UMTS Interfaces, Protocols, Message Flows and Procedures Analyzed and Explained

Ralf Kreher and Torsten Rüdebusch

Both of

Tektronix Berlin GmbH & Co. KG

Germany

John Wiley & Sons, Ltd

Published in 2005 by   John Wiley & Sons Ltd, The Atrium, Southern Gate, Chichester,
West Sussex PO19 8SQ, England

Telephone   (+44) 1243 779777

Email (for orders and customer service enquiries): cs-books@wiley.co.uk
Visit our Home Page on www.wiley.com

Reprinted with corrections September 2005
Reprinted December 2005, January and August 2006

Other Wiley Editorial Offices

John Wiley & Sons Inc., 111 River Street, Hoboken, NJ 07030, USA

Jossey-Bass, 989 Market Street, San Francisco, CA 94103-1741, USA

Wiley-VCH Verlag GmbH, Boschstr. 12, D-69469 Weinheim, Germany

John Wiley & Sons Australia Ltd, 33 Park Road, Milton, Queensland 4064, Australia

John Wiley & Sons (Asia) Pte Ltd, 2 Clementi Loop #02-01, Jin Xing Distripark, Singapore 129809

John Wiley & Sons Canada Ltd, 22 Worcester Road, Etobicoke, Ontario, Canada M9W 1L1

Wiley also publishes its books in a variety of electronic formats. Some content that appears
in print may not be available in electronic books.

British Library Cataloguing in Publication Data

A catalogue record for this book is available from the British Library

ISBN 10: 0-470-01351-6 (H/B)
ISBN 13: 978-0-470-01351-9 (H/B)

Typeset in 10/12pt Times by TechBooks, New Delhi, India.
Printed and bound in Great Britain by Antony Rowe Ltd, Chippenham, Wiltshire
This book is printed on acid-free paper responsibly manufactured from sustainable forestry
in which at least two trees are planted for each one used for paper production.

Contents

Preface

The successful trial, deployment, operation, and troubleshooting of 3G or UMTS infrastructures and applications is one of the most exciting, fascinating, and challenging tasks in today's mobile communications. Interoperability, roaming, and QoS awareness between multioperators and multitechnology network infrastructures are just a few of the problems that need to be met. In today's early deployments of UMTS networks, five main categories of problems can be differentiated:

1. Network Element Instability
2. Network Element Interworking
3. Multi-Vendor Interworking (MVI)
4. Configuration Faults
5. Network Planning Faults

To meet these challenges, it is vital to understand and analyze the message flows associated with UMTS.

UMTS Signaling focuses on providing an overview and reference to UMTS, details of the standards, the network architecture, and objectives and functions of the different interfaces and protocols. Furthermore, it comprehensively describes various procedures from Node B setup to different handover types in the UTRAN and the Core Network. The focus on wireline interfaces is unique in the market. All signaling sequences are based upon UMTS traces from various UMTS networks (trial and commercial networks) around the world. With this book the reader has access to the first universal UMTS protocol sequence reference, which allows you to quickly differentiate valid from invalid call control procedures. In addition, all main signaling stages are being explained – many of which had been left unclear in the standards so far – and valuable tips for protocol monitoring are provided.

What will you get out of *UMTS Signaling*?

- A comprehensive overview on UMTS UTRAN and Core Networks
  - Latest updates for Rel. 4, Rel. 5, and Rel. 6 features are included
  - Description of the real-world structure of ATM transport network on Iub and Iu interfaces
  - Valuable tips and tricks for practical interface monitoring
- In-depth description of the tasks and functions of UMTS interfaces and protocols
- A deep protocol knowledge improvement
- Potential to analyze specific protocol messages
- Support to reduce time and effort to detect and analyze problems

- Explanations of how to locate problems in the network
- Comprehensive descriptions and documentation of UMTS reference scenarios for different UMTS procedures
  - UTRAN signaling procedures
    - Description of RRC measurement procedures for radio network optimization
    - Analysis and explanation of PS calls with so-called channel-type switching, which is one of the most common performance problems of packet-switched services in today's 3G networks
    - SRNS Relocation scenarios – including full description of RANAP and RRC containers
    - More than 35 decoded message examples using Tektronix' protocol testers give a deep insight into control plane protocols on different layers
  - Core Network signaling procedures
    - In-depth evaluations on mobility management, session management, and call control procedures
    - Example call flows of the CS domain including practical ideas for troubleshooting
    - Tunnel management on Gn interfaces
    - Mobility management using optional Gs interface
    - Discussion on core network switch (MSC, SGSN) and database (HLR, VLR) information exchange over Mobile Application Part (MAP)
    - Short introduction to 3G intelligent services with CAMEL Application Part (CAP) protocol
    - Comprehensive description of Inter-MSC Handover procedures for 3G-3G, 3G-GSM, and GSM-3G handovers
    - Detailed description of RANAP, BSSAP, and RRC information

UMTS Signaling readers should be rather familiar with UMTS technology at a fairly detailed level as the book is directed to UMTS experts, who need to analyze UMTS signaling procedures at the most detailed level. This is why only an introductory overview section discusses the UMTS network architecture, the objectives and functions of the different interfaces, and the various UMTS protocols. Then the book leads right into the main part – the analysis of all main signaling processes in a UMTS networks, the so-called UMTS scenarios. All main procedures – from Node B Setup to Hard Handover – are described and explained comprehensively.

The combination of a network of UMTS experts from many different companies around the world with Tektronix' many years of experience in protocol analysis has resulted in this unique book, compendium, and reference. I hope it will prove helpful for the successful implementation and deployment of UMTS.

Alois Hauk
General Manager
Monitoring and Protocol Test
Tektronix, Inc.

If you have any kind of feedback or questions feel free to send us an e-mail to umts-signaling@tektronix.com.

For help with acronyms or abbreviations, refer to the glossary at the end of this book.

Acknowledgments

The Tektronix Network Diagnostics Academy has already trained hundreds of students in UMTS and other mobile technologies and in testing mobile networks. The experience from these trainings and our close customer relations pointed out that a book on UMTS Signaling is desperately needed.

We collected all the material that was available at Tektronix and that was provided by our partners at network equipment vendors and network operators to add it to this unique selection.

The authors would like to acknowledge the effort and time invested by all our colleagues at Tektronix who have contributed to this book.

Special thanks go to Jens Irrgang and Christian Villwock, Tektronix MPT, Berlin, for their coauthorship and their valuable advice and input for Section 1.6, "UMTS Security."

Without Juergen Placht (Sanchar GmbH) this book would not have existed. His unbelievable knowledge, experience, and efforts in preparing the very first slide sets for UMTS scenarios laid the basis for the material you have now in front of you.

Additionally, the material that Magnar Norderhus, Hummingbird, Duesseldorf, prepared for the first UMTS Training for Tektronix was the very first source that we have "blown up" for part one of this book.

Many thanks also go to Joerg Nestle Product Design, Munich, for doing a great job in the creation of all the graphics.

We would like to express thanks to Othmar Kyas, Marketing Manager of Tektronix Monitor & Protocol Test, for his strong belief in the Tektronix Network Diagnostics Academy and in *UMTS Signaling*, and for challenging us to make this book become real.

Of course, we must not forget to thank Mark Hammond and the team at Wiley. Mark wanted us to do the book and kept us moving, even though it took so much time to get all the permissions aligned with Tektronix.

Last but not least, a special "thank you" to our families and friends for their infinite patience and support throughout this project.

About the Authors

Ralf Kreher
Manager for Customer Training, Mobile Protocol Test, Tektronix, Inc.

Ralf Kreher leads the Customer Training Department for Tektronix' Mobile Protocol Test (MPT) business. He is responsible for the world-class seminar portfolio for mobile technologies and measurement products. Before joining Tektronix, he held a trainer assignment for switching equipment at Teles AG. He holds a Communication Engineering Degree of the Technical College Deutsche Telekom Leipzig. He currently resides in Germany.

Torsten Rüdebusch
Head of Knowledgeware and Training Department, Mobile Protocol Test, Tektronix, Inc.

Torsten Rüdebusch is the head of the Knowledgeware and Training Department for Tektronix' Mobile Protocol Test (MPT) business. He is responsible for providing leading edge technology and product seminars and the creation of knowledgeware products using the extensive Tektronix' expertise. Before joining Tektronix, he held an application engineer assignment at Siemens CTE. He holds a Communication Engineering Degree of the Technical College Deutsche Telekom Berlin. He currently resides in Germany.

1

UMTS Basics

UMTS is real. In several parts of the world we can walk in the stores of mobile network operators or resellers and take UMTS PC Cards or even third generation (3G) phones home and use them instantly. Every day the number of equipments and their feature sets gets broader. The "dream" of multimedia on mobile connections, online gaming, video conferencing or even real-time video becomes reality.

With rapid technical innovation the mobile telecommunication sector has continued to grow and evolve strongly.

The technologies used to provide wireless voice and data services to subscribers, such as Time Division Multiple Access (TDMA), Universal Mobile Telecommunications System (UMTS), and Code Division Multiple Access (CDMA), continue to grow in their complexity. This complexity continues to impart a time-consuming hurdle to overcome when moving from 2G to 2.5G and then to 3G networks.

GSM (Global System for Mobile Communication) is the most widely installed wireless technology in the world. Some estimates put GSM market share at up to 80 %. Long dominant in Europe, GSM is now gaining a foothold in Brazil and is expanding its penetration in the North American market.

One reason for this trend is the emergence of reliable, profitable 2.5G GPRS elements and services. Adding a 2.5G layer to the existing GSM foundation has been a cost-effective solution to current barriers while still bringing desired data services to market. The enhancement to EGPRS (EDGE; Enhanced Data Rates for GSM Evolution) allows a speed of 384 kbps. This is the maximum limit. Now EDGE goes under pressure, because High Speed Downlink Packet Access (HSDPA; see Section 1.2.3) and its speed of 2 Mbps will take huge parts of the market share once it is largely available.

So, the EGPRS operators will sooner or later switch to 3G UMTS services (Figure 1.1), the latest of which is UMTS Release 6 (Rel. 6). This transition brings new opportunities and new testing challenges, both in terms of revenue potential and addressing interoperability issues to ensure QoS (Quality of Service).

With 3G mobile networks, the revolution of mobile communication has begun. 4G and 5G networks will make the network transparent to the user's applications. In addition to horizontal handovers (for example between Node Bs), handovers will occur vertically between

UMTS Signaling   Ralf Kreher and Torsten Rüdebusch
© 2005 Tektronix, Inc.   ISBN: 0-470-01351-6.

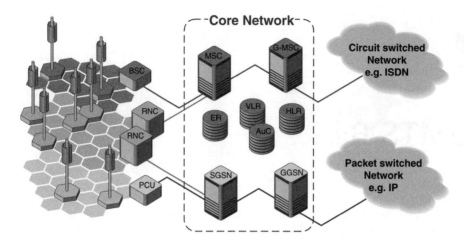

Figure 1.1 Component overview of a UMTS network.

applications, and the UTRAN (UMTS Terrestrial Radio Access Network) will be extended by a satellite-based RAN (Radio Access Network), ensuring global coverage.

Every day the number of commercial networks in different parts of the world increases rapidly. Therefore, network operators and equipment suppliers are desperate to understand how to handle and analyze UMTS signaling procedures in order to get the network into operation, detect errors, and troubleshoot faults.

Those experienced with GSM will recognize many similarities with UMTS, especially in Non-Access Stratum (or NAS) messaging. However, in the lower layers within the UTRAN and Core Network (CN), UMTS introduces a set of new protocols, which deserve close understanding and attention.

The philosophy of UMTS is to separate the user plane from the control plane, the radio network from the transport network, the access network from the CN, and the Access Stratum from the Non-Access Stratum.

The first part of this book is a refresher on UMTS basics, and the second part continues with in-depth message flow scenarios of all kinds.

1.1 Standards

ITU (the International Telecommunication Union) solicited several international organizations for descriptions of their ideas for a 3G mobile network:

CWTS   China Wireless Telecommunication Standard group
ARIB   Association of Radio Industries and Businesses, Japan
T1   Standards Committee T1 Telecommunications, United States
TTA   Telecommunications Technology Association, Korea
TTC   Telecommunication Technology Committee, Japan
ETSI   European Telecommunications Standards Institute

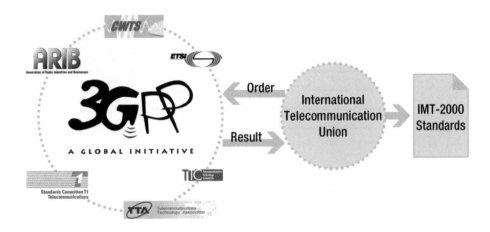

Figure 1.2   IMT-2000.

ITU decided which standards would be used for "International Mobile Telecommunications at 2000 MHz." Many different technologies were combined in IMT-2000 standards (Figure 1.2).

The main advantage of IMT-2000 is that it specifies international standards and also the interworking with existing PLMN (Public Land Mobile Network) standards, such as GSM.

In general, the quality of transmission will be improved. The data transfer rate will increase dramatically. Transfer rates of 384 kbps are already available; 2 Mbps (with HSDPA technology) is under test and almost ready to go live in certain parts of Asia. New service offerings will help UMTS to become financially successful for operators and attractive to users.

The improvement for the user will be the worldwide access available with a mobile phone, and the look and feel of services will be the same wherever he or she may be (Figure 1.3).

There is a migration path from 2G to 3G systems that may include an intermediate step, the so-called 2.5G network. Packet switches – GGSN or SGSN in case of a GSM

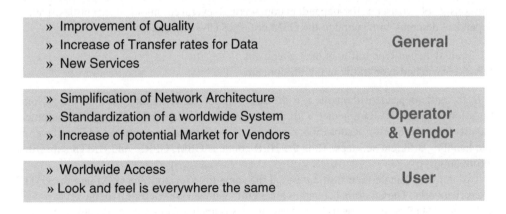

Figure 1.3   IMT-2000 standards benefit users, operators, and vendors.

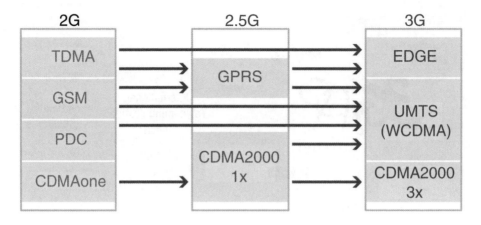

Figure 1.4   Possible migration paths from 2G to 3G.

network – are implemented in the already existing CN while the RAN is not changed significantly (Figure 1.4).

In case of a migration from GSM to UMTS a new Radio Access Technology (RAT; W-CDMA instead of TDMA) is introduced. This means the networks will be equipped with completely new RANs that replace the 2G network elements in the RAN. However, EDGE opens a different way to offer high-speed IP services to GSM subscribers without introducing W-CDMA.

The already existing CDMA cellular networks, which are especially popular in the Americas, will undergo an evolution to become CDMA2000 networks with larger bandwidth and higher data transmission rates.

1.2 Network Architecture

UMTS maintains a strict separation between the radio subsystem and the network subsystem, allowing the network subsystem to be reused with other RAT. The CN is adopted from GSM and consists of two user traffic-dependent domains and several commonly used entities. Traffic-dependent domains correspond to the GSM or GPRS CNs and handle:

• Circuit-switched-type traffic in the CS domain
• Packet-switched-type traffic in the PS domain

Both traffic-dependent domains use the functions of the remaining entities – the Home Location Register (HLR) together with the Authentication Center (AuC), or the Equipment Identity Register (EIR) – for subscriber management, mobile station roaming and identification, and handling different services. Thus the HLR contains GSM, GPRS, and UMTS subscriber information.

Two domains handle their traffic types at the same time for both the GSM and the UMTS access networks. The CS domain handles all circuit-switched type of traffic for the GSM as well as for the UMTS access network; similarly, the PS domain takes care of all packet-switched traffic for both the access networks.

1.2.1 GSM

The second generation of PLMN is represented Subsystem by a GSM network consisting of Network Switching Subsystem (NSS) and a Base Station (BSS) (Figure 1.5).

The first evolution step (2.5G) is a GPRS PLMN connected to a GSM PLMN for packet-oriented transmission.

The main element in the NSS is the Mobile Switching Center (MSC), which contains the Visitor Location Register (VLR). The MSC represents the edge toward the BSS and on the other side as Gateway MSC (GMSC), the connection point to all external networks, such as the Public Switched Telephone Network (PSTN) or Integrated Services Digital Network (ISDN). GSM is a circuit-switched network, which means that there are two different types of physical links to transport control information (signaling) and traffic data (circuit). The signaling links are connected to Signaling Transfer Points (STP) for centralized routing whereas circuits are connected to special switching equipment.

HLR   Home Location Register
SGSN   Serving GPRS Support Node with Location Register Function
GGSN   Gateway GPRS Support Node
AuC   Authentication Center
SCP   Service Control Point
SMSC   Short Message Service Center
CSE   CAMEL Service Entity (*C*ustomized *A*pplication for *M*obile network *E*nhanced *L*ogic)

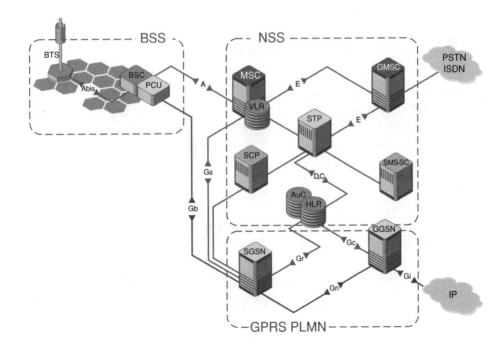

Figure 1.5 GSM network architecture.

The most important entity in BSS is the Base Station Controller (BSC), which, along with the Packet Control Unit (PCU), serves as the interface with the GPRS PLMN. Several Base Transceiver Stations (BTS) can be connected to the BSC.

1.2.2 UMTS Release 99

Figure 1.6 shows the basic structure of a UMTS Rel. 99 network. It consists of two different radio access parts BSS and UTRAN and the CN parts for circuit-switched (e.g. voice) and packet-switched (e.g. e-mail download) applications.

To implement UMTS means to set up a UTRAN, which is connected to a circuit-switched CN (GSM with MSC/VLR) and to a packet-switched CN (GPRS with SGSN). The interfaces are named Iu, whereas IuCS goes to the MSC and IuPS goes to the SGSN. Alternatively, the circuit and packet network connections could also be realized with an UMSC (UMTS MSC) that combines MSC and SGSN functionalities in one network element.

The corresponding edge within UTRAN is the Radio Network Controller (RNC). Other than in the BSS the RNCs of one UTRAN are connected with each other via the Iur interface.

The base stations in UMTS are called *Node B*, which is just its working name and has no other meaning. The interface between Node B and RNC is the Iub interface.

Release 99 (sometimes also named Release 3) specifies the basic requirements to roll out a 3G UMTS RAN. All following releases (4, 5, 6, etc.) introduce a number of features that allow operators to optimize their networks and to offer new services. A real network environment in the future will never be designed strictly following any defined release standard. Rather it must be seen as a kind of patchwork that is structured following the requirements of network

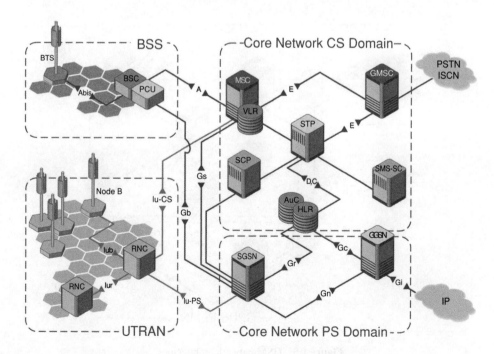

Figure 1.6   UMTS Rel. 99 network architecture.

operators and service providers. So it is possible to introduce, e.g., HSDPA, which is a feature clearly defined in Rel. 5 in combination with a Rel. 99 RAN.

In addition, it must be kept in mind that owing to changing needs of operators and growing experience of equipment manufacturers, every three months (four times per year!) all standard documents of all releases are revised and published with a new version. So development of Rel. 99 standards is not even finished yet.

It might also be possible that in later standard versions introduction of features promised in earlier version is delayed. This became true, for instance, for definition of Home Subscriber Server (HSS) that was originally introduced in early Rel. 4 standards, but then delayed to be defined in detail in Rel. 5.

The feature descriptions for higher releases in the following sections are based on documents not older than 2004–06 revision.

1.2.3 UMTS Release 4

3GPP Release 4 introduces some major changes and new features in the CN domains and the GERAN (GPRS/EDGE Radio Access Network), which replaces GSM BSS (Figure 1.7). Some of the major changes are:

- Separation of transport bearer and bearer control in the CS CN
- Introduction of new interfaces in CS CN

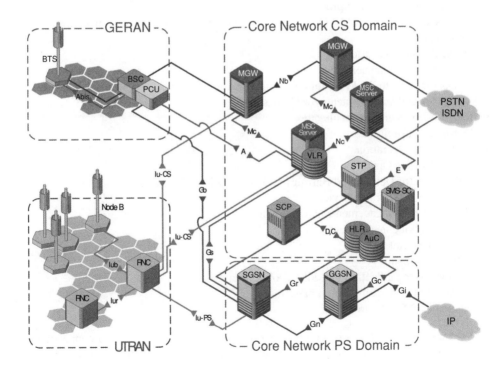

Figure 1.7   UMTS Rel. 4 network architecture.

- ATM (AAL2, ATM Adaptation Layer Type 2) or IP can now be used as data transport bearer in the CS domain
- Introduction of low chiprate (also called narrowband) TDD Describes the RAT behind the Chinese TD-SCDMA standard while UMTS TDD (wideband TDD, TD-CDMA) is seen as dominating TDD technology in European and Asian standards outside China. It is expected that interference in low chiprate TDD has less impact on cell capacity compared to same effect in wideband TDD. In addition, low chiprate TDD equipment shall support advanced radio transmission technologies like "smart antennas" and beamforming, which means to point a single antenna or a set of antennas at the signal source to reduce interference and improve communication quality
- IP-based Gb interface
- IPv6 support (optional)

The new features and services are[1]:

- Multimedia services in the CS domain
- Handover of real-time application in the PS domain
- UTRAN Transport Evolutions
  - AAL2 connection QoS optimization over Iub and Iur interfaces
  - Transport bearer modification procedure on Iub, Iur, and Iu
- IP transport of CN protocols
- Radio Interface Improvements
  - UTRA repeater specification
  - DSCH power control improvement
- Radio Access Bearer (RAB) QoS Negotiation over Iu interface during Relocation
- RAN improvements
  - Node B synchronization for TDD
  - RAB support enhancement
- Transparent End-to-End PS Mobile Streaming Applications
- Emergency call enhancements for CS-based calls
- Bearer independent CS architecture
- Real-time Facsimile
- Tandem Free Operation
- Transcoder Free Operation
- ODB (Operator Determined Barring) for Packet-Oriented Services
- Multimedia Messaging Service
- UICC/(U)SIM enhancements and interworking
- (U)SIM toolkit enhancements
  - USAT local link
  - UICC API testing
  - Protocol standardization of a SIM Toolkit Interpreter
- Advanced Speech Call Items enhancements
- Reliable QoS for PS domain

The main trend in Rel. 4 is the separation of control and services of CS connections and at the same time the conversation of the network to be completely IP-based. In CS CN the

---

[1] No specific order.

user data flow will go through Media Gateway (MGW), which are elements maintaining the connection and performing switching functions when required (bearer switching functions of the MSC are provided by the MGW). The process is controlled by a separate element evolved from MSC/VLR called MSC Server (control functions of the MSC are provided by the MSC Server and also contains the VLR functionality), which is in terms of voice over IP networks a signaling gateway. One MSC Server controls numerous MGWs. To increment control capacities, a new MSC Server will be added. To increase the switching capacity, one has to add MGWs.

1.2.4 UMTS Release 5

In 3GPP Release 5, the UMTS evolution continues. The shift to an all IP environment will be realized: all traffic coming from UTRAN is supposed to be IP-based (Figure 1.8). By changing GERAN, the BSC will be able to generate IP-based application packets. That is why the circuit-switched CN will not be part of UMTS Rel. 5 anymore. All interfaces will be IP-based rather than ATM-based.

The databases known from GSM/GPRS will be centralized in an HSS. Together with value-added services and CAMEL, it represents the Home Environment (HE). CAMEL could perform the communication with the HE completely. When the network has moved toward IP, the relationship between circuit- and packet-switched traffic will change. The majority of traffic will be packet-oriented because some traditionally circuit-switched services, including speech, will become packet-switched (VoIP). To offer uniform methods of IP application transport,

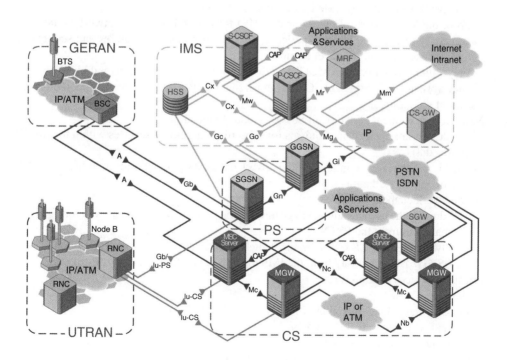

Figure 1.8   UMTS Rel. 5 basic architecture.

Rel. 5 will contain an IP Multimedia Subsystem (IMS), which efficiently supports multiple media components, e.g. video, audio, shared whiteboards, etc.

HSDPA will provide data rates of up to 10 Mbps in downlink direction and lower rates in uplink (e.g. Internet browsing or video on demand) through the new High Speed Downlink Shared Channel (HS-DSCH) (for details see *3GPP 25.855*).

New in Release 5

- All network node interfaces connected to IP network
- HSS replaces HLR/AuC/EIR
- IMS
  - Optional IPv6 implementation
  - Session Initiation Protocol (SIP) for CS signaling and management of IP multimedia sessions
  - SIP supports addressing formats for voice and packet calls and number translation requirements for SIP <-> E.164
- HSDPA integration
  - Data rates of up to 10 Mbps in downlink direction; lower rates in uplink (e.g. Internet browsing or video on demand)
  - New HS-DSCH
- All voice traffic is voice over packet
- MGW required at Point of Interconnection (POI)
- SGW (Signaling Gateway; MSC Server) translates signaling to "legacy" (SS7) networks
- AMR-WB, an enhanced Adaptive Multirate (Wideband) codec for voice services
- New network element MRF (Media Resource Function)
  - Part of the Virtual Home Environment (VHE) for portability across network boundaries and between terminals. Users experience the same personalized features and services in whatever network and whatever terminal
  - Very similar in function to an MGCF (Media Gateway Control Function) and MGW (Media Gateway) using H.248/MEGACO to establish suitable IP or SS7 bearers to support different kinds of media streams
- New network element CSCF (Call Session Control Function)
  - Provides session control mechanisms for subscribers accessing services within the IM (IP Multimedia) CN
  - CSCF is a SIP Server to interact with network databases [e.g. HSS for mobility and AAA (Authorization, Authentication, and Accounting) for security]
- New network element SGW
  - In CS domain the user signaling will go through the SGW, which is the gateway for signaling information to/from the PSTN
- New network element CS-GW (Circuit-Switched Gateway)
  - The CS-GW is the gateway from the IMS to/from the PSTN (e.g. for VoIP calls)
- Location services for PS/GPRS
- IuFlex
  - Breaking hierarchical mapping of RNCs to SGSNs (MSCs)
- Wideband AMR (new 16-kHz codec)
- End-to-end QoS in the PS domain
- GTT: Global Text Telephony (service for handicapped users)

- Messaging and security enhancements
- CAMEL Phase 4
  - New functions such as mid call procedures, interaction with optimal routing, etc.
- Load sharing
  - UTRAN (Radio Network for W-CDMA)
  - GERAN (radio network for GSM/EDGE)
  - W-CDMA in 1800/1900-MHz frequency spectrums
  - Mobile Execution Environment (MExE) support for Java and WAP applications

IMS
The Proxy-Call State Control Function (P-CSCF) is located together with the GGSN in the same network. Its main task is to select the I-CSCF in the user's home network and do some basic local analysis, e.g. QoS surveillance or number translation.

The Interrogating-CSCF (I-CSCF) provides access to the user's home network and selects the S-CSCF (in the home network, too).

The Serving-CSCF (S-CSCF) is responsible for the Session Control, handles SIP requests, and takes care of all necessary procedures, such as bearer establishment between home and visited network.

The HSS is the former HLR. It was renamed to emphasize that the database does not only contain location-related, but subscription-related data (subscribed services and their parameters, etc.) too (Figure 1.9).

HSDPA
HSDPA is a packet-based data service with data speed of up to 1.2–14.4 Mbps (and 20 Mbps for MIMO systems) over a 5-MHz bandwidth in downlink. HSDPA implementations include Adaptive Modulation and Coding (AMC), Multiple-Input Multiple-Output (MIMO), Hybrid Automatic Repeat Request (HARQ), fast cell search, and advanced receiver design (Figure 1.10).

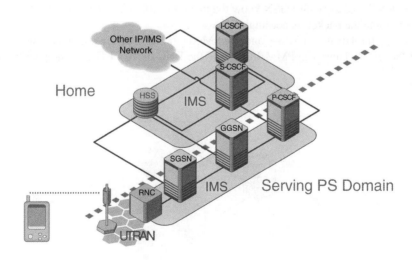

Figure 1.9 Overview of IMS architecture.

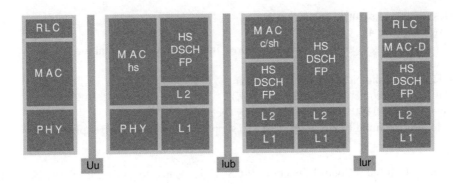

Figure 1.10 HSDPA protocol architecture.

IuFlex

Before UMTS Rel. 5 the RNC <-> SGSN relation was hierarchical: Each RNC was assigned to exactly one SGSN; each SGSN served one or more RNCs (Figure 1.11).

With Rel. 5, IuFlex allows "many-to-many" relations of RNCs, SGSNs, or MSCs, where RNCs and SGSNs are belonging to "Pool Areas" (can be served by one or more SGSNs/MSCs in parallel). All cells controlled by an RNC belong to one or more Pool Areas so that a UE (User Equipment) may roam in Pool Areas without changing the SGSN/MSC (Figure 1.12).

The integration of IuFlex now offers load balancing between SGSNs/MSCs in one Pool Area, reduction of SGSN relocations, and reduced signaling and access to HLR/HSS. An overlap of Pool Areas might allow mapping mobility patterns onto Pool Areas (e.g. cover certain residential zones plus city center).

When the UE performs a GPRS Attach, the RNC selects a suitable SGSN and establishes the connection. The SGSN encodes its NRI (Network Resource Identification) into the P-TMSI. Now UE, RNC, and Serving SGSN know the mapping IMSI <-> NRI, and RNC and SGSN are able to route the packets accordingly.

As long as the UE is in PMM-Connected Mode the RNC retains the mapping IMSI <-> NRI. If the status changes to PMM-Idle Mode the RNC deletes UE data (no packets from/to

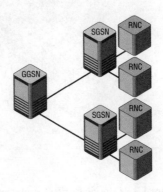

Figure 1.11 Hierarchical RNC <-> SGSN relation.

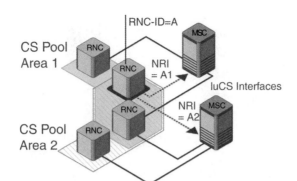

Figure 1.12   IuFlex basic description.

UE need to be routed). If the UE reenters PMM-Connected Mode, it again provides NRI of its Serving SGSN to the RNC.

1.2.5 UMTS Release 6

UMTS Release 6 is still under massive development; however, major improvements are already very clear: a clear path toward UMTS/WLAN Interworking, IMS "Phase 2," Push-to-Talk service, Packet-Switched Streaming Services (PSS), Multimedia Broadcast and Multicast Service, (MBMS), Network Sharing, Presence Service, and the definition of various other new multimedia services. Figure 1.13 describes the basic Rel. 6 architecture. The following paragraphs give a more detailed description of the new features and services that Rel. 6 will have to offer.

The P-CSCF is the first contact point for the GGSN to the IMS after PDP Context Activation. The S-CSCF is responsible for the Session Control for the UE and maintains and stores session states to support the services.

The Breakout-CSCF (B-CSCF) selects the IMS CN (if within the same IMS CN) or forwards the request (if breakout is within another IMS CN) for the PSTN breakout and the MGCF for PSTN interworking. Protocol mapping functionality is provided by the MGCF (e.g. handling of SIP and ISUP) while bearer channel mapping is being handled by the MGW. Signaling between MGW and MGCF follows H.248 protocol standard and handles signaling and session management. The MRF provides specific functions (e.g. conferencing or multiparty calls), including bearer and service validation.

New in Release 6
UMTS/WLAN Interworking (Figure 1.14)

- WLAN could be used at hotspots as access network for IMS instead of the UMTS PS domain (saves expensive 3G spectrum and cell space)
- Access through (more expensive) PS domain allows broadest coverage outside hotspots
- Handovers between 3G (even GPRS) and WLAN shall be supported (roaming)
- WLANs might be operated either by mobile operators or by third party
- Architecture definition for supporting authentication, authorization, and charging (standard IETF AAA Server) included:
  – AAA Server receives data from HSS/HLR

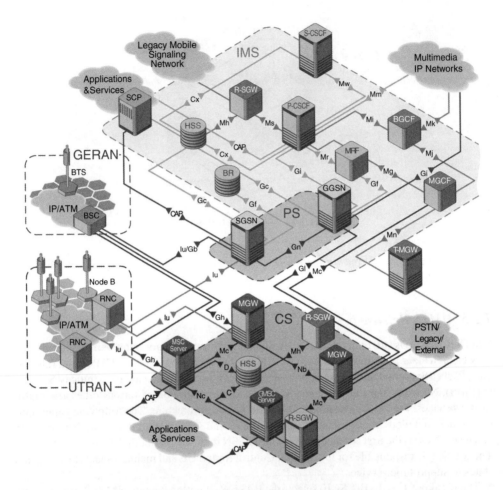

Figure 1.13   3GPP UMTS Rel. 6 network model.

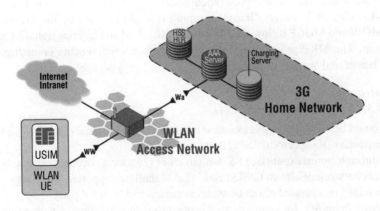

Figure 1.14   WLAN/UMTS support architecture.

Push-to-Talk over Cellular (PoC) Service

- Push-to-Talk is a real-time one-to-one or one-to-many voice communication (like with a walkie-talkie, half duplex only) over data networks
- Instead of dialing a number a subscriber might be selected, e.g. from a buddy list

Packet-Switched Streaming Services (PSS)

- PSS is used to transmit streaming content (subscriber can start to view, listen in real time, even though the entire content has not been downloaded)
- Support of End-to-End-Bitrate-Adaptation to meet the different conditions in mobile networks (allows to offer QoS from "best effort" to "guaranteed")
- Digital Rights Management (DRM) is supported
- Different codecs will be supported (e.g. MPEG-4 or Windows Media Video 9)

Network Sharing

- Allows cost-efficient sharing of network resources such as Network Equipment (Node B, RNC, etc.) or Spectrum (Antenna Sites), reduces time to market and deployment, and finally lets operators get earlier into profit generation
- Sharing can be realized with different models:
  - Multiple CNs share common RANs [each operator maintains individual cells with separate frequencies and separate MNC (Mobile Network Code); BTSs and RNCs are shared, but the MSCs and HLRs are still separated]
  - Sharing of a common CN with separated RANs (like above)
  - Operators agree on a geographical split of networks in defined territories with roaming contracts so that all the mobile users have full coverage over the territory

Presence Service

- User will have the option to make themselves "visible" or "invisible" to other parties and allow or decline services to be offered
- Users can create "buddy lists" and be informed about state changes
- Subscriber own "user profiles" that make service delivery independent of the type of UE or access to the network

Multimedia Broadcast and Multicast Service (MBMS)

- MBMS is an unidirectional point-to-multipoint bearer service (push service)
- Data is transmitted from a single source to multiple subscribers over a common radio channel
- Service could transmit, e.g., text, audio, picture, video
- User shall be able to enable/disable the service
- Broadcast mode sends to every user within reach (typically not charged, e.g. advertisement)
- Multicast mode selectively transmits only to subscribed users (typically charged service)
- Application examples
  - Multicast of, e.g., sport events
  - Multiparty conferencing
  - Broadcast of emergency information
  - Software download
  - Push-to-Talk

IMS "Phase 2"

- The IMS architecture of Rel. 5 was improved and enhanced for Rel. 6
- The main purpose is to integrate all the CNs to provide IP multimedia sessions on the basis of IP multimedia sessions, support real-time interactive services, to provide flexibility to the user, and to reduce cost
- QoS needed for voice and multimedia services is integrated
- Examples of supported services
  - Voice Telephony (VoIP)
  - Call conferencing
  - Group management
    - Setting up and maintaining user groups
    - Supporting service for other services (multiparty conferencing, Push-to-Talk)
  - Messaging
    - SIP-based messaging
    - Instant messaging
    - "Chat room"
    - Deferred messaging (equivalent to MMS)
    - Interworks with Presence Service to determine whether addressee is available
  - Location-based services
    - UE indicates local service request
    - S-CSCF routes request back to visited network
    - Mechanism for UE to retrieve/receive information about locally available services
  - IP <-> IMS Interworking functions
  - IMS <-> CS Interworking functions
  - Lawful interception integration

1.3 UMTS Interfaces

Figure 1.15 shows a basic overview of the different interfaces in an UMTS Rel. 99 network. A detailed description of objectives and functions follows in this chapter.

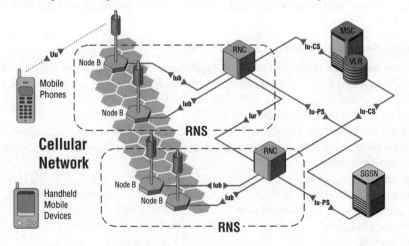

Figure 1.15   UMTS interface overview.

1.3.1 Iu Interface

The Iu interface is located between RNC and MSC for circuit-switched traffic and between RNC and SGSN for packet-switched traffic. Iu provides the connection to "classic" voice services at the same time as the connection for all kinds of packet services. It plays a vital role for the handover procedures in the UMTS network.

Objectives and Functions of the Iu Interface
The Iu interface shall take care of the interconnection of RNCs with the CN Access Points within a single PLMN and the interconnection of RNCs with the CN Access Points irrespective of the manufacturer of any of the elements. Other tasks are the interworking toward GSM, the support of all UMTS services, the support of independent evolution of Core, Radio Access, and Transport Networks, and finally the migration of services from CS to PS.

The Iu interface is split into two types of interfaces:

- IuPS (Packet Switched), corresponding interface towards the PS domain
- IuCS (Circuit Switched), corresponding interface towards the CS domain

The Iu interface supports the following functions:

- Establishing, maintaining, and releasing RABs
- Performing intra- and intersystem handover and SRNS relocation
- A set of general procedures, not related to a specific UE
- Separation of each UE on the protocol level for user-specific signaling management
- Transfer of NAS signaling messages between UE and CN
- Location services by transferring requests from the CN to UTRAN, and location information from UTRAN to CN
- Simultaneous access to multiple CN domains for a single UE
- Mechanisms for resource reservation for packet data streams

1.3.2 Iub Interface

The Iub interface is located between an RNC and a Node B. Via the Iub interface, the RNC controls the Node B. For example, the RNC allows the negotiating of radio resources, the adding and deleting of cells controlled by the individual Node B, or the supporting of the different communication and control links. One Node B can serve one or multiple cells.

Objectives and Functions of the Iub Interface
The Iub interface enables continuous transmission sharing between the GSM/GPRS Abis interface and the Iub interface and minimizes the number of options available in the functional division between RNC and Node B. It controls – through Node B – a number of cells and adds or removes radio links in those cells. Another task is the logical O&M support of the Node B and to avoid complex functionality as far as possible over the Iub. Finally, it accommodates the probability of frequent switching between different channel types.

The Iub interface supports the functions described in Table 1.1.

1.3.3 Iur Interface

The Iur interface connects RNCs inside one UTRAN.

Table 1.1 Iub function overview

Function	Description
Relocating SRNC	Changes the SRNC functionality as well as the related Iu resources [RAB(s) and signaling connection] from one RNC to another
Overall RAB management	Sets up, modifies, and releases RAB
Queuing the setup of RAB	Allows placing some requested RABs into a queue and indicates the peer entity about the queuing
Requesting RAB release	Requests the release of RAB (overall RAB management is a function of the CN)
Release of all Iu connection resources	Explicitly releases all resources related to one Iu connection
Requesting the release of all Iu connection resources	Requests release of all Iu connection resources from the corresponding Iu connection (Iu release is managed from the CN)
Management of Iub Transport Resources	
Logical O&M of Node B	Iub Link Management
	Cell Configuration Management
	Radio Network Performance Measurements
	Resource Event Management
	Common Transport Channel Management
	Radio Resource Management
	Radio Network Configuration Alignment
Implementation-specific O&M Transport	
System Information Management	
Traffic Management of Common Channels	Admission Control
	Power Management
	Data Transfer
Traffic Management of dedicated channels	Radio Link Management, Radio Link Supervision
	Channel Allocation/Deallocation
	Power Management
	Measurement Reporting
	Dedicated Transport Channel Management
	Data Transfer
Traffic Management of Shared Channels	Channel Allocation/Deallocation
	Power Management
	Transport Channel Management
	Data Transfer
Timing and Synchronization Management	Transport Channel Synchronization (Frame synchronization)
	Node B–RNC node synchronization
	Inter-Node B node synchronization

Objectives and Functions of the Iur Interface

The Iur interface provides an open interface architecture and supports signaling and data streams between RNCs, allows point-to-point connection and the addition or deletion of radio links supported by cells belonging to any RNS (Radio Network Subsystem) within the UTRAN. Additionally, it allows an RNC to address any other RNC within the UTRAN so as to establish signaling bearer or user data bearers for Iur data streams.

The Iur interface supports the following functions:

- Transport Network Management
- Traffic management of Common Transport Channels
- Preparation of Common Transport Channel resources
  - Paging
- Traffic Management of Dedicated Transport Channels
  - Radio link setup/addition/deletion
  - Measurement reporting
- Measurement reporting for common and dedicated measurement objects

1.4 UMTS Domain Architecture

From the beginning it was tried that UMTS be very modular in its structure. This is the base of becoming an international standard even though certain modules will be national-specific.

The two important big modules are the Access Stratum (Mobile and UTRAN) and the Non-Access Stratum [containing serving CN, Access Stratum, and USIM (Universal Subscriber Identity Module)] (Figure 1.16).

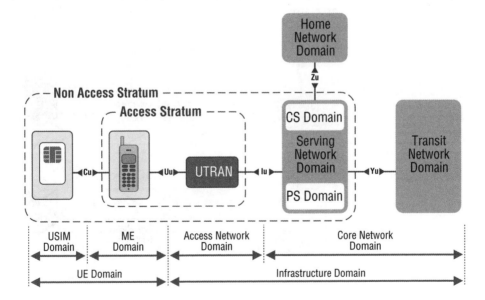

Figure 1.16 UMTS domain architecture.

1.5 UTRAN

Two new network elements are introduced in UTRAN: RNC and Node B. UTRAN is subdivided into individual RNS, where an RNC controls each RNS.

The RNC is connected to a set of Node B elements, each of which can serve one or several cells.

Existing network elements, such as MSC, SGSN, and HLR, can be extended to adopt the UMTS requirements, but RNC and Node B require completely new designs. RNC will become the replacement for BSC, and Node B fulfills nearly the same functionality as BTS. GSM and GPRS networks will be extended and new services will be integrated into an overall network that contains both existing interfaces, such as A, Gb, and Abis, and new interfaces that include Iu, Iub, and Iur (Figure 1.17).

The main UTRAN tasks are:

Admission Control (AC): Admits or denies new users, new radio access bearers, or new radio links. The AC should try to avoid overload situations and will not deteriorate the quality of the existing radio links. Decisions are based on interference and resource measurements (power or on the throughput measurements). Together with Packet Scheduler it allocates the bitrates sets (transmission powers) for Non-Realtime connections. The AC is employed at, for example, the initial UE access, the RAB assignment/reconfiguration, and at handover. The functionality is located in the RNC.

Power-based AC needs the reliable Received Total Wideband Power measurements from the Node B and assures the coverage stability. In the power-based case, the upper boundary for the AC operation is defined by the maximum allowed deterioration of the quality for the existing links (=the maximum allowed deterioration of the path loss). This limit is usually defined as P_{RX} Target [dB] (Figure 1.18).

Throughput-based AC assures the constant maximum cell throughput in every moment of the operation, but allows excessive cell breathing. On the linear scale the received power

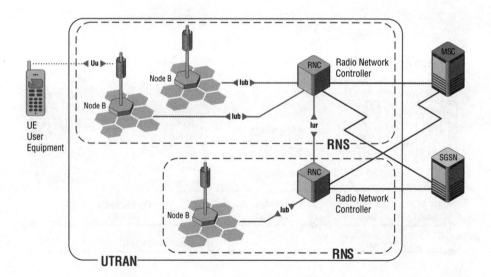

Figure 1.17   UTRAN.

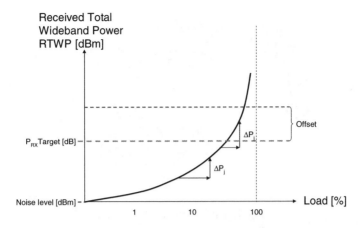

Figure 1.18   Throughput-based Admission Control.

changes [dB] can be expressed as the cell loading [%]. Via a simple equation the cell loading
[%] is bounded with the cell throughput [kbps] and call quality [E_b/N_0].

Congestion Control: Monitors, detects, and handles situations when the system is reaching a
near overload or an overload situation with the already connected users.

System Information Broadcasting: Provides the UE with the Access Stratum and Non-Access
Stratum information, which are needed by the UE for its operation within the network.

Ciphering: Encrypts information exchange and is located between UE and RNC.

Handover (HO): Manages the mobility of the radio interface. It is based on radio measurements
and for Soft/Softer HO it is used to maintain the QoS requested by the CN. An Intersystem
HO (IS-HO) is necessary to avoid losing the UE's network connection. In that case even
a lower QoS might be accepted. Handover may be directed to/from another system (for
example, UMTS to GSM HO).

Further functions of UTRAN are configuration and maintenance of the radio interface, power
control, paging, and macrodiversity.

1.5.1 RNC

The RNC is the main element in the RNS and controls usage and reliability of radio resources.
There are three types of RNCs: SRNC (Serving RNC), DRNC (Drift RNC), and CRNC (Controlling RNC). Tasks of the RNC are:

Call Admission Control: Provides resource check procedures before new users access the
network, as required by the CDMA air interface technology.

Radio Bearer Management: Sets up and disconnects radio bearers and manages their QoS.

Code Allocation: Manages the code planning that the CDMA technology requires.

Power Control: Performs the outer loop power control 10–100 times per second and defines
the SIR (Signal-to-Interference Ratio) for a given QoS.

Congestion Control: Schedules packets for PS CN data transmission.

O&M Tasks: Performs general management functions and connection to OMC.

Additionally, the RNC can act as a macrodiversity point, for example a collection of data from
one UE that is received via several Node Bs.

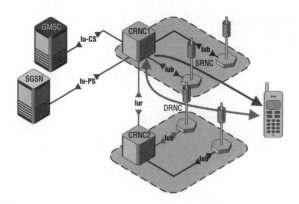

Figure 1.19   Different RNC types.

Different RNC Types

Controlling RNC (CRNC)
The CRNC controls, configures, and manages an RNS and communicates with NBAP (Node B Application Part) with the physical resources of all Node Bs connected via the Iub interfaces. Access requests of UEs will be forwarded from the related Node B to the CRNC (Figure 1.19).

Drift RNC (DRNC)
The DRNC receives connected UEs that are handed over (drifted) from an SRNC cell connected to a different RNS because, e.g., the received level of that cell became critical (mobility). The Radio Resource Control (RRC) however still terminates with the SRNC. The DRNC exchanges then routing information between SRNC and UE.

DRNC in Inter-RNC Soft HO situation is the only DRNC from SRNC point of view. It lends radio resources to SRNC to allow Soft HO. However, radio resources are controlled by CRNC function of the same physical RNC machine. Functions can be distinguished by protocol used: DRNC "speaks" RNSAP with SRNC via Iur, CRNC "speaks" NBAP with cells via Iub.

Serving RNC (SRNC)
The SRNC controls a user's mobility within a UTRAN and is the connection point to the CN toward MSC or SGSN, too. The RNC that has an RRC connection with an UE is its SRNC. The SRNC "speaks" RRC with UE via Iub, Uu and – if necessary – via Iur and "foreign" Iub (controlled by DRNC).

1.5.2 Node B

The Node B provides the physical radio link between the UE and the network. It organizes transmission and reception of data across the radio interface and also applies codes that are necessary to describe channels in CDMA systems. The tasks of a Node B are similar to those of a BTS. The Node B is responsible for:

Power Control: Performs the inner loop power control, which measures the actual SIR, compares it with the specific defined value, and may trigger changes in the TX power of a UE.
Measurement Report: Gives the measured values to the RNC.

Microdiversity: Combines signals (from the multiple sectors of the antenna that a UE is connected to) into one data stream before transmitting the sum-signal to the RNC. (The UE is connected to more than one sector of an antenna to allow for a Softer HO.)

The Node B is the physical unit to carry one or more cells (1 cell = 1 antenna).

There are three types of Node Bs:

- UTRA-FDD Node B
- UTRA-TDD Node B
- Dual Mode Node B (UTRA-TDD and UTRA-FDD)

Note: It is not expected to have 3.84 TDD and 1.28 TDD cells in the same network, but operators in same area are expected to work with different TDD versions.

So, three-band-Node Bs are not necessary.

1.5.3 Area Concept

The areas of 2G will be continuously used in UMTS.

UMTS will add a new group of locations specifying the UTRAN Registration Areas (URAs). These areas will be smaller Routing or Location Areas and will be maintained by UTRAN itself, covered by a number of cells The URA is configured in the UTRAN, and broadcast in relevant cells (Figure 1.20).

The different areas are used for Mobility Management, e.g. Location Update and Paging procedures.

Location Area (LA)
The LA is a set of cells (defined by the mobile operator) throughout which a mobile will be paged. The LA is identified by the LAI (Location Area Identity) within a PLMN (Public Land Mobile Network) and consists of MCC (Mobile Country Code), MNC (Mobile Network Code), and LAC (Location Area Code).

$$LAI = MCC + MNC + LAC$$

Routing Area (RA)
One or more RA is controlled by the SGSN. Each UE informs the SGSN about the current RA. RAs can consist of on one or more cells. Each RA is identified by a RAI (Routing Area

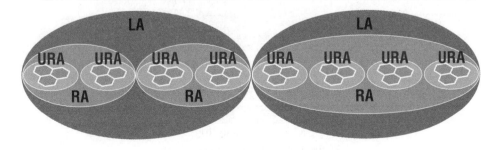

Figure 1.20 UMTS areas.

Identity). The RAI is used for paging and registration purposes and consists of LAC and RAC (Routing Area Code). The RAC (length: 1 octet fixed) identifies an RA within an LA and is part of the RAI.

$$RAI = LAI + RAC$$

Service Area (SA)
The SA identifies an area of one or more cells of the same LA, and is used to indicate the location of a UE to the CN.

The combination of SAC (Service Area Code), PLMN-ID (Public Land Mobile Network Identifier), and LAC (Location Area Code) is the Service Area Identifier.

$$SAI = PLMN\text{-}ID + LAC + SAC$$

UTRAN Registration Area (URA)
The URA is configured in the UTRAN is broadcast in relevant cells, and covers an area of a number of cells.

1.5.4 UMTS User Equipment and USIM

In UMTS the Mobile Station (MS) is called *User Equipment* (UE) and is constructed in a very modular way (Figure 1.21). It consists of following parts.

Mobile Termination (MT)
Represents the termination of the radio interface and, by that, the termination of an IMT-2000 family-specific unit. There are different MT messages for UMTS in Europe as opposed to in the United States.

Terminal Adapter
Represents the termination of the application-specific service protocols, for example, AMR for speech. This function will perform all necessary modifications to the data.

Terminal Equipment
Represents the termination of the service.

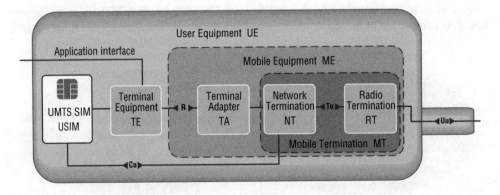

Figure 1.21 UMTS User Equipment.

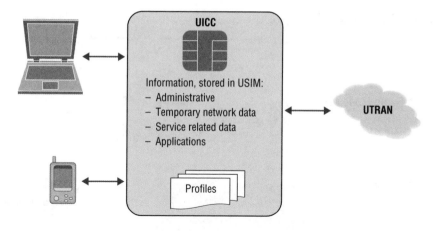

Figure 1.22   UMTS Service Identity Module (USIM).

USIM
Is a user subscription to the UMTS mobile network and contains all relevant data that enables access onto the subscribed network (Figure 1.22). Every UE may contain one or more USIM simultaneously (100 % flexibility). Higher layer standards like MM/CC/SM address 1 UE + 1 (of the several) USIM when they mention an MS.

The main difference between a USIM and a GSM SIM is that the USIM is downloadable (by default), can be accessed via the air interface, and can be modified by the network.

The USIM is a Universal Integrated Circuit Card (UICC), which has much more capacity than a GSM SIM. It can store Java applications. It can also store profiles containing user management and rights information and descriptions of the way applications can be used.

1.5.5 Mobiles

Mobile Terminations
The Mobile Terminations are divided into different groups (Figure 1.23).

Single Radio Mode MT
The UE can work with only one type of network because only one RAT is implemented.

Multiradio Mode MT
More than one RAT is supported. 3GPP specifies handover between different RATs in great detail.

Single Network MT
Independent of the Radio Mode, the Single Network MT is capable of using only one type of CN; for example only the packet-switched CN (PC Card).

Multinetwork MT
Independent of the Radio Mode, the Multinetwork MT can work with different types of CNs. At the beginning of UMTS, the multinetwork operations will have to be performed sequentially,

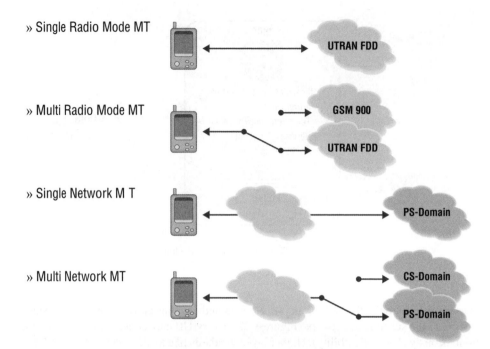

» Single Radio Mode MT

» Multi Radio Mode MT

» Single Network M T

» Multi Network MT

Figure 1.23 Types of Mobile Terminations.

but, at a later stage, parallel operations could also be possible. This ability will depend heavily on the overall performance of the UE and the network capacity.

The first UMTS mobiles should be Multiradio–Multinetwork mobiles.

Mobile Capabilities

The possible features of UTRAN and CN will be transmitted via System Information on the radio interface via broadcast channels. A UE can, by listening on these channels, configure its own settings to work with the actual network (Figure 1.24).

On the other hand, the UE will also indicate its own capabilities to the network by sending MS Classmark and MS Radio Access Capability information to the network.

Below is an extract of possible capabilities:

- Available W-CDMA modes, FDD or/and TDD
- Dual mode capabilities, support of different GSM frequencies
- Support of GSM PS features, GPRS or/and HSCSD

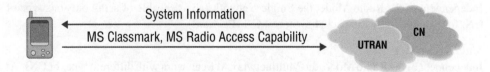

Figure 1.24 Mobile capabilities.

- Available encryption algorithms
- Properties of measurement functions, timing
- Ability of positioning methods
- Ability to use universal character set 2 (16-bit characters)

In GSM, MS Classmark 1 and 2 were used. In UMTS, MS Classmark 2 and the new MS Classmark 3 are used. The difference is the number of parameter for different features can be transmitted.

1.5.6 QoS Architecture

There is 1:1 relation between Bearer Services and QoS in UMTS networks.

Other than in 2G systems where a bearer was a traffic channel in 3G the bearer represents a selected QoS for a specific service. Only from the point of view of the physical layer a bearer is a type of channel.

A Bearer Service is a service that guarantees a QoS between two endpoints of communication. Several parameters will have to be defined from operators.

A Bearer Service is classified by a set of values for these parameters:

- Traffic class
- Maximum bitrate
- Guaranteed bitrate
- Delivery order
- Maximum SDU (Service Data Unit) size
- SDU format information
- SDU error ratio
- Residual bit error ratio
- Delivery of erroneous SDUs
- Transfer delay
- Traffic handling priority
- Allocation/retention priority

The End-to-End Service will define the constraints for the QoS. These constraints will be given to the lower Bearer Services, translated into their configuration parameters and again passed to the lower layer. By that, UMTS sets up a connection through its own layer architecture fulfilling the requested QoS (Figure 1.25).

Problems are foreseen in the External Bearer Services because they are outside of UMTS and the responsibility of the UMTS network operator.

QoS classes with QoS attributes have been specified to meet the needs of different End-to-End Services (Figure 1.26).

Conversational Class
Real-time applications with short predictable response time. Symmetric transmission without buffering of data and with a guaranteed data rate.

Streaming Class
Real-time applications with short predictable response time. Asymmetric transmission with possible buffering of data and with a guaranteed data rate.

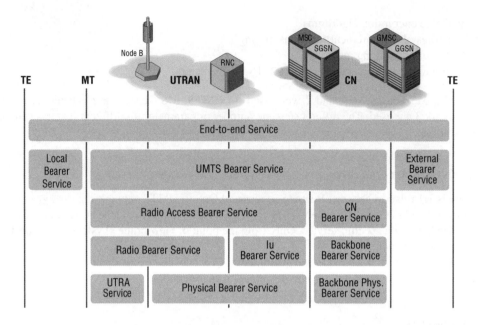

Figure 1.25 UMTS Bearer/QoS architecture.

Interactive Class
Non-real-time applications with acceptable variable response time. Asymmetric transmission with possible buffering of data but without guaranteed data rate.

Background Class
Non-real-time applications with long response times. Asymmetric transmission with possible buffering of data but without a guaranteed data rate.

1.6 UMTS SECURITY

After experiencing GSM, the 3GPP creators wanted to improve the security aspects for UMTS.
   For example, UMTS addresses the "Man-in-the-Middle" Fake BTS problem by introducing a signaling integrity function.

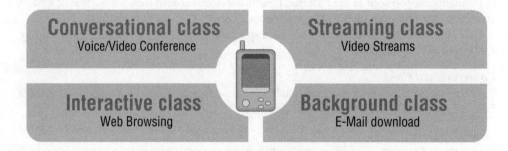

Figure 1.26 UMTS Bearer/QoS classes.

Figure 1.27   Ciphering in ancient Greece.

The most important security features in the access security of UMTS are:

* Use of *temporary identities* (TMSI, P-TMSI)
* Mutual *authentication* of the user and the network
* Radio access network *encryption*
* Protection of *signaling integrity* inside UTRAN

1.6.1 Historic Development

Although ciphering and cryptanalysis became a hot topic accelerated by the current geopolitic environment, information security is not a new issue.

Four hundred years B.C. the ancient Greeks used the so-called *skytals* (Gr. *Skytale*) for encryption. A skytal is a wooden stick of fixed diameter with a long paper strip winded around the stick. The sender wrote a message on the paper in longitudinal direction. The unwinded paper strip gave no meaningful information to the courier or other unauthorized person. Only a receiver who owns a stick with the same diameter was able to decipher the message (Figure 1.27).

Caesar was ciphering secret information simply by replacing every character with another one that was in the alphabet three places behind it. The word "cryptology" would be ciphered as "fubswrorjb." Code books were widely used in the twelfth century. Certain key words of a text were replaced by other predefined words with completely different meaning. A receiver who owns an identical code book is able to derive the original message.

Kasiski's and William F. Friedman's fundamental research about statistical methods in the nineteenth century are the foundation of modern methods for ciphering and cryptanalysis. The Second World War gave another boost for ciphering technologies. The Enigma was an example of advanced ciphering machines used by the German military. Great Britain under Alan Turing with his "bomb" was able to crack Enigma (Figure 1.28).

Another milestone was Claude E. Shannon's article "Communication Theory of Secret Systems" published in 1949. It gives the information-theoretic basis for cryptology and proves Vernam's "One-Time-Pad" as a secure cryptosystem.

In the last century several ciphering technologies have been developed, which can be divided in symmetric and asymmetric methods. Symmetric methods are less secure because the same key is used for ciphering and deciphering. Examples are the Data Encryption Standard (DES) developed by IBM and the International Data Encrypted Algorithm (IDEA) proposed by Lai and Massey.

Asymmetric technologies use one encryption key (public key) and another decryption key (private key). It is not possible to calculate the decryption key by knowing only the encryption

Figure 1.28   Enigma and Bomb as examples for decryption and encryption.

key. The most common asymmetric ciphering method is RSA, developed by *R*ivest, *S*hamir, and *A*dleman in 1978. The method is based on the principle of big prime numbers: It is relatively easy to detect two prime numbers x and y with 1000 and more digits. However, even today it is not possible to calculate the factors of the product "x * y" in reasonable time. Kasumi from Mitsubishi developed an algorithm for ciphering and integrity protection used in UMTS networks. The 3GPP standard is open for other ciphering methods, but today Kasumi is the first and only ciphering algorithm used in UMTS.

Security Threats and Protection in Mobile Networks
In a digital mobile network the subscriber is exposed to several basic attacks as described below (Figure 1.29):

- Eavesdropping (theft of voice and data information)
- Unauthorized identification
- Unauthorized usage of services
- Offending the data integrity (data falsification by an intruder)
- Observation
  - Detection of the current location
  - Observation of communication relations (who is communicating with whom?)
  - Generation of behavior profiles

As an example for unlawful observation, Figure 1.30 shows a part of a Measurement Report Message captured on the GSM Abis interface. An active mobile permanently measures the power level and the bit error rate of its serving cell and up to six neighbor cells. This information

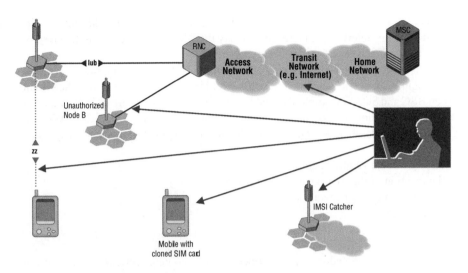

Figure 1.29 Potential attack points of intruders.

is transmitted from the mobile over the BTS to the BSC. In addition, the BTS sends the Timing Advance Information to the mobile. The Timing Advance is a value in the range of 0–63. The Timing Advance is an indicator of the distance between BTS and mobile. Assuming that the maximum cell size in GSM is 30 km, the Timing Advance value allows estimating the distance with 500 m precision. In urban places however, the cell size is much smaller. Combining that information, a potential intruder can relatively exactly determine the location of the mobile subscriber. GSM was originally designed as a circuit-switched voice network. In contradiction

```
000001--   Actual Timing Advance                    1
L3 Information
00001011   IE Name                          L3 Information
00000000   Spare                            0
00010010   LLSDU Length                     18
**B18***   DTAP LLSDU                       06 15 2a 2a 01 25 06 a7 97 63 85...
E-GSM 04.08 (DTAP) 5.3.0) (DTAP)  MEASREP (= Measurement report)
Measurement report
----0110   Protocol Discriminator           radio resources management msg
0000----   Skip Indicator                   0
-0010101   Message Type                     21
0-------   Extension bit                    0
Measurement Results
0-------   BA-USED                          0
-0------   DTX-USED                         not used
--101010   RXLEV-FULL-SERVING-CELL          -69 dBm to -68 dBm
0-------   Spare                            0
-0------   Measurement results valid        Valid
--101010   RXLEV-SUB-SERVING-CELL           -69 dBm to -68 dBm
0-------   Spare                            0
-000----   RXQUAL-FULL-SERVING-CELL         BER less than 0.2%
----000-   RXQUAL-SUB-SERVING-CELL          BER less than 0.2%
***b3***   NO-NCELL-M                       4 NCELL measurement result
--100101   RXLEV-NCELL 1                    -74 dBm to -73 dBm
00000---   BCCH-FREQ-NCELL 1                0
-----110   BSIC-NCC-NCELL 1                 6
101-----   BSIC-BCC-NCELL 1                 5
```

Figure 1.30 Measurement result message sent unciphered via GSM radio channels.

to the voice data, controlling information is never ciphered in GSM. In addition, the ciphering is limited to the air interface. It is needless to say that short messages are transferred over the signaling network and therefore are never ciphered.

GPRS as extension to GSM already offers significant security improvements. User and controlling information are ciphered not only over air interface but also over the Gb interface between BSC and SGSN. Commonly used in commercial networks are GEA1[2] and GEA2, and recently under development is GEA3. The most secure mobile network is the UMTS network.

UMTS actively combats prior mentioned threats offering the following security procedures:

- Ciphering of control information and user data
- Authentication of the user toward the network
- Authentication of the network toward the user
- Integrity protection
- Anonymity

The UMTS security procedures are described in the following sections. Security mechanism over transport networks (Tunneling, IPsec) are not part of this book.

Principles of GSM Security and the Evolution to UMTS Security

As UMTS can be seen as an evolution of the 2G (GSM) communication mobile systems, the security features for UMTS are based on the GSM security features and are enhanced. When UMTS was defined from the Third Generation Partnership Project, better known as 3GPP, there was the basic requirement to adopt the security features from GSM that have proved to be needed and robust and to be as compatible with the 2G security architecture as possible. UMTS should correct the problems with GSM by addressing its real and perceived security weaknesses and to add new security features to secure the new services offered by 3G.

The limitations and weaknesses of the GSM security architecture stem by large from designing limitations rather than on defects in the security mechanisms themselves. GSM has the following specific weaknesses that are corrected within UMTS.

- Active attacks using a false base station
  - Used as "IMSI catcher" (collect "real" IMSIs of MSs that try to connect with the base stations) → cloning risk
  - Used to intercept mobile originated calls – encryption is controlled by network, so user is unaware if it is not on
- Cipher keys and authentication data are transmitted in clear between and within networks
  - Signaling system vulnerable to interception and impersonation
- Encryption of the user and signaling data does not carry far enough through the network to prevent being sent over microwave links (BTS to BSC) – encryption terminated too soon
- Possibility of channel hijack in networks that does not offer confidentiality
- Data integrity is not provided, except traditional noncryptographic link-layer checksums
- IMEI (International Mobile Equipment Identity – unique) is an unsecured identity and should be treated as such – as the Terminal is an unsecured environment, trust in the terminal identity is misplaced
- Fraud and lawful interception was not considered in the design phase of 2G

---

[2] GPRS Encryption Algorithm.

- There is no HE knowledge or control of how an SN (Serving Network) uses authentication parameters for HE subscribers roaming in that SN
- Systems do not have the flexibility to upgrade and improve security functionality over time
- Confidence in strength of algorithms
  - Failure to choose best authentication algorithm
  - Improvements in cryptanalysis of A5/1
    - Key length too short
    - Lack of openness in design and publication

Furthermore, there are challenges that security services will have to cope within 3G systems, and these will probably be:

- Totally new services are likely to be introduced
- There will be new and different providers of services
- Mobile systems will be positioned as preferable to fixed line systems for users
- Users will typically have more control over their service profile
- Data services will be more important than voice services
- The Terminal will be used as a platform for e-commerce and other sensitive applications

The following features of GSM security are reused for UMTS:

- User authentication and radio interface encryption
- Subscriber identity confidentiality on the radio interface
- SIM as a removable, hardware security module in UMTS, called USIM
  - Terminal independent
  - Management of all customer parameter
- Operation without user assistance
- Minimized trust of the SN by the HE

1.6.2 UMTS Security Architecture

Based on Figure 1.31, showing the order of all transactions of a connection, the next section will cover the Authentication and Security Control part and explain the overall security functions for the connection.

The 3G security architecture (Figures 1.32 and 1.33) is a set of security features and enhancements that are fully described in *3GPP 33.102* and is based on the three security principles.

Authentication and Key Agreement (AKA)
Authentication is provided to assure the claimed identity between the user and the network. It is divided into two parts:

- Authentication of the user toward the network
- Authentication of the network toward the user (new in UMTS)

This is done in so-called one-pass authentication, reducing messages sent back and forth. After these procedures the user will be sure that he is connected to his served/trusted network and the network is sure that the claimed identity of the user is true. Authentication is needed for the other security mechanisms as confidentiality and integrity.

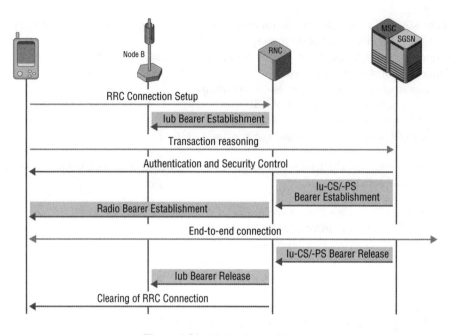

Figure 1.31   Network transitions.

Integrity

Integrity protection is used to secure that the content of a signaling message between the user
and the network has not been manipulated, even if the message might not be confidential. This
is done by generating "stamps" individually from the user and the network that are added to
the transferred signaling messages. The stamps are generated based on a pre-shared secret key
K, which is stored in the USIM and the AuC. At transport level, the integrity is checked by

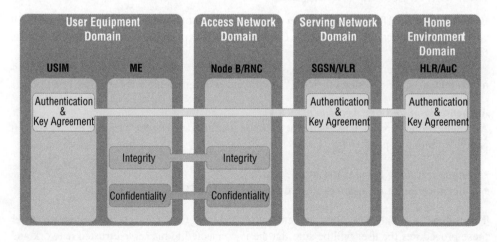

Figure 1.32   UMTS security architecture.

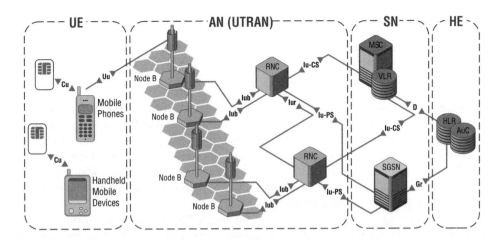

Figure 1.33   UMTS interface and domain architecture overview.

CRC checksum, but these measures are only to achieve bit-error-free communication and are not equivalent to transport level integrity.

Confidentiality
Confidentiality is used to keep information secured from unwanted parties. This is achieved by ciphering of the user/signaling data between the subscriber and the network and by referring to the subscriber by temporary identities (TMSI/P-TMSI) instead of using the global identity, IMSI. Ciphering is carried out between the users terminal (USIM) and the RNC. User confidentiality is between the subscriber and the VLR/SGSN. If the network does not provide user data confidentiality, the subscriber is informed and has the opportunity to refuse connections.

Parts that are confidential are:

- Subscriber identity
- Subscriber's current location
- User data (voice and data)
- Signaling data

1.6.3 Authentication and Key Agreement (AKA)

UMTS security starts with the AKA, the most important feature in the UMTS system. All other services depend on them since no higher level services can be used without authentication of the user.

Mutual Authentication

- Identifying the user to the network
- Identifying the network to the user

Key Agreement

- Generating the Cipher Key
- Generation the Integrity Key

After Authentication and Key Agreement

- Integrity protection of messages
- Confidentiality protection of signaling data
- Confidentiality protection of user data

The mechanism of mutual authentication is achieved by the user and the network showing knowledge of a secret key (K) which is shared between and available only to the USIM and the AuC in the user's HE. The method was chosen in such a way as to achieve maximum compatibility with the current GSM security architecture and facilitate migration from GSM to UMTS. The method is composed of a challenge/response protocol identical to the GSM subscriber authentication and key establishment protocol combined with a sequence number-based one-pass protocol for network authentication.

The authenticating parties are the AuC of the user's HE (HLR/AuC) and the USIM in the user's MS. The mechanism consists of the distribution of authentication data from the HLR/AuC to the VLR/SGSN and a procedure to authenticate and establish new cipher and integrity keys between the VLR/SGSN and the MS.

AKA Procedure

Once the HE/AuC has received a request from the VLR/SGSN, it sends an ordered array of *n* Authentication Vectors (AVs) to the VLR/SGSN. (Figure 1.34). Each AV consists of the

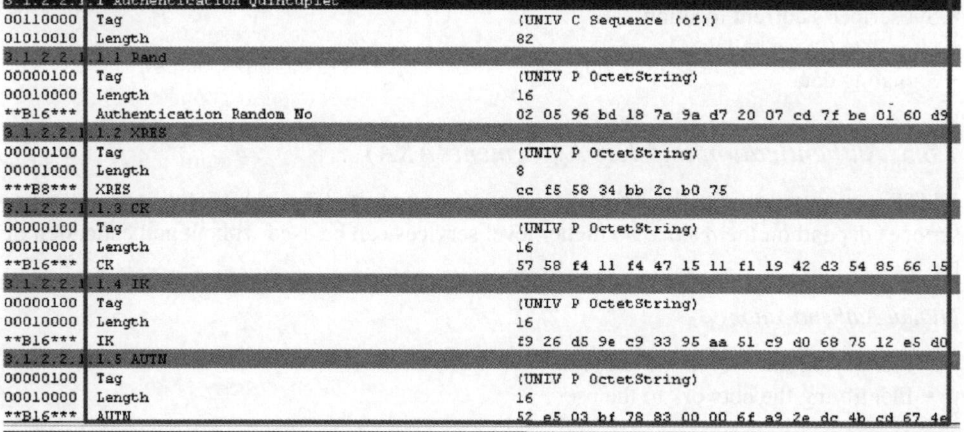

From	2. MSG	3. Prot	3. MSG		Procedure Code	Last Prot	Last MSG
G62 -..	SD	RL	RL			SCCP	CC
Gr G62	MSU	SCCP	UDT			MAP	BEG
Gr G62	MSU	SCCP	UDT			MAP	END
G62	SD	RL	RL		id Direct Transfer	CWM DMTAP	ACRO

		Frame View	
BITMASK	ID Name		Comment or Value
01010110	Length		86
3.1.2.2.1	Quintuplet List		
10100001	Tag		(CONT C [1])
01010100	Length		84
3.1.2.2.1.1 Authentication Quintuplet			
00110000	Tag		(UNIV C Sequence (of))
01010010	Length		82
3.1.2.2.1.1.1 Rand			
00000100	Tag		(UNIV P OctetString)
00010000	Length		16
B16*	Authentication Random No		02 05 96 bd 18 7a 9a d7 20 07 cd 7f be 01 60 d9
3.1.2.2.1.1.2 XRES			
00000100	Tag		(UNIV P OctetString)
00001000	Length		8
B8	XRES		cc f5 58 34 bb 2c b0 75
3.1.2.2.1.1.3 CK			
00000100	Tag		(UNIV P OctetString)
00010000	Length		16
B16*	CK		57 58 f4 11 f4 47 15 11 f1 19 42 d3 54 85 66 15
3.1.2.2.1.1.4 IK			
00000100	Tag		(UNIV P OctetString)
00010000	Length		16
B16*	IK		f9 26 d5 9e c9 33 95 aa 51 c9 d0 68 75 12 e5 d0
3.1.2.2.1.1.5 AUTN			
00000100	Tag		(UNIV P OctetString)
00010000	Length		16
B16*	AUTN		52 e5 03 bf 78 83 00 00 6f a9 2e dc 4b cd 67 4e

Figure 1.34 Example for AV (Authentication Vector) sending from HE to SN in authentication data response.

following components: a random number RAND, an expected response XRES, a Cipher Key CK, an Integrity Key IK, and an Authentication Token AUTN. Each AV is valid only for one AKA between the VLR/SGSN and the USIM and are ordered based on sequence number. The VLR/SGSN initiates an AKA by selecting the next AV from the ordered array and sending the parameters RAND and AUTN to the user. If the AUTN is accepted by the USIM, it produces a response RES that is sent back to the VLR/SGSN. AVs in a particular node are used on a first-in/first-out basis. The USIM also computes CK and IK. The VLR/SGSN compares the received RES with XRES. If they match, the VLR/SGSN considers the AKA exchange to be successfully completed. The established keys CK and IK will then be transferred by the USIM and the VLR/SGSN to the entities that perform ciphering and integrity functions. VLR/SGSNs can offer secure service even when HE/AuC links are unavailable by allowing them to use previously derived cipher and integrity keys for a user so that a secure connection can still be set up without the need for an AKA. Authentication is in that case based on a shared integrity key, by means of data integrity protection of signaling messages (Figure 1.35).

AKA is performed when the following events happen:

- Registration of a user in an SN
- After a Service Request
- Location Update Request
- Attach Request
- Detach Request
- Connection reestablishment request

Registration of a subscriber in an SN typically occurs when the user goes to another country. The coverage area of an operator is nationwide, and roaming between national operators will therefore be limited. The first time the subscriber connects to the SN, he gets registered in the SN.

Service Request is the possibility for higher level protocols/applications to ask for AKA to be performed, e.g. performing AKA to increase security before an online banking transaction. The terminal updates the HLR regularly with its position in Location Update Requests.

Attach request and detach request are procedures to connect and disconnect the subscriber to/from the network.

Connection re-establishment request is performed when the maximum number of local authentications has been conducted.

A weakness of the AKA is that the HLR/AuC does not check whether the information sent from the VLR/SGSN (Authentication information) is correct or not.

Algorithms Used for AKA (Tables 1.2 and 1.3)
The security features of UMTS are fulfilled with a set of cryptographic functions and algorithms. A total of 10 functions are needed to perform all the necessary features, f0–f5, f8, and f9.

f0 is the random challenge generating function, the next seven are key generating functions, and so they are all operator-specific. The keys used for authentication are generated only in the USIM and the AuC, the two domains that the same operator is always in charge of.

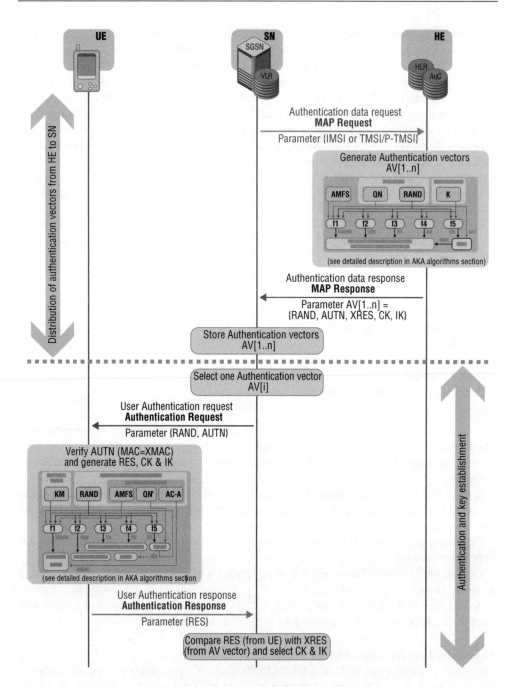

Figure 1.35 AKA procedure – sequence diagram.

Table 1.2 AKA function overview

Function	Description	Input parameter	Output Parameter
f0	The random challenge generating function	RAND	RAND
f1	The network authentication function	AMF, K, RAND	MAC-A (AuC side)/XMAC-A (UE side)
f2	The user authentication function	K, RAND	RES (UE side)/XRES (AuC side)
f3	The cipher key derivation function	K, RAND	CK
f4	The integrity key derivation function	K, RAND	IK
f5	The anonymity key derivation function	K, RAND	AK
f8	The confidentiality key stream generating function	COUNT-C, BEARER, DIRECTION, LENGTH, CK	\<Keystream block\>
f9	The integrity stamp generating function	IK, FRESH, DIRECTION, COUNT-I, MESSAGE	MAC-I (UE side)/XMAC-I (RNC side)

Table 1.3 AKA parameter overview

Parameter	Definition	Bit size
K	Pre-shared secret key stored in the USIM and AuC	128
RAND	The random challenge to be sent to the USIM	128
SQN	Sequence number	48
AK	Anonymity Key	48
AMF	Authentication Management Field	16
MAC	Message Authentication Code	64
MAC-A/XMAC-A	MAC used for AKA	64
MAC-I/XMAC-I	Message authentication code for data integrity	64
CK	Cipher key for confidentiality	128
IK	Integrity key for integrity checking	128
RES	Response	32–128
XRES	The expected result from the USIM	32–128
AUTN	Authentication token that authenticates the AuC toward the USIM (AMF, MAC-A, SQN')	128 (16 + 64 + 48)
COUNT-I	The integrity sequence number	32
FRESH	The network-side random value	32
DIRECTION	Either 0 (UE→RNC→uplink) or 1 (RNC->UE=downlink)	1
MESSAGE	The message themselves	Variant

Functions f8 and f9 are used in USIM and RNC, and since these two domains may be of different operators, they cannot be operator-specific. The functions use the pre-shared secret key (K) indirectly. This is to keep from distributing K in the network, and keep it safe in the USIM and AuC.

The functions f1–f5 are called key generating functions and are used in the initial AKA procedure. The lifetime of Key is depended on how long the keys have been used. The maximum limits for use of same keys are defended by the operator, and whenever the USIM finds the keys being used for as long as allowed, it will trigger the VLR/SGSN to use a new AV.

The functions f1–f5 shall be designed so that they can be implemented with a 8-bit microprocessor running at 3.25 MHz with 8-kbyte ROM and 300-kbytes RAM and produce AK, XMAC-A, RES, CK, and IK in less than 500-ms execution time.

When generating a new AV the AuC reads the stored value of the sequence number SQN and then generates a new SQN' and a random challenge RAND. Together with the stored AV and Authentication Management Field (AMF) and the pre-shared secret key (K), these four input parameters are ready to be used. The functions f1–f5 use these inputs and generate the values for the Message Authentication Code, MAC-A, the expected result, XRES, the CK, the IK, and the AK. With the SQN xor'ed AK, AMF and MAC, the AUTN is made. The AV is send to the SGSN/VLR and stored there, while the parameter pair AUTN and RAND are then send from the SGSN/VLR to the User. The CK and IK are used, after a successful authentication, for confidentiality (ciphering) and integrity (Figure 1.36).

Only one of the four parameters that the AuC has is stored in the USIM, the pre-shared secret key (K). The rest of the parameters it has to receive from the network (RAND and AUTN). The secret key K is then used with the received AMF, SQN', and RAND to generate the Expected Message Authentication Code (XMAC-A). This is then compared with the MAC-A. If the XMAC and MAC matches, the USIM have authenticated that the message is originated in its HE and thereby connected to an SN that is trusted by the HE. With a successful network authentication, the USIM verifies if the sequence number received is within

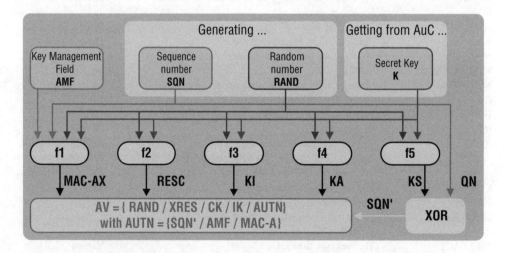

Figure 1.36 Authentication Vector generation on the AuC side (HE).

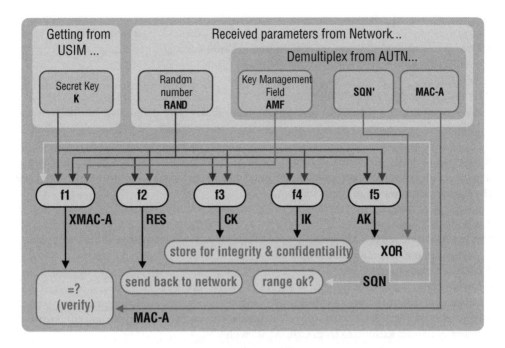

Figure 1.37   User Authentication Response on the user side.

the correct range. With a sequence number within the correct range, the USIM continues to generate the RES, which is send back to the network to verify a successful user authentication (Figure 1.37).

1.6.4 Kasumi/Misty

The Kasumi algorithm is the core algorithm used in functions f8 (Confidentiality) and f9 (Integrity). Kasumi is based on the block cipher "Misty" proposed by Mitsuru Matsui (Mitsubishi) and first published in 1996. Misty where translated from English to Japanese means Kasumi.
    Misty was designed to fulfill the following design criteria:

High security

- Provable security against differential and linear cryptanalysis

Multiplatform

- High speed in both software and hardware implementations
    - Pentium III (800 MHz) (Assembly Language Program)
      - Encryption speed   230 Mbps
    - ASIC H/W (Mitsubishi 0.35 micron CMOS Design Library)
      - Encryption speed   800 Mbps
      - Gate size   50 kgates

Compact

- Low gate count and low power consumption in hardware
  - ASIC (Mitsubishi 0.35 micron CMOS Design Library)
    - Gate size   7.6 kgates
    - Encryption speed   72 Mbps
- A requirement for W-CDMA encryption algorithm: "gate size must be smaller than 10 kgates"

Kasumi is a variant of Misty1 designed for W-CDMA systems and has been adopted as a mandatory algorithm for data confidentiality and data integrity in W-CDMA by 3GPP in 1999. Here are some examples of improvement:

- Simpler key schedule
- Additional functions to complicate cryptanalysis without affection provable security aspects
- Changes to improve statistical properties
- Minor changes to speed up
- Stream ciphering f8 uses Kasumi in a from of output feedback, but with:
  - BLKCNT added to prevent cycling
  - Initial extra encryption added to protect against chosen plaintext attack and collision
- Integrity f9 uses Kasumi to form CBC MAC with
  - Nonstandard addition of second feedforward

  Mitsubishi Electric Corporation, Japan, holds the rights on essential patents on the algorithms.

  Therefore, the Beneficiary must get a separate royalty-free IPR License Agreement from Mitsubishi.

  Basically Kasumi is a block cipher that produces a 64-bit output from a 64-bit input under the control of a 128-bit key. A detailed description can be found in the *3GPP Specification TS 35.202*. Misty1 and Kasumi have been widely studied since its publication, but no serious flaws have been found.

1.6.5 Integrity – Air Interface Integrity Mechanism

Most control signaling information elements that are sent between the UE and the network are considered sensitive and must be integrity-protected. Integrity protection shall apply at the RRC layer. A message integrity function (f9) shall be applied on the signaling information transmitted between the UE and the RNC. User data are, on the other hand, not integrity-protected and it is up to higher level protocols to add this if needed. Integrity protection is required, not optional, in UMTS for signaling messages.

  After the RRC connection has been established and the security mode set up procedure has been performed, all dedicated control signaling messages between UE and the network shall be integrity-protected (Figure 1.38).

Threats Against Integrity

Manipulation of messages is the one generic threat against integrity. This includes deliberate or accidental modification, insertion, replaying, or deletion by an intruder.

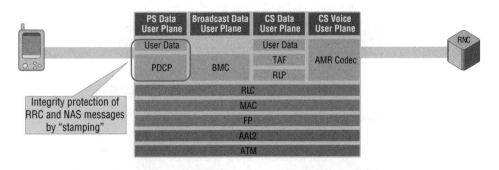

Figure 1.38 Iub control plane.

Both user data and signaling/control data are venerable to manipulation, and the attacks may be conducted on the radio interface, in the fixed network, or on the terminal and the USIM/UICC.

The threats against integrity can be summarized to:

- *Manipulation of transmitted data*: Intruders may manipulate data transmitted over all reachable interfaces.
- *Manipulation of stored data*: Intruders may manipulate data that are stored on system entities, in the terminal, or stored by the USIM. These data include the IMEI stored on the terminal, and data and applications downloaded to the terminal or USIM. Only the risks associated with the threats to data stored on the terminal or USIM are regarded to be significant, and only the risk for manipulation of the IMEI is regarded as being of major importance.
- *Manipulation by masquerading*: Intruders may masquerade as a communication participant and thereby manipulate data on any interface. It is also possible to manipulate the USIM behavior by masquerading as the originator of malicious applications or data downloaded to the terminal or USIM.

On the radio interface this is considered to be a major threat, whereas manipulation of the terminal or USIM behavior by masquerading as the originator of applications and/or data is considered to be of medium significance. Masquerading could be done both to fake a legal user and to fake an SN.

Distribution of Keys
The integrity protection in UMTS is implemented between the RNC and the UE. Therefore, IK must be distributed from the AuC to the RNC. The IK is part of an AV which is sent to the SN (VLR/SGSN) from the AuC following an authentication data request. To facilitate subsequent authentications, up to five AVs are sent for each request. The IK is sent from the VLR/SGSN to the RNC as part of a RANAP message called security mode command.

Integrity Function f 9
The function f 9 is used in a similar way as the AUTN. It adds a "stamp" to messages to ensure that the message is generated at the claimed identity, either the USIM or the SN, on behalf of the HE. It also makes sure that the message has not been tampered with.

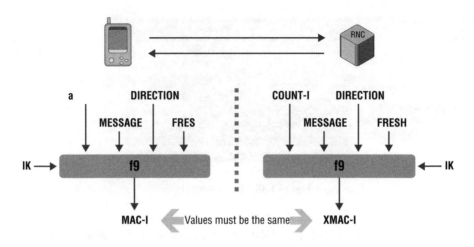

Figure 1.39   Integrity check procedure.

The input parameters to the algorithm are the Integrity Key (IK), the integrity sequence number (COUNT-I), a random value generated by the network side (FRESH), the direction bit DIRECTION, and the signaling data MESSAGE. On the basis of these input parameters the user computes MAC for data integrity MAC-I using the integrity algorithm f9. The MAC-I is then appended to the message when sent over the radio access link. The receiver computes XMAC-I on the message received in the same way as the sender computed MAC-I on the message sent and verifies the data integrity of the message by comparing it to the received MAC-I (Figure 1.39).

Protection against replay is important and guaranteed with:

- The value of COUNT-I is incremented for each message, while the generation of a new FRESH value and initialization of COUNT-I take place at connection setup.
- The COUNT-I value is initialized in the UE and therefore primarily protects the user side from replay attacks. Likewise the FRESH value primarily provides replay protection for the network side.

Integrity Initiation – Security Mode Setup Procedure

The VLR/SGSN initiates integrity protection (and encryption) by sending the RANAP message security mode control to the SRNC. This message contains a list of allowed integrity algorithms and the IK to be used. Since the UE can have two ciphering and integrity key sets (for the PS and CS domains, respectively), the network includes a CN type indicator in the security mode command message.

The security mode command to UE starts the downlink integrity protection; i.e., all subsequent downlink messages sent to the UE are integrity-protected. The security mode complete from UE starts the uplink integrity protection; i.e., all subsequent messages sent from the UE are integrity-protected. The network must have the "UE security capability" information

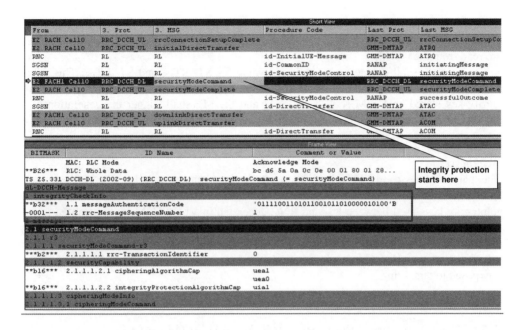

Figure 1.40   Example of "stamped" message for integrity check.

before the integrity protection can start; i.e., the "UE security capability" must be sent to the network in an UMTS security–integrity protection unprotected message. Returning the "UE security capability" to the UE in a protected message later will allow UE to verify that it was the correct "UE security capability" that reached the network.

Some messages does not include integrity protection (Figure 1.40); these messages are:

- HANDOVER TO UTRAN COMPLETE
- PAGING TYPE 1
- PUSCH CAPACITY REQUEST
- PHYSICAL SHARED CHANNEL ALLOCATION
- RRC CONNECTION REQUEST
- RRC CONNECTION SETUP
- RRC CONNECTION SETUP COMPLETE
- RRC CONNECTION REJECT
- RRC CONNECTION RELEASE (CCCH only)
- SYSTEM INFORMATION (BROADCAST INFORMATION)
- SYSTEM INFORMATION CHANGE INDICATION
- TRANSPORT FORMAT COMBINATION CONTROL (TM DCCH only)

Key Lifetime
To avoid attacks using compromised keys, a mechanism is needed to ensure that a particular integrity key set is not used for an unlimited period of time. Each time an RRC connection is

released, the values START$_{CS}$ and START$_{PS}$ of the bearers that were protected in that RRC connection are stored in the USIM. When the next RRC connection is established these values are read from the USIM.

The operator shall decide on a maximum value for START$_{CS}$ and START$_{PS}$. This value is stored in the USIM. When the maximum value has been reached, the cipher key and integrity key stored on USIM shall be deleted, and the ME shall trigger the generation of a new access link key set (a cipher key and integrity key) at the next RRC connection request message.

Weaknesses

The main weaknesses in UMTS integrity protection mechanisms are:

- Integrity keys used between UE and RNC generated in VLR/SGSN are transmitted unencrypted to the RNC (and sometimes between RNCs)
- Integrity of user data is not offered
- For a short time during signaling procedures, signaling data are unprotected and hence exposed to tampering.

1.6.6 Confidentiality – Encryption (Ciphering) on Uu and Iub

Threats Against Confidentiality

There are several different threats against confidentiality-protected data in UMTS. The most important threats are:

- *Eavesdropping* on user traffic, signaling, or control data on the radio interface
- *Passive traffic analysis*: Intruders may observe the time, rate, length, sources, or destinations of messages on the radio interface to obtain access to information
- *Confidentiality of authentication data in the UICC/USIM*: Intruders may obtain access to authentication data stored by the service provider in the UICC/USIM

The radio interface is the easiest interface to eavesdrop, and should therefore always be encrypted. If there is a penetration of the cryptographic mechanism, the confidential data would be accessible on any interface between the UE and the RNC. Passive traffic analysis is considered as a major threat. Initiating a call and observing the response, active traffic analysis, is not considered as a major threat. Disclosure of important authentication data in the USIM, i.e. the long-term secret K, is considered a major threat. The risk of eavesdropping on the links between RNCs and the UICC-terminal interface is not considered a major threat, since these links are less accessible for intruders than the radio access link. Eavesdropping of signaling or control data, however, may be used to access security management data or other information, which may be useful in conducting active attacks on the system.

Ciphering Procedure

Ciphering in UMTS is performed between UE and RNC over air and Iub Interfaces. Figure 1.41 shows the protocol stack of the Iub interface for Rel. 99.

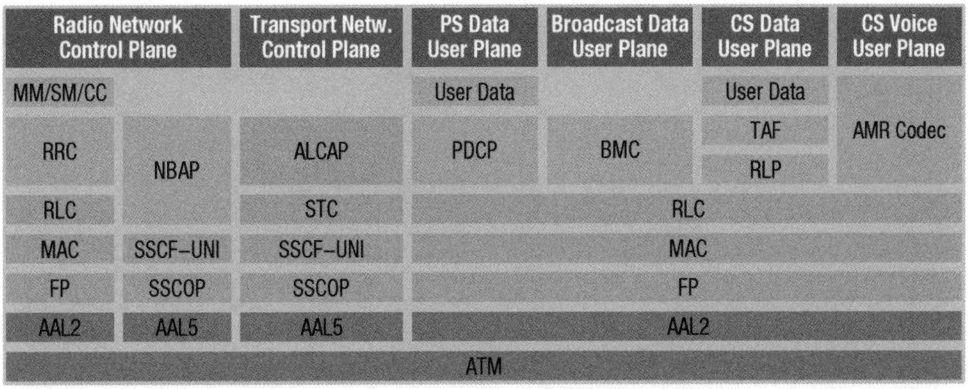

Radio Network Control Plane		Transport Netw. Control Plane		PS Data User Plane	Broadcast Data User Plane	CS Data User Plane	CS Voice User Plane
MM/SM/CC				User Data		User Data	
RRC	NBAP	ALCAP		PDCP	BMC	TAF	AMR Codec
						RLP	
RLC		STC		RLC			
MAC	SSCF–UNI	SSCF–UNI		MAC			
FP	SSCOP	SSCOP		FP			
AAL2	AAL5	AAL5		AAL2			
ATM							

Figure 1.41   Iub Protocol stack.

The Iub protocol stack contains a Radio Network Control Plane, a Transport Network Control Plane, and a User Plane for AMR coded voice, IP packages, video streaming, etc. The Radio Network Control Plane is splitted into two parts, the NAS and the NBAP. The NAS contains Mobility Management (MM), Session Management (SM), and Call Control Management (CC) for communication between UE and CN.

Before UE and RNC are able to exchange NAS messages and user data, one or more transport channel is required. All information related to the establishment, modification, and release of transport channels are exchanged between RNC and Node B over NBAP and ALCAP. Transport channels are based on AAL2 connections. The concept of those transport channels is very important for the understanding of ciphering and integrity protection.

Task of the transport channel is an optimal propagation of signaling information and user data over the air interface. In order to do so, a transport channel is composed of several RBs. The characteristic of every RB is defined during establishment by the NBAP and RRC layer. This is done by a list of attributes, so-called Transport Format Set (TFS). The TFS describes the way of data transmission using different parameters, like block size, transmission time interval (TTI), and channel coding type.

The UTRAN selects for the communication between mobile and network these RABs, which use the radio resources in the most efficient way. Every RAB has its own identifier and every transport block has its own sequence number. This technique allows from one side a fast switchover between Radio Bearers and from the other one a parallel communication over several RABs. This technique requires a bearer-independent ciphering mechanism.

Ciphering will be activated with the messages flow shown in Figure 1.42. Ciphering is always related to a certain transport channel. Therefore, ciphering will be activated independently for control and user planes and independently for packet-switched and circuit-switched planes. In other words, if a mobile subscriber has two independent sessions (voice calls and IP packet transfer) activated, UE and RNC need to exchange the ciphering activation procedure two times. It is important to note that NAS messages exchanged prior ciphering activation (typically the Authentication procedure) are not ciphered.

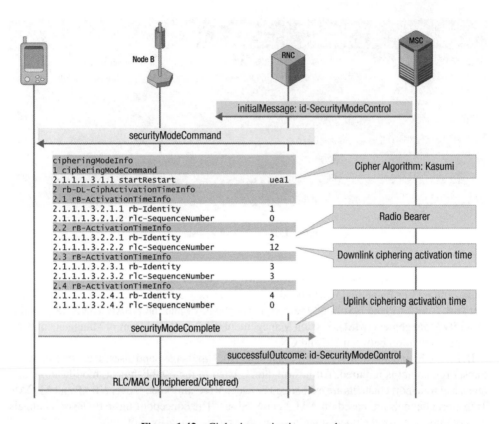

Figure 1.42   Ciphering activation procedure.

Message **securityModeCommand** establishes the Activation Time for the RABs in downlink direction and the message **securityModeComplete** determines the Activation Time in uplink direction. Ciphering for a certain RAB starts for that RLC (Radio Link Control) block where Sequence Number is equal to Activation Time.

The ciphering depth depends on the RLC mode. The RLC protocol contains Control PDUs (Packet Data Units) (never ciphered) and Data PDUs. For Data PDU's, the RLC protocol works in three different modes:

- Unacknowledged Mode (UM)
- Acknowledged Mode (AM)
- Transparent Mode (TM)

UM and AM messages (e.g. Data) are secured against bit errors with a check sequence, while TM information (e.g. AMR voice) isn't. Therefore, RLC UM and RLC AM are ciphered beginning with RLC layer and above, while ciphering for RLC TM already starts with the MAC layer (Figure 1.43).

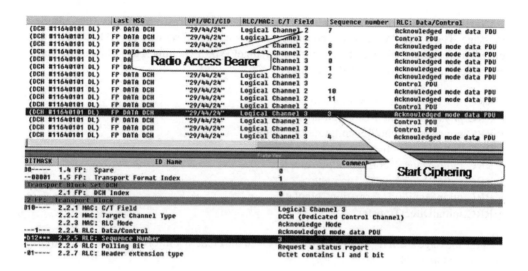

Figure 1.43 RLC: Ciphering Activation Time.

The Kasumi algorithm itself needs the following parameters (Figure 1.44):

- Cipher Sequence Number COUNT
- Direction (uplink or downlink)
- RB Identifier
- Block Length
- Ciphering Key CK

CK is never sent over the Uu and Iub interfaces. The RNC receives this value from MSC or SGSN and the USIM calculates CK as described before (Figure 1.44).

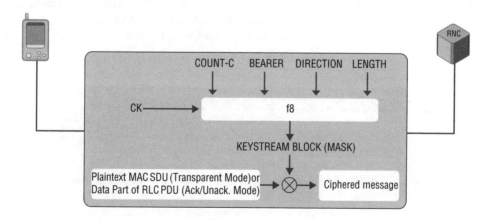

Figure 1.44 UTRAN encryption.

COUNT is initially derived from the START value of the **rrcConnectionSetupComplete** message. The START value is not constant during a ciphering session. It can be modified by different procedures, like Cell Reselection or Channel Type Switching. The following messages can trigger an update of the COUNT value:

- RRC_rrcConnectionSetupComplete
- RRC_physicalChannelReconfigurationComplete
- RRC_transportChannelReconfigurationComplete
- RRC_radioBearerSetupComplete
- RRC_radioBearerReconfigurationComplete
- RRC_radioBearerReleaseComplete
- RRC_utranMobilityInformationComplete
- RRC_initialDirectTransfer

If the message **securityModeFailure** is received the ciphering information shall be removed from USIM and RNC.

Advantages of this method:
1. The key can be generated even before the message is available to the algorithm.
2. To decipher, the receiving side generates the same Keystream Block (Mask) and adds it, bit-by-bit, to the received encrypted message. This second addition of the mask cancels out the mask that was previously added and thereby decrypts the message.

A second bit-by-bit addition negates the first addition = successful deciphering!

Testing UMTS Networks when Ciphering Is Active

As described earlier, ciphering in UMTS networks is also performed between the UE and RNC over the Uu (air) and the Iub interface.

Ciphering causes the RRC and NAS messages to be encrypted (Figure 1.45).

RRC and NAS messages contain key information to perform network optimization and troubleshooting, which results in the fact that when ciphering is active, traditional protocol analyzer and network monitoring systems cannot be used to carry out these two very important tasks.

Radio Network Control Plane	Transport Netw. Control Plane	PS Data User Plane	Broadcast Data User Plane	CS Data User Plane	CS Voice User Plane
Ciphered	NBAP	ALCAP	Ciphered		
		STC			
MAC	SSCF-UN	SSCF-UN	MAC		
FP	SSCOP	SSCOP	FP		
AAL2	AAL5	AAL5	AAL2		
ATM			ATM		

Figure 1.45   Ciphered Iub protocol stack.

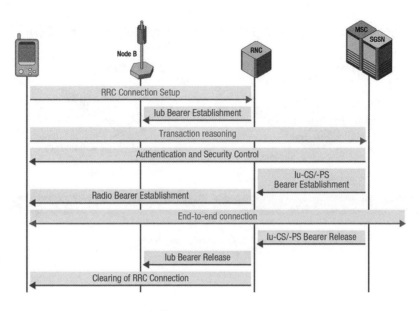

Figure 1.46 UMTS network transactions.

In UMTS networks, in order to perform network optimization and troubleshooting, protocol test equipment would need the ability to decipher the messages. As shown here for the Iub interface, connected to the Iu and Iub, protocol analyzers collect the ciphering parameters, feed them to the deciphering algorithm, and allow full access to the content of the protocol messages. In addition to network optimization and troubleshooting, the equipment also enables the testing of the impact of Iub ciphering/deciphering on network element/network behavior and performance.

Please see Chapter 2 for a short introduction to network monitoring, troubleshooting, and network optimization.

1.6.7 UMTS Network Transactions

Figure 1.46 shows the order of the necessary transactions of a connection. It further indicates the interworking of pure signaling exchange and Radio Bearer procedures.

The procedures running between UE, Node B, and RNC will exchange Access Stratum messages whereas procedures going through to the CN, MSC and SGSN, will exchange NAS messages.

1.7 Radio Interface Basics

To understand the relation between UTRAN signaling messages and the UE, it is also necessary to discuss some procedures and methods used on UMTS air interface, which is done in this section.

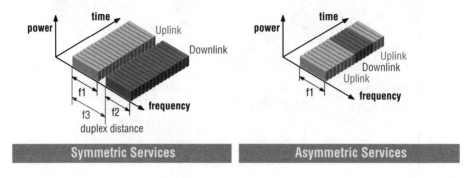

Figure 1.47   Duplex methods.

1.7.1 Duplex Methods

Duplex methods are used to separate transmit and receive signals, for example, speak and listen signals.

Two different methods of duplex control are used on the radio interface. By these methods it is guaranteed that TX and RX data can be separated from each other. These methods have no limits for parallel usage of the radio interface (Figure 1.47).

One method is Frequency Division Duplex (FDD). It provides an uplink and downlink radio channel between network and user, and frequencies are separated by a duplex spacing. Users tune in between uplink and downlink frequencies to transmit and receive signals, respectively. FDD is also used in GSM, where the unidirectional frequency is 200 kHz.

The other method is Time Division Duplex (TDD). A common carrier is shared between uplink and downlink and resources are switched in time. Users are allocated to one or more time slots for uplink and downlink transmission. The main advantage of TDD operation is that it allows an asymmetric flow which is more suitable for data transmission.

In UMTS, these methods will be used as UTRA-FDD and later as UTRA-TDD. The bandwidth of f1 and f2 will be 5 MHz, and the duplex distance will be 190 MHz.

1.7.2 Multiple Access Methods

The Multiple Access Methods feature specifies how user signals can be separated from each other. Again, there is no overall capacity of a cell or a radio access system that could be derived from this method (Figure 1.48).

Multiplex methods are used to divide the limited resources of a cell between the different MSs in a cell.

FDMA   Uses different frequencies to separate the users. This technique is used in analogue systems.

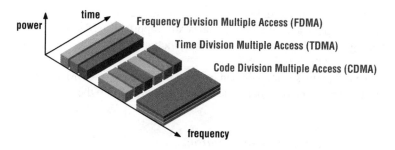

Figure 1.48   Multiple Access Method.

TDMA   Uses different time slots over the whole frequency to separate the users. In this case, different users use the air interface resources at different times. This technique is used in GSM.

CDMA   Uses the whole frequency bandwidth over the whole time. Using different codes applied to their data separates different users. This will be used in UMTS.

For network operators, the difference in planning is that for FDMA and TDMA, frequency planning is the major task, whereas for CDMA, code planning is the major task.

1.7.3 UMTS CDMA

The tasks that result from the CDMA technique are mainly implemented in Node B and in the UE (Figure 1.49).

The following work steps must be performed before the signal can be transmitted via the antenna:

- Spreading of the data with OVSF codes
- Scrambling of the spread stream with scrambling codes
- Modulation of the digital signal onto the air interface

OVSF   Orthogonal Codes with Variable Spreading Factor

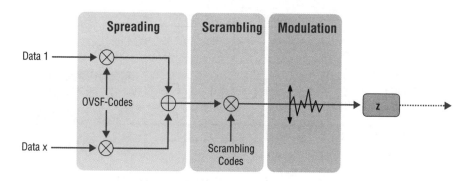

Figure 1.49   UMTS CDMA.

The receiver will have to perform these steps in reverse order.

Since spreading codes and scrambling codes are important to identify UTRAN signaling messages belonging to a defined user, a short introduction into these techniques is given, while modulation is outside the scope of this book.

However, the following section will demonstrate the process for CDMA-FDD only, because TDD is close to implementation, but typically not introduced into the networks yet.

1.7.4 CDMA Spreading/Channelization

CDMA can use different methods of spreading channelization:

- Direct Sequence CDMA (DS-CDMA)
- Frequency Hopping CDMA (FH-CDMA)
- Time Hopping CDMA (TH-CDMA)
- Hybrid Modulation CDMA (HM-CDMA)
- Multi-Carrier CDMA (MC-CDMA)

UMTS will use, in the first stage, the DS-CDMA technique (Figure 1.50).

Every bit of the data (*symbol*) stream will be spread (coded) by a number of code bits (*chips*). By this, the data stream becomes a chip stream with the length:

data bits × code chips

The input data rate is also called symbol rate.

For the spreading, the data bit values have to be turned to nonreturn to zero (NRZ) codes: for example, +1 or −1.

Binary zero is presented as +1 and binary one is presented as −1.

Multiplying the code to the bit using the XOR function performs the spreading. As can be seen, the chip stream is a picture of the code; i.e., if a binary zero has to be spread, the chip stream is the code. If a binary one has to be spread, the chip stream is the inverted code.

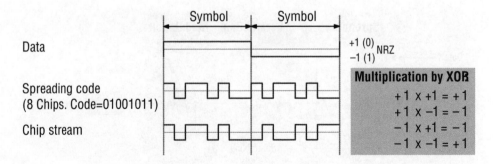

Figure 1.50   Spreading using Direct Sequence CDMA.

One of the main reasons for spreading is to convert a narrowband signal to a wideband signal, nearly as wide as the radio interface frequency band.

In UMTS, the chip stream has always the size of 3,840,000 chips/s, for example 3.84 Mcps, equal to a frequency of 3.84 MHz.

Depending on the data stream variable, spreading codes have to be used. First of all, the value of the code is not important, but its length is.

Secondly, the used codes should be orthogonal; they differ completely from each other. In the uplink direction, the UE separates different data channels from each other by using different codes for each data channel.

1.7.5 Microdiversity – Multipath

The transmission of a radio wave is not straight. Because of reflection, diffraction, and scattering of the radio wave, the received signal appears as a multiple of the sent signal, different in time. This phenomenon is called *Multipath* (Figure 1.51).

In UMTS, it means that the UE and the Node B receive multiple signals from each other. A special *RAKE receiver* is implemented in both units to overcome this problem. It receives each of the parallel signals in a finger and combines them to one strong output signal, which will be given to the higher layer.

Microdiversity stands for the small diversity the receiver has to deal with.

1.7.6 Microdiversity – Softer Handover

A special case where microdiversity is used is the *Softer Handover*. In this situation the UE is connected to more than one sectors of a Node B (Figure 1.52). The advantage is to get a stronger RX signal. The disadvantage is that more radio resources are in use than necessary. It is up to the network planning if and when this feature is used.

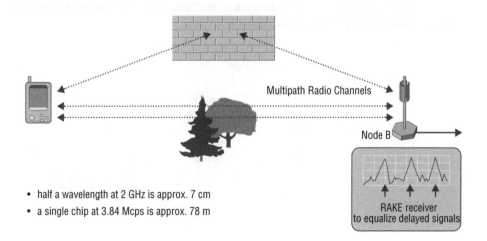

- half a wavelength at 2 GHz is approx. 7 cm
- a single chip at 3.84 Mcps is approx. 78 m

Figure 1.51   Multipath.

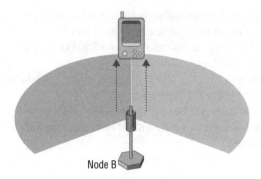

Figure 1.52 Softer Handover.

1.7.7 Macrodiversity – Soft Handover

The function of *Macrodiversity* is to collect data from one UE coming into the network via different Node Bs. Macrodiversity is implemented in the SRNC. The maximum number of parallel serving Node Bs in Rel. 99 is 3, but maybe increased in further releases of UMTS standards.

The described situation is called *Soft Handover* (Figure 1.53). It will again use more resources than necessary for a single connection not only on the radio interface, but also in the UTRAN on the different Iub and Iur interfaces. The advantage is that in case of transmission errors on one radio link there is a high chance to get the same frame error-free on a different link. The SRNC compares the incoming messages from all links and selects the error-free frames. So it is prevented that the Node Bs needs to change their transmission power to keep contact with the UE that is close to the cell border. A change of transmission power could cause interference of the neighborhood cells or cell breathing effects.

In the downlink direction, several Node Bs may send data to the UE, but the UE will only receive the data of the sender with the strongest RX signal.

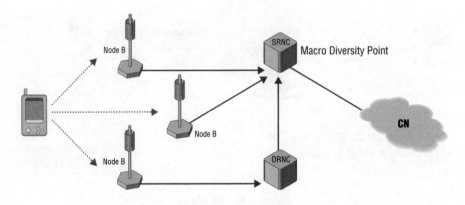

Figure 1.53 Soft Handover.

Data rate (After channel coding)	SF	Chip rate
960 kbit/s	4	3.84 Mcps
480 kbit/s	8	3.84 Mcps
240 kbit/s	16	3.84 Mcps
120 kbit/s	32	3.84 Mcps
60 kbit/s	64	3.84 Mcps
30 kbit/s	128	3.84 Mcps
15 kbit/s	256	3.84 Mcps
7.5 kbit/s	512	3.84 Mcps

FDD Example:

A Call requires a 12.2 kbit/s voice channel. With special channel coding it will increase up to 30 kbit/s

Looking into the table will indicate to use SF=128 (C_{128}).

Figure 1.54   UMTS spreading.

1.7.8 UMTS Spreading

Figure 1.54 lists possible Spreading Factor (SF) values both for CDMA forms and for the uplink and downlink directions. The table within this figure also shows the SFs that should apply for certain data rates to reach 3.84 Mcps.

Possible SF:

- FDD UL: $4 - 8 - 16 - 32 - 64 - 128 - 256$
- FDD DL: $4 - 8 - 16 - 32 - 64 - 128 - 256 - 512$
- TDD : $1 - 2 - 4 - 8 - 16$

Transmission of pure signaling information should always use SF=256.

1.7.9 Scrambling

Scrambling describes the multiplication of another code to the chip stream without changing its length and is done to remove the quasi-orthogonal signals from different users and to identify different sources:

Scrambling in Uplink

- Short scrambling codes (256 bits) are used in Node B if there is advanced multiuser detection or an interference cancellation receiver.
- Long scrambling codes (38,400 bits) are used if the RAKE receiver implemented in the Node B.

Scrambling in Downlink

- Long scrambling codes (38,400 bits) are used.

Note: Scrambling does not spread the chip stream.

Table 1.4 Channelization and scrambling

	Channelization	Scrambling
Usage	Uplink: Separation of physical data (DPDCH) and control channels (DPCCH) from same terminal Downlink: Separation of connections to different users within one cell	Uplink: Separation of terminals Downlink: Separation of sectors (cells)
Length	4-256 chips (1.0–66.7 μs) Downlink also 512 chips	38.400 chips (10 ms) Uplink also 256 chips (66.7 μs)
Number of codes	Spreading factor dependent	Uplink: several millions Downlink: 512
Code family	Orthogonal Variable Spreading Factor	Long 10 ms code: Gold code Short code: Extended S(2) code family
Spreading	Yes, increase transmission bandwidth	No

A *Scrambling code* is a random code called *Gold code*, and because of their random appearance, they are also called *Pseudo-Noise (PN) codes*.

Scrambled signals of different users are orthogonal to each other again. Scrambling codes are of length 38,400 bits (*long scrambling code*). With evolved Node Bs *short scrambling codes*, 256 bits will be used.

1.7.10 Coding Summary

Table 1.4 and Figure 1.55 gives an overview of channelization and scrambling. In uplink and downlink, these codes have different meaning as it is decribed in the figure.

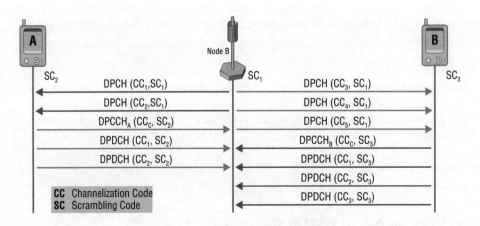

Figure 1.55 Channelization and scrambling.

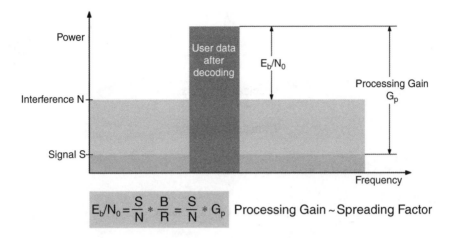

$$E_b/N_0 = \frac{S}{N} * \frac{B}{R} = \frac{S}{N} * G_p \quad \text{Processing Gain} \sim \text{Spreading Factor}$$

Figure 1.56 Signal to Interference. B: Bandwidth of radio interface; R: User data rate.

1.7.11 Signal to Interference

Every user is an interference source to all other users in one cell (also in neighboring cells). To guarantee the success of the request QoS, a special ratio has to be calculated: E_b/N_0. This value represents the ratio between the *energy of one signal* (bit) compared to the interference at the receiver.

The value is the SIR multiplied by the Processing Gain, which is more or less the SF (Figure 1.56).

If for any reason E_b/N_0 gets too low, one way of increasing the ratio is to increase the SF. With a fixed chip stream rate of 3.84 Mcps, the SF cannot just be increased. So the data rate also has to be changed; the data rate must be decreased and then the SF can be increased.

1.7.12 Cell Breathing

Cell breathing describes a constant change of the range of a geographical area covered by a Node B cell based on the amount of traffic currently using that transmitter. When a cell becomes heavily loaded, it shrinks. Subscriber traffic is then redirected to a neighboring cell that is more lightly loaded, which is called load balancing. Cell breathing is a common phenomenon of 2G and 3G wireless systems including CDMA (Figure 1.57).

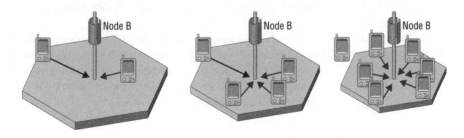

Figure 1.57 Cell breathing.

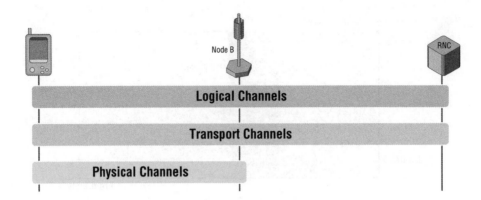

Figure 1.58   UMTS channels.

The cause of cell breathing is the given QoS. The QoS then defines/causes Eb/N0, limited bandwidth, and limited TX power.

Part of the cell breathing is also the *Near-Far-Effect*, where users who are closer to a Node B use less TX power than users who are further away from the Node B. The reason for this cell breathing effect is the fact that the RX power should, ideally, be the same for all users; for example, the SIR should be the same for all users.

Summary
- Every service requires a certain Eb/N0 ratio (QoS).
- Received SIR should be the same for all users in a cell.
- Users with longer distance to Node B than others must use higher transmit power.
- Users with higher data rates (smaller SF) must use higher transmit power.
- If interference increases, the signal power must be increased.
- Signal power can only be increased to a maximum ($\sim$0.5 W).
- Result: the "usable" area of a cell shrinks!

1.7.13 UMTS Channels

Three types of UMTS channel levels (Figure 1.58) are defined.[3]

Physical Channels
Each Physical Channel is identified by its frequency, spreading code, scrambling code, and phase of the signal. Physical Channels provide the bearers for the different transport channels (see overviews below).

Dedicated Physical Channels identify a destination UE by SF and scrambling code. One or more Dedicated Physical Data Channels (DPDCHs) can be configured in uplink or downlink direction. The DPCCH is used for radio interface related control information only. One Dedicated Physical Control Channel (DPCCH) always belongs to the set of DPDCHs and is used for RRC messages and other signaling between UE and network.

---

[3] *3GPP 25.301; 3GPP 25.302; 3GPP 25.211.*

Transport Channels

Transport Channels are unidirectional virtual channels, mapped onto physical channels. They provide bearers for information exchange between MAC protocol and physical layer. Only transport channels of one type (e.g. DCHs) are mapped.

Logical Channels

Logical Channels are uni- or bidirectional and provide bearers for information exchange between MAC protocol and RLC protocol. There are two types of Logical Channels:

- Control Channels for signaling information of the control planes
- Traffic Channels for user data of the user planes

Table 1.5 gives all physical channels available in a UMTS network.

Different Types of Physical Channels in UTRA-FDD

Dedicated Physical Data Channel (DPDCH): Transmission of user data and higher layer signaling (RRC, NAS) in uplink direction coming from higher layers.

Dedicated Physical Control Channel (DPCCH): Transmission of radio control information in uplink direction. This channel exists only once per radio connection.

Dedicated Physical Channel (DPCH): Transmission of user data and control information in downlink direction. Both types of information will be mapped onto the DPCH.

Synchronization Channel (SCH): Cell search and synchronization of the UE to the Node B signal. Subdivided into Primary Synchronization Channel (P-SCH) and Secondary Synchronization Channel (S-SCH).

Table 1.5 Physical channels in UMTS

Channel	Abbreviation	Direction	Duplex mode
Dedicated Physical Data Channel	DPDCH	Uplink	FDD
Dedicated Physical Control Channel	DPCCH	Uplink	FDD
Dedicated Physical Channel	DPCH	Downlink	FDD, TDD
		Uplink	TDD
Synchronization Channel	SCH	Downlink	FDD
Primary Synchronization Channel	P-SCH	Downlink	FDD
Secondary Synchronization Channel	S-SCH	Downlink	FDD
Common Control Physical Channel	CCPCH	Downlink	FDD, TDD
Primary Common Control Physical Channel	P-CCPCH	Downlink	FDD
Secondary Common Control Physical Channel	S-CCPCH	Downlink	FDD
Common Pilot Channel	CPICH	Downlink	FDD
Physical Random Access Channel	PRACH	Uplink	FDD, TDD
Physical Common Packet Channel	PCPCH	Uplink	FDD
Paging Indicator Channel	PICH	Downlink	FDD
Acquisition Indicator Channel	AICH	Downlink	FDD
Physical Downlink Shared Channel	PDSCH	Downlink	FDD, TDD
Physical Uplink Shared Channel	PUSCH	Uplink	TDD

Common Control Physical Channel (CCPCH): Transmission of common information and is divided into Primary Common Control Physical Channel (P-CCPCH) and Secondary Common Control Physical Channel (S-CCPCH). P-CCPCH transmits the broadcast channel (BCH) and S-CCPCH transports the Forward Access Channel (FACH) and the Paging Channel (PCH). FACHs and PCH can be mapped to the same or to separate S-CCPCHs.

Common Pilot Channel (CPICH): Supports channel estimation and allows estimations in terms of power control. It is subdivided into Primary Common Pilot Channel (P-CPICH) and Secondary Common Pilot Channel (S-CPICH), which differ in scrambling code and availability within a cell.

Physical Random Access Channel (PRACH): Transmission of the Random Access Channel (RACH), which is used for the random access of a UE and for transmission of a small amount of data in the uplink direction.

Physical Common Packet Channel (PCPCH): Common data transmission using the collision detection CSMA/CD method.

Paging Indicator Channel (PICH): Transmission of the Page Indicator (PI) to realize the paging in the downlink direction. One PICH is always related to an S-CCPCH, which transports the PCH.

Acquisition Indicator Channel (AICH): Transmits the positive acknowledgment of a random access of a UE via PRACH or PCPCH.

Physical Downlink Shared Channel (PDSCH): Common transmission of data in downlink direction. Parallel UEs will have different codes assigned.

Different Types of Physical Channels in UTRA-TDD

Dedicated Physical Channel (DPCH): Bidirectional transmission channel for user data and control information.

Common Control Physical Channel (CCPCH): Same as in FDD mode.

Physical Random Access Channel (PRACH): Same as in FDD mode.

Physical Uplink Shared Channel (PUSCH): Common transmission of data and control information in the uplink direction. Parallel UEs will have different codes assigned.

Physical Downlink Shared Channel (PDSCH): Common transmission of data in downlink direction.

Paging Indicator Channel (PICH): Same as in FDD mode.

1.7.14 Transport Channels

W-CDMA (*3GPP 25.302, 3GPP 25.211–25.215*) is interworking with the higher layer, Media Access Control (MAC) protocol.

It offers the transport channels to the MAC. To be flexible in data rates, etc., all information on the transport channel is described by transport formats and certain attributes.

1.7.15 Common Transport Channels

Common Transport Channels can be used by all UEs located in the same cell. A special identifier, the so-called RNTI (Radio Network Temporary Identity), is used to mark messages coming from or sent to a single UE on RACH, FACH or shared channels.

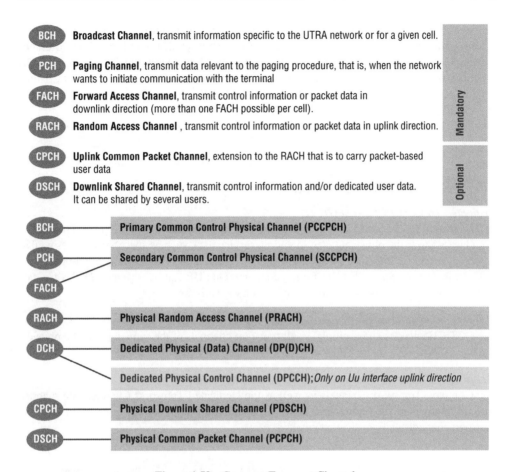

Figure 1.59   Common Transport Channels.

To the Common Transport Channels belong (Figure 1.59):

Broadcast Channel (BCH): Transmits system information.
(Mandatory)
Paging Channel (PCH): Calls a UE, which has no RRC connection.
(Mandatory)
Forward Access Channel (FACH): Transmits a small amount of data in the downlink direction.
   There can be multiple FACHs in one cell with different bandwidths.
(Mandatory)
Random Access Channel (RACH): Transmits the acknowledgment to a Paging Request and
   transmits a small amount of data in the uplink direction.
(Mandatory)
Uplink Common Packet Channel (CPCH): Transmits a small number of data packets in the
   uplink direction. The difference from RACH are fast power control, collision detection, and
   a status monitoring function.
(Optional)

Downlink Shared Channel (DSCH): Transmits a small number of user data packets or control information in the downlink direction. It is shared between different users. The difference from FACH are fast power control and a variable bitrate on a Frame-by-Frame base. DSCH is not mandatory in every cell, but, if it exists, it is related to a Dedicated Transport Channel [similar to GSM Associated Control Channel (ACCH)].

Shared Channels requires a parameter to identify a UE, the RNTI.

The DSCH is always related to DCHs: several DCHs can be mapped into one DSCH. (Optional)

1.7.16 Dedicated Transport Channels

Dedicated Transport Channel (DCH)

DCHs are used for the transport of user data and control information for a particular UE coming from layers above the physical layer, including service data, such as speech frames, as well as higher layer control information, such as handover commands or measurement reports.

There is no need for a UE identifying parameter. One UE can have several DCHs for data transmission but only one for control information transmission.

Coded Composite Transport Channel (CCTrCH)

The CCTrCH encodes and multiplexes all transport channels of the same type on the physical layer.

Mapping of Transport Channels onto Physical Channels

The Common Transport Channels as well as the Dedicated Transport Channels are mapped onto physical channels. Figure 1.60 shows an example of the relationship between different transport channels and physical channels.

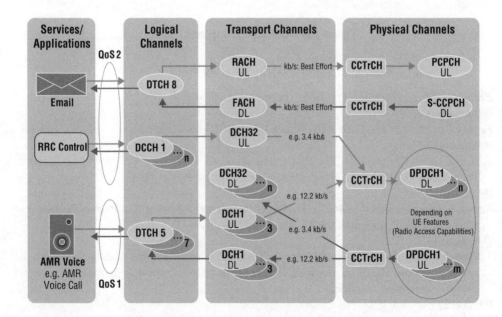

Figure 1.60   Mapping of transport channels.

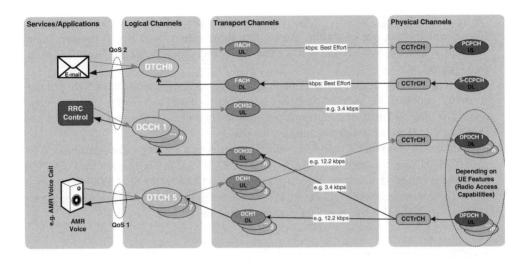

Figure 1.61   Channel mapping example.

1.7.17 Initial UE Radio Access

If a UE is switched on the first time in a cell of the UMTS network it starts to perform the following Initial UE Radio Access procedure that can be described in four steps (Figure 1.62):

1. UE reads the Primary Synchronization Channel, which is not scrambled and spread by a predefined spreading code (SF=256). By reading this, the UE become time synchronic with the Node B.

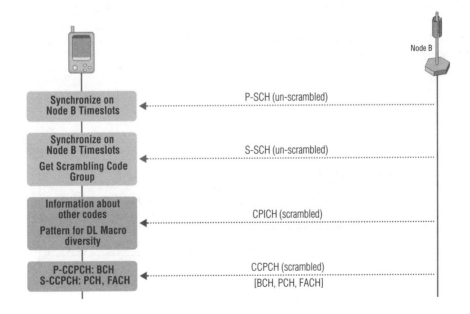

Figure 1.62   Initial UE Radio Access.

2. UE reads the S-SCH, which is also not scrambled. The S-SCH will transmit five hex values, which come out of a table. By reading these values the UE will become frame synchronic with the Node B and will get the scrambling group the actual Node B is using (see Figure 1.54).
3. UE can now read the Common Pilot Channel, which is scrambled with one of eight primary scrambling codes of the scrambling group. It is a matter of trial and error to find the correct code. The Pilot Channel will contain further information about other necessary codes and about the DL Macrodiversity synchronization pattern.
4. UE will read the Common Control Physical Channel, which uses the same scrambling code as the CPICH, to get detailed information about UTRAN and the CN, to allow the P-CCPCH to transport the BCH, and to be able to get paged, and to allow the S-CCPCH to transport PCH. The system information in the BCH will also indicate the secondary scrambling code of the actual Node B for further data transmission on the DCHs.

1.7.18 Power Control

Because of the fact that the SIR should be the same for all users in a cell, the demand for power control in UMTS is very high. Two forms of power control exist in UMTS (Figure 1.63).

Open Loop Power Control is a kind of one-way power control used before the UE is connected to the RRC and describes the ability of the UE transmitter to set the output power to a specific value for initial *uplink* and *downlink* transmission powers. The power control tolerance is ±9dB (normal conditions) or ±12dB (extreme conditions).

- SIR should be the same for all users in a cell
- Each user produces a signal, which, to other users, is just noise
- Received Signal = S (User 1) + $\sum$ N (User n − 1)
- The goal must be to keep Signal S at a minimum so that the noise will be low

Closed Loop Power Control is performed when the UE has an RRC connection. It contains an *Inner and Outer Loop Power Control* mechanism.

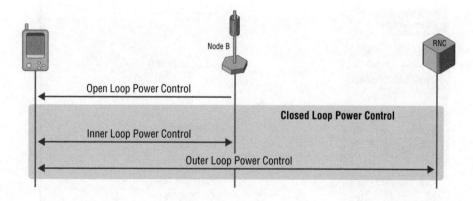

Figure 1.63   Power control.

Inner Loop Power Control (1500 Hz) runs in the uplink and describes the ability of an UE transmitter to adjust output power in accordance with one or more Transmit Power Control (TPC) commands of the downlink. The received uplink SIR shall be kept at a given SIR target. UE transmitters might change the output power (step size of 1, 2, and 3 dB) in the slot after TPC_cmd was derived. Serving cells estimate SIR of received uplink DPCH, generate TPC commands (TPC_cmd), and transmit the commands once per slot according to:

$$\text{If } SIR_{est} > SIR_{target} \rightarrow \text{TPC command is "0"}$$

$$\text{If } SIR_{est} < SIR_{target} \rightarrow \text{TPC command "1"}$$

After reception of TPC command, the UE derives a TPC command for each slot. The UE-specific higher layer parameter **PowerControlAlgorithm** determines which of the two algorithms is used for the evaluation.

Outer Loop Power Control maintains the quality of communication for bearer service quality requirements, using power as low as possible. Uplink outer loop power control takes care of setting a target SIR in Node Bs for individual uplink inner loop power control. This target SIR is updated for each UE according to the estimated uplink quality (BLock Error Ratio, Bit Error Ratio) for each RRC connection. Downlink outer loop power control describes the ability of the UE receiver to converge to required link quality (BLER) defined by the network (RNC) in downlink.

Open Loop Power Control
By receiving the CPICH and the BCCH information parameters on the BCH, the UE can estimate a TX power. The stronger the RX signal, the less the TX power will be (Figure 1.64).

Closed Loop Power Control
After finding out what type of service the UE wants to get, the SRNC will define a QoS target for the Radio Bearer (SIR). Node B will store the target SIR and will compare it with the actual measurements of that UE. The result of the comparison will be given to the SRNC, which in turn will send a new target. The communication between Node B and SRNC is called the

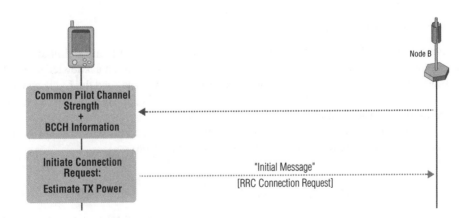

Figure 1.64   Open Loop Power Control.

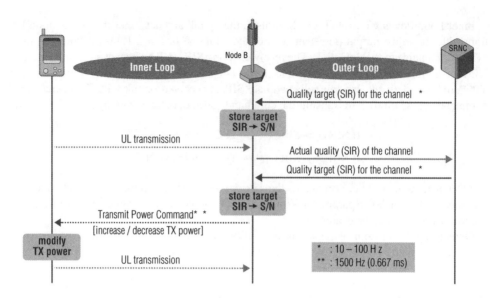

Figure 1.65   Closed Loop Power Control.

Outer Loop Power Control and will be performed between 10 and 100 times per second. This
is why this method is called *Slow Power Control*!

On the other side, the Node B must control the UE TX power to reach the given SIR. Node
B sends TPC commands to the UE to indicate either to increase or to decrease the TX power.
UE will have to modify its TX power immediately. This method is called the Inner Loop Power
Control and is performed up to 1500 times per second, and thus called *Fast Power Control*
(Figure 1.65).

Power control mechanisms will become a very important part of network optimization in
the future, but in current state of deployment there is still only little experience in this field for
network operators.

1.7.19 UE Random Access

After estimating the TX power (Open Loop Power Control) the UE will send an Initial Access
frame on the Physical Random Access Channel. It will then wait for an acknowledgment. If
there is no acknowledgment, then the UE will increase TX power and send the frame again. It
will perform this until it receives an Access Detected message via the AICH or until it reaches
the maximum value for TX power.

Now the UE knows about good TX power strength and will send the real Random Access
Information containing the RRC Connection Request (Figure 1.66).

1.7.20 Power Control in Soft Handover

In Soft Handover, the UE is connected to more than one Node B. All Node Bs will by default
transmit **Transmit Power Command** messages. The rule to follow is that the less the TX

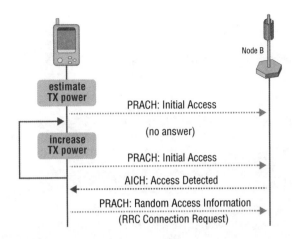

Figure 1.66 UE random access.

power the better! In the example, one Node B indicates to decrease the TX power. This would mean to the UE to decrease the TX power even if it would lose the contact to the other Node Bs. By this rule the Near-Far-Effect cannot become an endless problem.

A special alternative is the *Site Selection Diversity Transmission (SSDT)*. Using this UTRAN option the RNC will get the measurements of the actual radio interface connection toward one UE by several Node Bs and decide that some of the Node Bs should stop transmitting DCHs and also stop transmitting TPC commands to the UE. Only the Node B with the best radio contact will be the UE "server" in downlink (Figure 1.67).

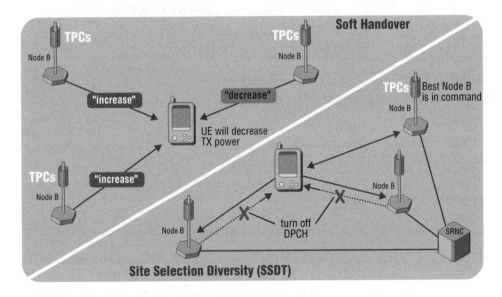

Figure 1.67 Power control in Soft Handover.

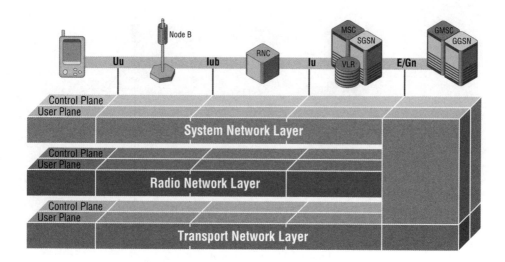

Figure 1.68   UMTS network protocol architecture.

1.8 UMTS Network Protocol Architecture

The protocol architecture of UTRAN (Figure 1.68) is subdivided into three layers:

1. *Transport Network Layer*: Physical and transport protocols and functions to provide AAL2 resources and allow communication within UTRAN and CN. The protocols are not UMTS-specific.
2. *Radio Network Layer*: Protocols and functions to allow management of radio interface and communication between UTRAN components and between UTRAN and UE.
3. *System Network Layer*: NAS protocols to allow communication between CN and UE.

   Each of the layers is divided into a control and a user plane.

Control plane: Transmission of control signaling information.
User plane: Transmission of user data traffic.

   The following sections give an overview about protocol stacks on the different interfaces in UTRAN and the CN. The description of functions of protocol layers, their messages and procedures follows in Chapter 2, which deals with all protocol details.

1.8.1 Iub – Control Plane

The protocol stacks of Uu and Iub interfaces – control plane – contain (Figure 1.69):

ATM   Asynchronous Transfer Mode is used in UMTS as the transmission form on all Iu interfaces. The physical layer is SDH over fiber. The smallest unit in ATM is the ATM cell. It will be transmitted in *Virtual Channel*. Many virtual channels are running within a *Virtual Path*.

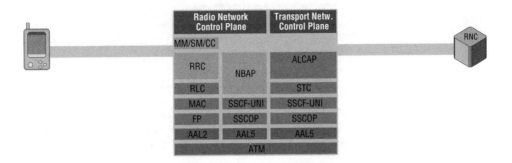

Figure 1.69 Iub – control plane.

AAL   ATM Adaptation Layer
            To transmit higher protocols via ATM, it is required to have adaptation sub-
            layers. These sublayers contain a common adaptation and a service-specific
            adaptation part.

UP FP   User Plane Framing Protocol
            Used on Iur and Iub interfaces to frame channels supported between SRNC
            and Node Bs

SSCOP   Service Specific Connection Oriented Protocol
            Provides mechanisms for establishment and release of connections and reliable
            exchange of signaling information between signaling entities

MAC   Medium Access Control Protocol
            Coordinates access to physical layer. Logical channels of higher layers are
            mapped onto transport channels of lower layers. MAC also selects appropri-
            ate TFSs depending on necessary transmission rate and organizes the priority
            handling between different data flows of one single UE

RLC   Radio Link Control Protocol
            Offers transport services to the higher layers called Radio Bearer Services; the
            three-work modes are transparent, acknowledged, and unacknowledged mode

SSCF   Service Specific Coordination Function (User-Network-I/F, Network-
            Network-I/F)
            Not a protocol but an internal coordination function, which does internal adapta-
            tion of the information coming or going to higher layers, for example, MTP3-B
            routing information

STC   Signaling Transport Converter
            Is an internal function, which has no own messages; it converts primitives
            from lower and higher layers (either MTP3 or MTP3-B primitives) and their
            parameters fitting the requirements of the other

RRC   Radio Resource Control Protocol
            Is a sublayer of Layer 3 on UMTS radio interface and exists in the control plane
            only. It provides information transfer service to the NAS and is responsible for
            controlling the configuration of UMTS radio interface layers 1 and 2

AAL2L3   AAL2 Layer 3 Protocol
            Generic name for transport signaling protocol to set up and release transport
            bearers. In UMTS the main ALCAP protocol is the AAL2 signaling protocol

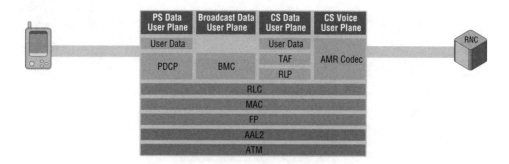

Figure 1.70   Iub – user plane.

NBAP   Node B Application Part
                Protocol is used between RNC and Node B to configure and manage the Node
                B and set up channels on Iub and Uu interfaces
MM   Mobility Management
                A generic term for the specific mobility functions provided by a PLMN in-
                cluding, e.g., tracking a mobile as it moves around a network and ensuring that
                communication is maintained
SM   Session Management protocol is used between UE and SGSN and creates,
                modifies, monitors, and terminates sessions with one or more participants,
                including multimedia and Internet telephone calls
CC   Call Control includes some basic procedures for mobile call control (no trans-
                port control!): Call Establishment, Call Clearing, Call Information Phase, and
                other miscellaneous procedures

1.8.2 Iub – User Plane

The user plane protocol stacks of Uu and Iub interfaces introduce some new layers (Figure 1.70):

PDCP   Packet Data Convergence Protocol
                Is used to format data into a suitable structure prior to transfer over the
                air interface and provides its services to the NAS at the UE or the relay
                at the RNC
BMC   Broadcast/Multicast Protocol
                Adapts broadcast and multicast services on the radio interface and is a
                sublayer of L2 that exists in the user plane only
Application Data IP-based packet protocols

Speech (AMR) will be transported transparently on AAL2.

1.8.3 Iur – User/Control Plane

The Iur interface between RNCs shows two alternative solutions on the transport network
layer: Either SCCP and RNSAP messages can be transported using MTP3-B running on top of
SSCOP, or it is possible to run SCCP on top of M3UA if the lower transport layer is IP-based
(Figure 1.71).

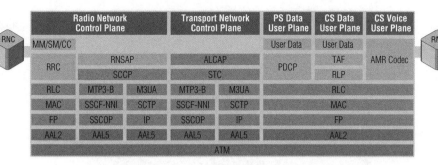

Figure 1.71 Iur – user/control plane.

IP		Internet Protocol
		Provides connectionless services between networks and includes features for addressing, type-of-service specification, fragmentation and reassembly, and security
SCTP		Stream Control Transmission Protocol
		Transport protocol that provides acknowledged error-free nonduplicated transfer of data. Data corruption, loss of data, and duplication of data are detected by checksums and sequence numbers. Retransmission mechanisms are applied to correct loss or corruption of data
MTP3-B		Message Transfer Part Level 3 Broadband
		Fulfills the same sort of work as the standard narrowband MTP; it provides identification and transport of higher layer messages (PDUs), routing, and load sharing
M3UA		MTP Level 3 User Adaptation Layer
		Provides equivalent primitives to MTP3 users as provided by MTP3. ISUP and/or SCCP are unaware that expected MTP3 services are offered remotely and not by local MTP3 layer. M3UA extends access to MTP3 layer services to a remote IP-based application
SCCP		Signaling Connection Control Part
		Provides a service for transfer of messages between any two signaling points in the same or different network
RNSAP		Radio Network Subsystem Application Part
		Communication protocol used on the Iur interface between RNCs and is specified using ASN.1 Packed Encoding Rules (PER)

Speech (AMR) will be transported transparently on AAL2.

1.8.4 IuCS – User/Control Plane

The protocol stack of IuCS interface – control/user plane – contains (Figure 1.72):

AMR	Adaptive Multirate Codec (speech)
	Offers a wide range of data rates and is used to lower codec rates as interference increases on the air interface

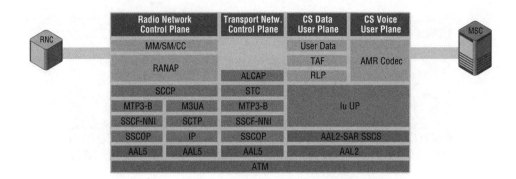

Figure 1.72   IuCS – user/control plane.

TAF   Terminal Adaptation Function (V. and X. series terminals)
        A converter protocol to support the connection of various kinds of TE to the MT
RLP   Radio Link Protocol
        Controls circuit-switched data transmission within the GSM and UMTS PLMN

The CS domain refers to the set of all entities handling the circuit-switched type of user traffic as well as entities supporting the related signaling. These are the MSC, the GMSC, the VLR, and the IWF [InterWorking Function(s)] towards the PSTN/ISDN networks.

1.8.5 IuPS – User/Control Plane

The PS domain includes the related entities for packet transmission, the SGSN, GGSN, and BG (Border Gateway) (Figure 1.73).

Note: The user plane payload (IP-traffic) is transported using AAL5. So there is no ALCAP layer necessary in the control plane to set up and delete switched virtual AAL2 ATM connections.

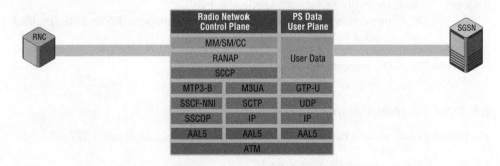

Figure 1.73   IuPS – user/control plane.

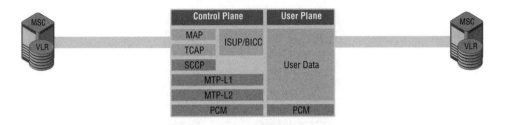

Figure 1.74   E – user/control plane.

1.8.6 E – User/Control Plane

The E interface protocol stack is well-known form GSM environment with both control and user planes (Figure 1.74).

PCM   Pulse Code Modulation
      An analogue signal is encoded into a digital bit stream by first sampling, then quantizing, and finally encoding into a bit stream. The most common version of PCM converts a voice circuit into a 64 kbps stream

TCAP Transaction Capability Application Part
      Enables deployment of advanced intelligence in networks by supporting noncircuit-related information exchange between signaling points using SCCP connectionless service

MAP   Mobile Application Part
      Enables real-time communication between nodes in mobile networks. Example: transfer of location information from VLR to the HLR

ISUP ISDN User Part
      Part of SS7 protocol layer, used for setting up, management, and release of voice calls and data between calling and called parties

MTP2 Message Transfer Part Level 2
      Takes care of reliable transmission through retransmission techniques of signaling units over signaling links

MTP3 Message Transfer Part Level 3
      Represents the highest level of MTP and takes care of the general MTP management and the discrimination, distribution, and routing of signaling messages

Note: The MAP is also able to carry containers with, for example, RANAP and BSSAP messages to exchange these messages between different MSCs in case of inter-MSC or intersystem handover procedures.

1.8.7 Gn – User/Control Plane

The protocol stack on GPRS Gn interface has not changed significantly in comparison with 2.5G networks (Figure 1.75).

GTP-C GPRS Tunneling Protocol – Control
       GTP-C messages are exchanged between GSNs to create, update, and delete GTP tunnels, for path management and to transfer GSN capability information between

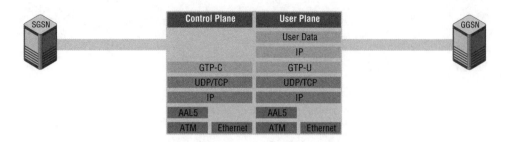

Figure 1.75   Gn – user/control plane.

	GSN pairs. GTP-C is also used for communication between GSNs and the Charging Gateways.
GTP-U	GPRS Tunneling Protocol – User
GTP-U	messages are exchanged between GSN pairs or GSN/RNC pairs for path management and error indication, to carry user data packets and signaling messages
UDP	User Datagram Protocol
	UDP is a connectionless, host-to-host protocol that is used on PS networks for real-time applications
TCP	Transmission Control Protocol provides reliable connection-oriented, full-duplex point-to-point services

1.9 ATM

Asynchronous Transfer Mode (ATM) is used in UMTS as the transmission form on all Iu interfaces. The physical layer is SDH over fiber.

The smallest unit in ATM is the *ATM cell*. It will be transmitted in a *Virtual Channel*. Many virtual channels are running within a *Virtual Path* (Figure 1.76).

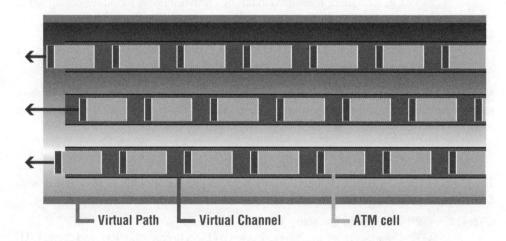

Figure 1.76   Asynchronous Transfer Mode.

A Virtual Path is, for example, the Permanent Virtual Connections (PVCs) for exchanging NBAP and ALCAP messages between RNC and Node B. This connection will be set up once and will run until it is changed or deleted by O&M operation. Over this PVC many user connections are running, which represents virtual channels.

1.9.1 ATM Cell

An ATM cell contains two address parameters, Virtual Path Identifier (VPI) and Virtual Channel Identifier (VCI); an identification of the type of payload; a cell loss priority; and a header CRC. This means that the transmission of the payload contents is not secured by a checksum. For transmission error detection and correction, the higher layers must have certain functions.

The header is 5 bytes and the payload is 48 bytes long.

The example in Figure 1.77 shows a possible configuration of the UMTS Iur interface. Certain signaling protocols are running over different VCIs as well as the user data traffic. The VCIs for traffic are *switched virtual connections (SVCs)*; that is, they will be set up only on request.

1.9.2 ATM Layer Architecture

To transmit higher protocols via ATM, it is required to have adaptation sublayers. These sublayers contain a common adaptation and a service-specific adaptation part (Figure 1.78).

The *Convergence Sublayer* is responsible for getting the PDUs of the higher layer and for modifying its size so that each of the PDUs fits into an SSCS and a CPCS message, respectively. Additionally, extra parameters will be inserted to guarantee that a receiver can allocate each message to a specific stream of information.

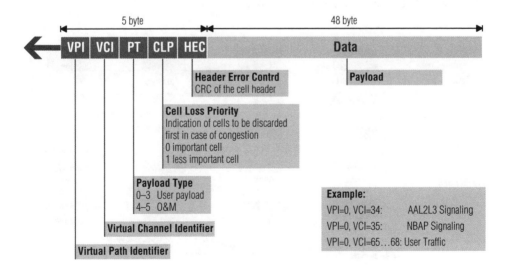

Figure 1.77 ATM cell.

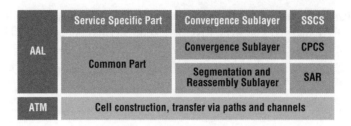

Figure 1.78   ATM layer architecture.

The *Segmentation Sublayer* is responsible for segmenting the SSCS or CPCS message so that each part of the original will fit into the ATM cell payload. Reassembly is the counterpart to segmentation and is performed at the receiver side.

1.9.3 ATM Adaption Layer (AAL)

The AAL is specified by four classes (A–D) that differ from each other in bitrate, synchronization method, and connection type (Figure 1.79):

A   Constant Bitrate Service (CBR)
B   Unspecified Bitrate Service (UBR)
C   Available Bitrate Service (ABR)
D   Variable Bitrate Service (VBR)

For each class a specific adaptation layer has been developed to support the specific use of it. These are ATM Adaptation Layers 1, 2, 3/4, and 5.

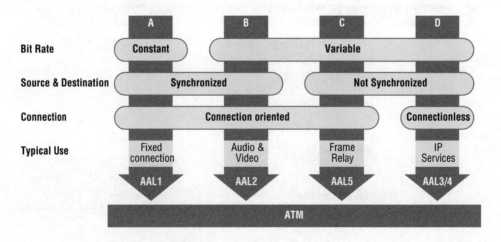

Figure 1.79   ATM Adaption Layer.

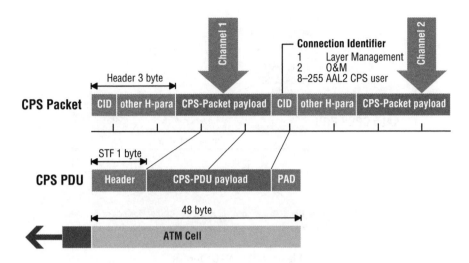

Figure 1.80   AAL2 format.

Each of the AALs contains a different frame structure which contains all necessary parameters to support the need. Part of every AAL frame is a data field in which the AAL-SDU message, or segment of a message of a higher protocol, will be placed and transmitted.

1.9.4 AAL2

The AAL2 (*ITU-T I.363.2*) provides for the bandwidth-efficient transmission of low-rate, short, and variable length packets in delay-sensitive applications (Figure 1.80). More than one AAL2 user information stream can be supported on a single ATM connection.

The AAL2 is subdivided into the Common Part Sublayer (CPS) and the Service Specific Convergence Sublayer (SSCS). Different SSCS protocols may be defined to support specific AAL2 user services, or groups of services. The SSCS may also be null, merely providing for the mapping of the equivalent AAL primitives to the AAL2 CPS primitives and vice versa.

AAL2 has been developed to transport multiple data streams. The *Connection Identifier (CID)* identifies every stream. The CID value can be found in the Layer 3 signaling as a reference (AAL2L3/ALCAP signaling protocol).

The *Start Field (STF)* is used to point to the payload or *Padding (PAD)* as well as for transmission error detection.

1.9.5 AAL5

The AAL5 (*ITU-T I.363.5*) enhances the service provided by the ATM layer to support functions required by the next higher layer (Figure 1.81). This AAL performs functions required by the user, control, and management planes, and supports the mapping between the ATM layer and the next higher layer.

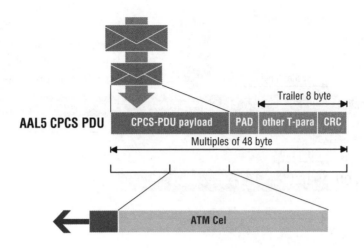

Figure 1.81   AAL5 format.

The AAL5 supports the nonassured transfer of user data frames. The data sequence integrity is maintained and transmission errors are detected. The AAL5 is characterized by transmitting in every ATM cell (but the last) of a PDU, 48 octets of user data. In most of the cells, there is no overhead encountered.

The AAL5 does not support input data streams; it supports frames. Maximum size of a frame is 64 kbytes. The higher layer message will be put in the *CPCS-PDU* payload field. One of the tail parameters will indicate the length of the payload. The *CRC* field will also protect the payload. In this way, the transmission is protected and is mainly used for transmission of control information, for example signaling.

1.10 User Plane Framing Protocol

The user plane Framing Protocol (FP, defined in *3GPP 25.427*) transports Transport Block Sets (TBSs) across the Iub and Iur interfaces. It is also responsible for transmission of outer loop power control information between Node B and SRNC and for transfer of radio interface parameters from SRNC to Node B. A set of FP signaling messages supports mechanisms for transport channel synchronization and node synchronization. In addition, FP also provides transport services for DSCH TFIs (Transport Format Indicators) from SRNC to Node B.

Transport channels on the Iur that lead from SRNC to DRNC must have the same configuration parameters as the same transport channels on Iub between DRNC and Node B.

The SRNC is responsible for the complete configuration of the transport channels. Appropriate signaling messages are exchanged between SRNC and Node B(s) via Iub and – if necessary – via Iur control plane. Transport channels in downlink direction are multiplexed by the Node B onto radio physical channels, and de-multiplexed in uplink direction from the radio physical channels.

Figure 1.82   UP FP frame architecture.

1.10.1 Frame Architecture

There are two different FP frame formats for data and control frames (Figure 1.82). In case of FP data frame the frame type field in the header is set to "0".

Header

Frame Type (FT): 0=data
Connection Frame Number (CFN): Reference to radio frame
Transport Format Indicator (TFI): Information about data block

Payload

Transport Block (TB): Block of data of DCH.
CFN: Indicator as to which radio frame the first data was received on uplink or will be transmitted on downlink.
TFI: The local number of the transport format used for the transmission time interval.
Transport Block (TB): A block of data to be transmitted or received over the air interface. The transport format indicated by the TFI describes the transport block length and transport block set size.

1.10.2 FP Control Frame Architecture

In case of a FP control frame, the frame type field is set to "1" and control frame type indicates the name of the FP signaling message (Figure 1.83).

Header

Frame Type (FT):   0=data, 1=control
Control Frame Type: OUTER LOOP POWER CONTROL
                           TIMING ADJUSTMENT
                           DL SYNCHRONIZATION
                           UL SYNCHRONIZATION
                           DSCH TFCI SIGNALING
                           DL NODE SYNCHRONIZATION
                           UL NODE SYNCHRONIZATION
                           RX TIMING DEVIATION
                           RADIO INTERFACE PARAMETER UPDATE
                           TIMING ADVANCE

Figure 1.83   UP FP control frame architecture.

Payload
Contains only parameters (CFN, Time of Arrival, UP SIR Target, other Timing Information, etc.)

1.11 Medium Access Protocol (MAC)

The MAC protocol (*3GPP 25.321*) coordinates the access of the physical layer. The logical channels of higher layers are mapped onto transport channels of lower layers. MAC also selects the appropriate TFSs, depending on necessary transmission rate, and organizes the priority handling between different data flows of one single UE.

If the UE uses Common Transport Channels, MAC provides a unique RNTI for each single UE, which is also known by the RRC.

In case of random access to the network via RACH, MAC defines a priority by assigning an Access Service Class. The values of ASC can be 0–7, whereas 0 is the highest priority. An Emergency Call would, for example, get the ASC=0. During Radio Bearer connection setup, MAC will receive a MAC Logical Link Priority (MLP). This corresponds with ASC.

1.11.1 MAC Architecture

The diagrams that describe the MAC architecture show the different MAC entities (Figure 1.84), which are:

MAC-b is the MAC entity that handles the following transport channel:

 • Broadcast Channel (BCH)

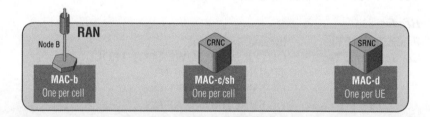

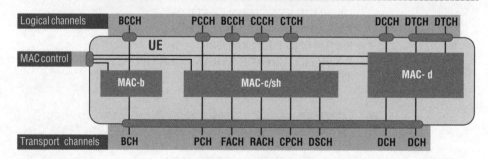

Figure 1.84 MAC Architecture.

MAC-c/sh is the MAC entity that handles the following transport channels:

- Paging Channel (PCH)
- Forward Access Channel (FACH)
- Random Access Channel (RACH)
- Common Packet Channel (CPCH). The CPCH exists only in FDD mode
- Downlink Shared Channel (DSCH)
- Uplink Shared Channel (USCH). The USCH exists only in TDD mode

MAC-d is the MAC entity that handles the following transport channel:

- Dedicated Transport Channel (DCH)

All entities are controlled via the MAC Control SAP, which is connected to the RRC unit.

1.11.2 MAC Data PDU

The MAC Data PDU contains the following information elements (definitions following *3GPP 25.321*; Figure 1.85).

Target Channel Type Field
The TCTF field is a flag that provides identification of the logical channel class on FACH and RACH transport channels. The flag tells whether the channel carries BCCH, CCCH, CTCH, SHCCH, or dedicated logical channel information.

UE-Id (Different RNTIs)
The UE-Id field provides an identifier of the UE on Common Transport Channels.
The following types of UE-Id used on MAC are defined:

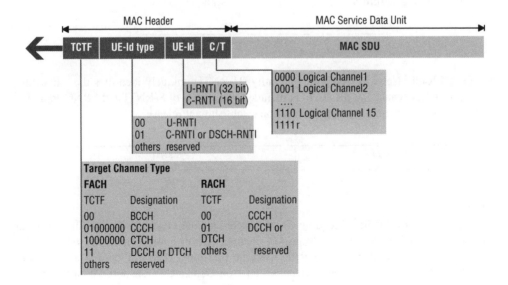

Figure 1.85   MAC Data PDU.

- The **Cell Radio Network Temporary Identity** (**C-RNTI**) uniquely identifies a UE within one cell and is assigned by the SRNC (=CRNC).

    It is used on DTCH and DCCH in uplink, may be used on DCCH in downlink, and is used on DTCH in downlink when mapped onto Common Transport Channels except when mapped onto the DSCH transport channel.

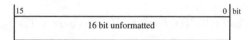

- The **DSCH Radio Network Temporary Identity** (**DSCH-RNTI**) uniquely identifies a UE within one cell, in case DSCH-TrCHs are used as bearers for DCCH/DTCH. The DSCH-RNTI is assigned by the CRNC. In FDD, the DSCH-RNTI is used on DTCH and DCCH in downlink when mapped onto the DSCH transport channel.

- The **SRNC Radio Network Temporary Identity** (**S-RNTI**) uniquely identifies a UE in the SRNS (e.g. in RNSAP messages) and is assigned by SRNC for a RRC-connection establishment. An S-RNTI is discarded, if the RRC connection is released or when the SRNC changes (e.g. during an SRNC relocation).

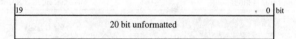

- The **DRNC Radio Network Temporary Identity** (**D-RNTI**) uniquely identifies a UE in RNSAP messages from SRNC to DRNC and is assigned by DRNC.

19	0	bit
20 bit unformatted		

- UTRAN Radio Network Temporary Identity (U-RNTI) uniquely identifies the UE within the UTRAN, because the SRNC-Id is included. It consists of S-RNTI and SRNC-Id and is assigned/released upon an RRC-connection establishment/release.

11	0	19	0	bit
SRNC-Id		S-RNTI		

C/T Field

The C/T field provides identification of the logical channel instance when multiple logical channels are carried on the same transport channel (for example it indicates which radio signaling bearer is used in case of RRC message transport). The C/T field is also used to provide identification of the logical channel type on Dedicated Transport Channels and on FACH and RACH when used for user data transmission.

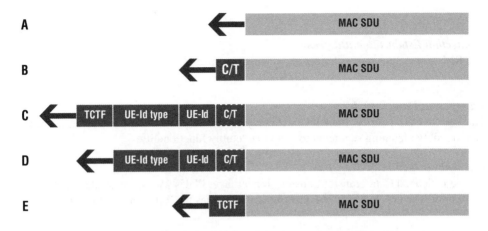

Figure 1.86   MAC header alternatives.

1.11.3 MAC Header Alternatives

Depending on the channel used, the MAC header can contain a different parameter (Figure 1.86):

A   DTCH or DCCH will be mapped on DCH; there is no multiplexing of dedicated channels in MAC. No header information is required.

B   DTCH or DCCH will be mapped on DCH. MAC performs multiplexing of dedicated channels. C/T is required.

C   DTCH or DCCH will be mapped on RACH/FACH. If multiplexing of dedicated channels is necessary, C/T is included.

D   DTCH or DCCH will be mapped on DSCH/USCH as long as DTCH or DCCH are the only logical channels. If multiplexing of dedicated channels is necessary, C/T is included.

E   Could be used if BCCH is mapped on FACH and must be used if CCCH is mapped on RACH/FACH and CTCH messages are used.

1.12 Radio Link Control (RLC)

The RLC protocol offers transport services to the higher layers called *Radio Bearer Services* and is specified in *3GPP 25.322*.

RLC supports segmentation and the transport of user and signaling information.

The RLC sublayer consists of RLC entities for the UE-UTRAN interface, of which there are three modes of operation:

- Transparent Mode (TM)
- Unacknowledged Mode (UM)
- Acknowledged Mode (AM)

1.12.1 RLC Services

Connection Establishment/Release
The RLC connection establishment and release organizes the setup or ending of RLC connections.

Transparent Data Transfer
Transparent Data Transfer transmits higher layer PDUs without adding any protocol information, possibly including segmentation and reassembly functionality.

Unacknowledged Data Transfer
Unacknowledged Data Transfer transmits higher layer PDUs without guaranteeing delivery to peer entity. The unacknowledged data transfer mode has the following characteristics:

- Detects erroneous data by using a sequence-number check function. The RLC sublayer delivers to the receiving higher layer only the SDUs that are free of transmission errors.

Unique Delivery
Using duplication detection, the RLC sublayer delivers each SDU to the receiving higher layer only once.

Immediate Delivery
The receiving RLC sublayer entity delivers an SDU to the higher layer receiving entity as soon as the SDU arrives at the receiver.

Acknowledged Data Transfer
Acknowledged Data Transfer transmits higher layer PDUs and guarantees delivery to peer entity. In case RLC is unable to deliver data correctly, the user of RLC at the transmitting side is notified. In-sequence delivery and out-of-sequence delivery are supported.
  Acknowledged Data Transfer mode has the following characteristics:

- *Error-free delivery*: Ensured by means of retransmission; the receiving entity delivers only error-free SDUs to the higher layer.
- *Unique delivery*: Using duplication detection, the RLC sublayer delivers each SDU to the receiving higher layer only once.
- *In-sequence delivery*: RLC sublayer provides support for a sequential delivery of SDUs. The RLC sublayer delivers SDUs to the receiving higher layer entity in the same order as the transmitting higher layer entity submits to RLC sublayer.
- *Out-of-sequence delivery*: As an alternative to in-sequence delivery, the receiving RLC entity delivers SDUs to the higher layer in a different order than they were submitted to the RLC sublayer at the transmitting side.

1.12.2 RLC Functions

Segmentation and Reassembly
Used for variable length of higher layer PDUs into or from smaller RLC Payload Units (PUs). The PDU size depends on the actual set of transport formats.

Concatenation
Concatenates the contents of RLC SDU with the first segment of the next RLC if they do not fill an integer number of RLC PUs. The SDU may be put into RLC PU in concatenation with the last segment of the previous RLC SDU.

Padding
When concatenation is not applicable and the remaining data to be transmitted does not fill the entire RLC PDU of a given size, then the remainder of the data field is filled with padding bits.

Transfer of User Data
RLC conveys data between users of RLC services and supports acknowledged, unacknowledged, and transparent data transfer. The QoS setting controls the transfer of user data.

Error Correction
Errors are corrected by retransmitting the data while in the acknowledged data transfer mode. Data can be retransmitted using commands such as Selective Repeat, Go Back N, or a Stop-and-Wait ARQ (Automatic Repeat Request).

In-Sequence Delivery of Higher Layer PDUs
Ensure transfer of higher layer PDUs (submitted for transfer by RLC) in the correct order, using acknowledged data transfer. If the function is not used, out-of-sequence delivery is provided.

Duplicate Detection
The RLC detects duplicated PDUs that it receives and ensures that the resultant higher layer PDU is delivered only once to the higher layer.

Flow Control
The RLC receiver controls the rate at which a peer RLC transmitting entity may send information.

Sequence Number Check (Unacknowledged Data Transfer Mode)
The RLC guarantees integrity of reassembled PDUs and provides a mechanism for detection of corrupted RLC SDUs through checking the sequence number in RLC PDUs when the PDUs are reassembled into an RLC SDU. All corrupted RLC SDUs will be discarded.

Protocol Error Detection and Recovery
The RLC detects and recovers from errors in the operation of its protocol.

Ciphering
Ciphering prevents unauthorized acquisition of data and is performed in the RLC layer for nontransparent RLC mode.

Suspend/Resume Function
Suspend/Resume function of data transfer works in the same way as in LAPDm (Ref. GSM 04. 05)

Transparent Mode
No RLC information will be added to the message. Erroneous messages will be detected, registered, and discarded. There is no sequence control function available.

Table 1.6 RLC function overview table

Transparent entity	Unacknowledged entity	Acknowledged entity
Segmentation/reassembly	Segmentation/reassembly	Segmentation/reassembly
Transfer of application data	Concatenation	Concatenation
	Padding	Padding
	Transfer of application data	Transfer of application data
	Ciphering	Ciphering
	Sequence number check	Error correction
		In-sequence delivery
		Flow control
		Duplicate detection
		Protocol error detection and recovery
		Suspend/resume functionality

This mode is used for streaming application data where the data does not have to be segmented.

Applications using transparent mode are video and audio data applications.

Unacknowledged Mode

By using the sequence number, the uniqueness of a data package can be checked, but there is no error correction method specified in this mode. Certain RLC information will be added to the message, and segmentation and ciphering will be performed.

Applications using unacknowledged mode are certain RRC procedures, where the RRC layer is responsible for the receive acknowledgment, the Cell Broadcast Service (CBS), and the VoIP.

Acknowledged Mode

Supports ARQ with all the necessary parameters, and performs segmentation and ciphering.

Applications using acknowledged mode are secure transmission and packet-oriented data transfer.

Table 1.6 explains the functions in combination with the different modes.

Note: Ciphering is not part of RLC in the transparent entity.

1.12.3 RLC Architecture

A UM and a Tr RLC entity can be configured to be a transmitting RLC entity or a receiving RLC entity. The transmitting RLC entity transmits RLC PDUs; the receiving RLC entity receives RLC PDUs (Figure 1.87).

An AM RLC entity consists of a transmitting side and a receiving side, where the transmitting side of the AM RLC entity transmits RLC PDUs and the receiving side of the AM RLC entity receives RLC PDUs.

Elementary procedures are defined between a "Sender" and a "Receiver." In UM and Tr, the transmitting RLC entity acts as a Sender and the peer RLC entity acts as a Receiver. An AM RLC entity acts either as a Sender or as a Receiver depending on the elementary procedure. The Sender is the transmitter of AMD PDUs and the Receiver is the receiver of AMD PDUs. A Sender or a Receiver can reside either at the UE or at the UTRAN.

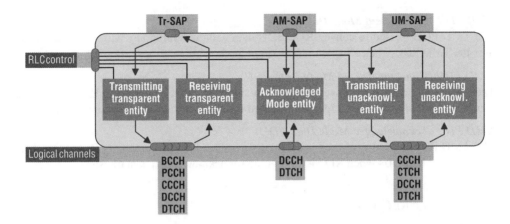

Figure 1.87   RLC architecture.

There is one transmitting and one receiving RLC entity for each TM and UM service. There is one combined, transmitting and receiving entity for the AM service.

Each RLC UM and TM entity uses one logical channel to send or receive Data PDUs. An AM RLC entity can be configured to use one or two logical channels to send or receive data and control PDUs. If two logical channels are configured, they are of the same type – DCCH or DTCH.

1.12.4 RLC Data PDUs

Figure 1.88 shows the three different types of RLC Data PDUs.

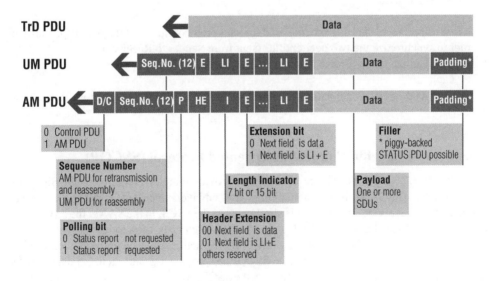

Figure 1.88   RLC Data PDUs.

TMD PDU (Transparent Mode Data PDU)
The TMD PDU is used to convey RLC SDU data without adding any RLC overhead. RLC uses the TMD PDU when RLC is in transparent mode.

UMD PDU (Unacknowledged Mode Data PDU)
The UMD PDU is used to convey sequentially numbered PDUs containing RLC SDU data. RLC uses UMD PDUs when RLC is configured for unacknowledged data transfer.

AMD PDU (Acknowledged Mode Data PDU)
The AMD PDU is used to convey sequentially numbered PDUs containing RLC SDU data. RLC uses AMD PDUs when RLC is configured for acknowledged data transfer.

1.12.5 Other RLC PDUs

Other RLC PDUs are:

- RESET PDU – to reset RLC protocol entities and all their system variables
- RESET ACK PDU – acknowledgment to RESET PDU

Control PDUs are used only in acknowledged mode.
STATUS PDU and **Piggybacked STATUS PDU** are used:

- By the Receiver to inform the Sender about missing and received AMD PDUs in the Receiver; selective and group acknowledgment is possible
- By the Receiver to inform the Sender about the size of the allowed transmission window
- By the Sender to request that the Receiver move the reception window
- By the Receiver to acknowledge to the Sender the receipt of the request to move the reception window

RESET PDU is used:

- To reset all protocol states, protocol variables, and protocol timers of the peer RLC entity in order to synchronize the two peer entities (sent from Sender to Receiver)
- To increment the Hyper Frame Number (Ciphering)

RESET ACK PDU is an acknowledgment of the **RESET PDU** (sent from Receiver to Sender).

1.13 Service Specific Connection Oriented Protocol (SSCOP)

The SSCOP (*ITU-T Q.2110*) has been defined to provide functions required in the Signaling AAL (SAAL). The SAAL is a combination of two sublayers: a common part and a service-specific part. The service-specific part is also known as the SSCS. In the SAAL, the SSCS itself is functionally divided into the SSCOP and an SSCF which maps the services provided by the SSCOP to the needs of the user of the SAAL. This structure allows a common connection-oriented protocol with error recovery (the SSCOP) to provide a generic reliable data transfer service for different AAL interfaces defined by the SSCF. Two such SSCFs, one for signaling at the User-Network Interface (UNI) and one for signaling at the Network-to-Network Interface (NNI), have been defined. It is also possible to define additional SSCFs over the common SSCOP to provide different AAL services.

Sequence Integrity
Preserves the order of SSCOP SDUs that were submitted for transfer by SSCOP.

Error Correction by Selective Retransmission
Through a sequencing mechanism, the receiving SSCOP entity can detect missing SSCOP SDUs. This function corrects sequence errors through retransmission.

Flow Control
Allows an SSCOP receiver to control the rate at which the peer SSCOP transmitter entity may send information.

Error Reporting to Layer Management
Indicates to the layer which management errors have occurred.

Keep Alive
Verifies that the two peer SSCOP entities participating in a connection are remaining in a link-connection-established state even in the case of a prolonged absence of data transfer.

Local Data Retrieval
Allows the local SSCOP user to retrieve in-sequence SDUs that have not yet been released by the SSCOP entity.

Connection Control
Performs the establishment, release, and resynchronization of an SSCOP connection. It also allows the transmission of variable length user-to-user information without a guarantee of delivery.

Transfer of User Data
Conveys user data between users of the SSCOP. SSCOP supports both assured and unassured data transfer.

Protocol Error Detection and Recovery
Detects and recovers from errors in the operation of the protocol.

Status Reporting
Allows the transmitter and receiver peer entities to exchange status information.

1.13.1 Example SSCOP

The example in Figure 1.89 shows the setup, connection, and release phase of an SSCOP connection on the IuPS.

BGN (Begin) and BGAK (Begin Ack) represent the connection setup.

During connection, data of higher layers will be transmitted with Sequenced Data PDUs, SD. Every SD contains a sequence number, N(S). After the internal time-out of Timer-POLL, an acknowledgment will be requested, POLL-PDU. The acknowledgment is then the STAT

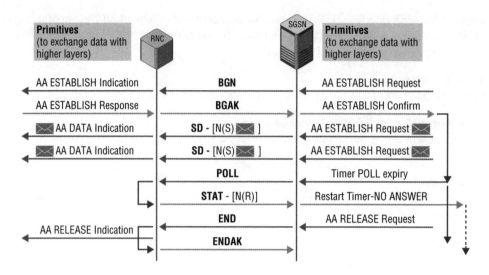

Figure 1.89 SSCOP (message flow).

message containing a receive sequence number, N(R). If there are no SD messages on the link in the meantime, then the POLL-STAT procedure will also run. The POLL-STAT procedure confirms that the SSCOP connection (link integrity) is established.

END and *ENDAK* represent the disconnect procedure.

Note: An SSCOP connection is a permanent connection; it is not user-dependent. The connection is set up for each signaling link, for example VPI/VCI for signaling information. All user signaling will be transferred via an SSCOP connection.

1.14 Service Specific Coordination Function (SSCF)

SSCF (*ITU-T Q.2140*) is not a protocol but an internal coordination function, which does internal adaptation of the information coming or going to higher layers, for example MTP3-B routing information. SSCF provides the following mapping functions:

- Mapping of primitives from Layer 3 to signals of the SSCOP
- Mapping of destination address (Signaling Point Code, SPC) to SSCOP connection

Because of this modular concept, SSCOP can work with many different higher layer protocols.

1.15 Message Transfer Part Level 3 – Broadband (MTP3-B)

MTP3-B (*ITU.T Q.2210*) fulfills the same sort of work as the standard narrowband MTP; it provides identification and transport of higher layer messages (PDUs), routing, and load sharing (Figure 1.90).

The main address parameters are Originating and Destination Point Codes (ODC and DPC). Their unique value represents the SPC of a network component.

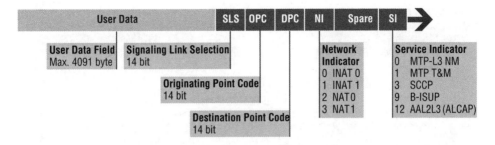

Figure 1.90 MTP3-B.

Network Indicator

Is used on Points of Interconnection (POI) to build virtual interconnection networks on or between national and international network level.

Service Indicator

Identifies the contents of the user data field (for example the higher layer protocol).

1.16 Internet Protocol (IP)

The Internet Protocol (*RFC 791*), actual version 4, IPv4, provides connectionless services between networks and includes features for addressing, type-of-service specification, fragmentation and reassembly, and security.

IP transmits data without a connection and without protection of the data, such as ciphering, authentication, flow control, or any other error correction mechanism.

The addressing is symmetrical; the source and destination addresses are always in the header.

The data contained in an IP message can be 64 kbytes maximum, whereas every IP node must be able to handle packet sizes of 576 bytes minimum.

The next generation is IP version 6. The goal was to improve these negative features of IPv4. The address range has been enhanced to 128 bits. This version now includes protection mechanisms, including ciphering and integrity check of data and address. It also now supports real QoS.

- IP version 4
  - No error control or correction
  - No sequence or flow control
  - Fragmentation and reassembly of data; header minimum of 20 bytes
  - Address size: 32 bits, source and destination included
- IP version 6
  - Qos parameter included and used
  - Remote configuration of IP users
  - Authentication and ciphering mechanism included
  - Signature of address and contents
  - Fragmentation and reassembly of data; header minimum of 40 bytes
  - Address size: 128 bits, source and destination included

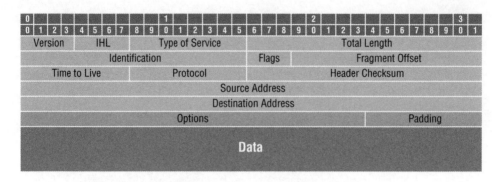

Figure 1.91 IPv4 frame architecture.

1.16.1 IPv4 Frame Architecture

Flags and *Fragment Offset* will indicate if the data part contains a full message or just a segment of one. The offset value will indicate a multiple of 8 bytes (Figure 1.91).

Time to Live is not a timer but a hop counter which has to be decremented by every IP node. If it reaches zero, the packet will be deleted.

Protocol identifies the next higher layer protocol. Table 1.7 gives some examples.

Source and Destination Address contains the 32-bit node address, for example 192.168.1.17.

Options are optional and usually not included.

1.17 Signaling Transport Converter (STC)

STC (*ITU-T Q.2150.1*) is an internal function, which has none of its own messages. It converts primitives from lower and higher layers (either MTP-3 or MTP3-B primitives) and their parameters fitting the requirements of the other. Other functions are:

• Provision of OPC, DPC and SIO value
• UMTS: Service Indicator (part of SIO) = 12 (AAL2-L3)

In UMTS the (AAL2L3) Signaling protocol could be on STC. To allow AAL2L3 to set up user connections in the network, it has to send certain messages to the partner instances. The routing and selection of SPCs is the responsibility of STC.

Table 1.7 Protocol parameter meaning

Value	Meaning
1	ICMP, Internet Control Message Protocol
2	IGMP, Internet Group Management Protocol
6	TCP, Transmission Control Protocol
17	UDP, User Datagram Protocol
41	IPv6, IP version 6
132	SCTP, Stream Control Transmission Protocol

1.18 Signaling Connection Control Part (SCCP)

SCCP (*ITU-T Q.711–716*) provides a service for transfer of messages between any two signaling points in the same or different network and is used in the same way as it is known from SS7 and GSM. It can act as connectionless or connection-oriented transport protocol and provides these connection types:

- Connectionless (Class 0&1) and connection oriented (Class 2&3)
- Class 0
  – Addressing purposes, DPC or Global Title (GT)
- Class 2
  – Used on IuCS, IuPS, and Iur interfaces to organize connections Class 3
  – Planned to be used on Iur interface by some switch manufacturerso
  – SCCP user is called Subsystem and identified by a subsystem number (SSN)[QA11]
- Class 3
  – Flow control connection-oriented class
  – (Probably) not used on Iur

The SCCP (Signaling Connection Control Part) user is called Subsystem and identified by an SSN that has the same function as protocol field described in the IP protocol part (1.16) before. For instance on Iu interface RANAP is an SCCP subsystem. However, a subsystem can be both a higher layer protocol or a network element/function.

Table 1.8 gives an overview of SSN used in live network environments.

The differences between connectionless and connection-oriented data exchange are as follows.

Connectionless
SCCP is responsible for the end-to-end addressing. SCCP creates addresses by either giving the DPC or a GT of the endpoint. A GT has to be translated (GTT – Global Title Translation) on the way to the destination. On the last link to the destination point, the GT will be replaced with a DPC.

Table 1.8 SSN overview

Value	Meaning
6	HLR
7	VLR
8	MSC
12	INAP/MAP operator defined
142	RANAP
143	RNSAP
146	CAP
147	gsmSCF
149	SGSN
150	GGSN
192	BSSAP+ on Gs interface
254	BSSAP

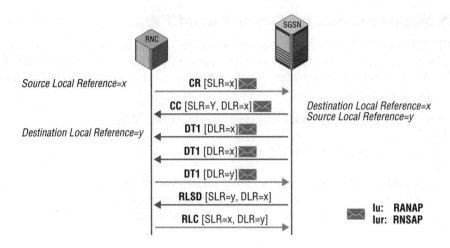

Figure 1.92   SCCP CO (message flow).

Connection Oriented
SCCP is responsible for the user connection running on one interface. It is not controlling an end-to-end connection. The connection is identified by a Source and a Destination Local Reference Number (SLR, DLR).

To identify the transported higher protocol, SCCP uses an SSN.

1.18.1 Example SCCP

Connection-Oriented Example
Connection Request (CR) and Connection Confirm (CC) represent the setup phase. In the set-up phase, the two sides exchange the local reference numbers. A negative response would be the Connection Refused (CREF) message which would contain a cause, explaining the problem (Figure 1.92).

Some procedures use the CREF message as a fast method to release the procedure, for example, if all necessary information has been already send in the CC.

During the connection Data Form 1 (**DT1**) message will transport higher layer messages.

To release the SCCP connection, a Released (**RLSD**) and a Release Complete (**RLC**) message will be exchanged.

Every main user procedure has its own SCCP connection on the Iu interfaces.

1.19 Abstract Syntax Notation One (ASN.1) in UMTS

The protocols RANAP, RNSAP, NBAP, RRC (Basic-PER, octet aligned), MAP, and CAP (BER) are specified by using ASN.1 (ISO/IEC 8824-1), a protocol description language. ASN.1 provides these functions:

* Automatic generation of network component protocol software
* Fast access of information by receiving entity

- Compact form of information transmission by using special encoding rules
  - Basic Encoding Rules (BER; ISO/IEC 8825 and 8825-1): used for MAP and CAP, easy-to-read raw data contents, messages are quite large, data clearly structured, big messages
  - Packed Encoding Rules (PER; ISO/IEC 8825-2): used for the other named protocols (running within UTRAN), very compact raw data contents, and messages gets small
    - Data structure is known by sender and receiver –> omits extra data-specific information, compact and smaller message

1.19.1 ASN.1 BER

Every protocol element is represented by an identifying TAG, a length field, and a contents field.

In case of primitive contents form, the contents field consists of one value.

In case of constructor contents forms, the content field consists of one or more other TAG-Length–Contents constructions. In this case, the first byte of the contents field is a TAG.

ASN.1 BER is used by MAP and CAP in the CN (Figure 1.93).

1.19.2 ASN.1 PER

The packed encoding rules have the goal to transmit as little data as possible.

That is why a preamble (similar to a tag field) and the length field can be missing. Sender and Receiver have to use the same protocol versions; otherwise the Receiver will not understand the contents of a received message.

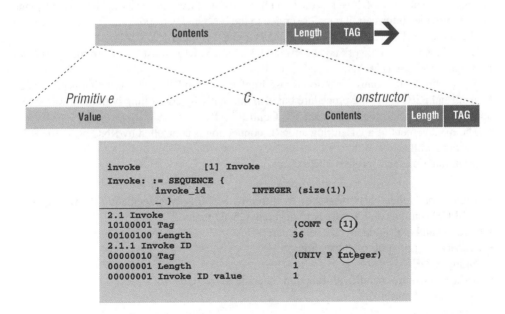

Figure 1.93 ASN.1 BER.

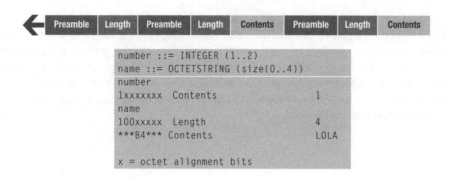

Figure 1.94   ASN.1 PER.

There are two alternative PER: octet aligned and not octet aligned.

In UMTS the octet aligned version will be used, with the exception of the RRC, which uses unaligned. This means that even if a field requires just 1 bit (see the number in Figure 1.94), the field would be of size 1 byte, 7 bits filled with zeros (x).

ASN.1 PER is used by NBAP, RNSAP, RRC, and RANAP in the UTRAN.

1.20 Radio Resource Control (RRC)

The RRC (*3GPP 25.331*) protocol is the most complex one in UMTS. It reflects the tasks of the RNC.

RRC is a sublayer of Layer 3 on UMTS radio interface and exists in the control plane only. It provides information transfer service to the NAS and is responsible for controlling the configuration of UMTS radio interface Layers 1 and 2 (Figure 1.95).

Because RNC is a network node between UE and CN, RRC must be able to transport NAS message across the UTRAN.

The Radio Bearer control function is also implemented in RRC. This function is performed by a special RRC signaling procedure but also internally by controlling both the lower layers and the user plane protocols via the RRC Control SAP.

The management of a UE during an RRC connection is controlled by RNC using the RRC protocol as well.

Main functions and services of RRC are:

- Routing of higher layer messages to different Mobility Management/Call Management (MM/CM) entities on UE side or to different CN domains
- Creation and management of Radio Bearers
- Broadcasting of system information
- Paging of UEs
- Dedicated Control handles all functions specific to one UE
  - Location Management
  - Handover

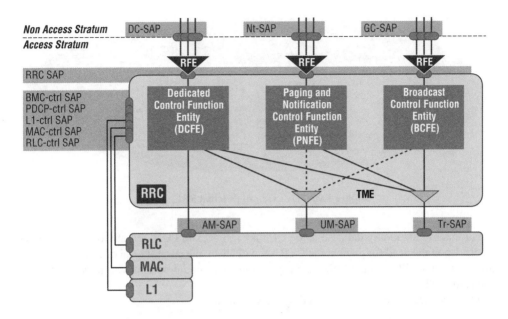

Figure 1.95 RRC architecture.

- SMS Routing
- Power Management (outer loop power control)
- Configuration of lower layer protocols
- Setup of RRC measurement settings
- Management of measurement report

1.20.1 RRC States[4]

To understand some of the signaling procedures later in this book, for example cases of physical channel reconfiguration (channel-type switching), it is necessary to have a closer look at the RRC state machine (Figure 1.95). The RRC state machine (Figure 1.96) provides further details.

After power is on, the UE stays in Idle Mode (RRC Idle) until it transmits a request to establish an RRC connection. In Idle Mode the connection of the UE is closed on all layers of the access stratum. In Idle Mode the UE is identified by NAS identities such as IMSI, TMSI, and P-TMSI. In addition, the UTRAN has no information about the individual Idle Mode UEs, and it can only address all UEs in a cell or all UEs monitoring a paging occasion. The UTRAN Connected Mode is entered when the RRC connection is established. The UE is assigned an RNTI to be used as UE identity on Common Transport Channels (Figure 1.97).

The RRC states within the UTRAN Connected Mode reflect the level of the UE connection and which transport channels can be used by the UE.

---

[4] *3GPP 25.331.*

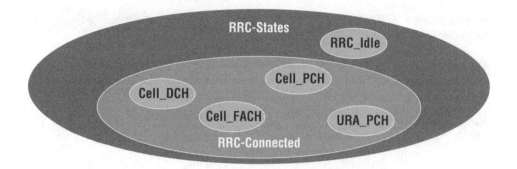

Figure 1.96   Overview of different RRC states.

For inactive stationary data users, the UE may fall back to PCH on both the Cell and URA levels. Upon the need for paging, the UTRAN will check the current level of connection of the given UE, and will decide whether the paging message should be sent within the URA or via a specific cell.

The *RRC_Idle* state is characterized by the following:

- UE is unknown in UTRAN, and no RNTIs have been assigned; TMSI or P-TMSI might be allocated if UE was registered into the network previously
- UE monitors downlink PICH/PCH (paging must be detected to change into RRC Connected Mode)
- If UE is moving, it will perform Routing and Location Area Update procedures

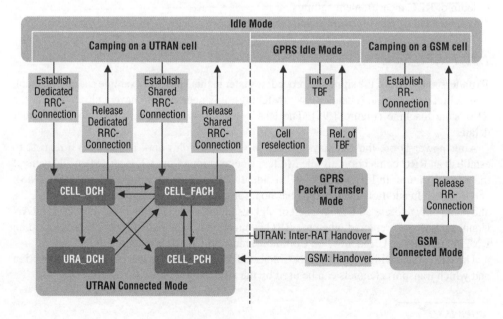

Figure 1.97   RRC state overview.

- Cell Reselection will be performed depending on the radio conditions but no Cell Updates nor URA Updates will happen
- DCCHs or DPCHs do not exist
- UE sends RRC_CONN_REQ on RACH to change into RRC Connected Mode

Note: Not all states may be applicable for all UE connections. For a given QoS requirement on the UE connection, only a subset of the states may be relevant.

The transition to the UTRAN Connected Mode from the Idle Mode can only be initiated by the UE by transmitting a request for an RRC connection. The event is triggered either by a paging request from the network or by a request from higher layers in the UE. When the UE receives a message from the network that confirms the RRC connection establishment, the UE enters the CELL_FACH or CELL_ DCH state of UTRAN Connected Mode.

In case of a failure, to establish the RRC Connection, the UE goes back to Idle Mode. The possible causes are radio link failure, a received reject response from the network, or lack of response from the network (time-out).

Connected Mode States (Figure 1.98)
The *CELL_DCH* state is characterized by the following:

- A dedicated physical channel is allocated to the UE in uplink and downlink
- Common/shared channels might be configured
- The UE is known on cell level according to its current active set
- Soft and Hard HO might be initiated
- No Cell Update nor URA Update is initiated by the UE
- The UE sends Measurement Reports to RNC according to the RNC setup
- The UE can use Dedicated Transport Channels (DCH), downlink and uplink (TDD) shared transport channels (TCH), and a combination of these transport channels.

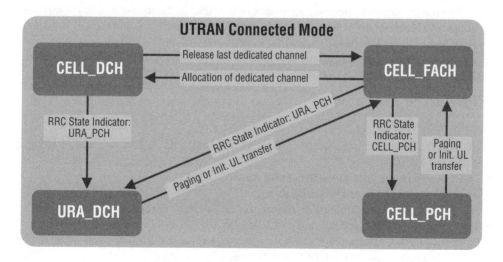

Figure 1.98   UTRAN – Connected Mode States.

The CELL_DCH state is entered from the Idle Mode through the setup of an RRC connection, or by establishing a dedicated physical channel from the CELL_FACH state.

A PDSCH may be assigned to the UE in this state, to be used for a DSCH. In TDD a PUSCH may also be assigned to the UE in this state, to be used for a USCH. If PDSCH or PUSCH are used for TDD, a FACH transport channel may be assigned to the UE for reception of physical shared channel allocation messages.

The *CELL_FACH* state is characterized by the following:

- No dedicated physical channel is allocated to the UE
- The UE continuously monitors a FACH in downlink
- The UE is assigned a default common or shared transport channel in the uplink (e.g. RACH or CPCH) that it can use anytime according to the access procedure for that transport channel
- No Soft or Hard HO might be initiated.
- UTRAN knows the position of the UE on the cell level according to the cell where the UE last made a cell update
- The UE performs Cell Updates, but no URA updates
- In TDD mode, one or several USCH or DSCH transport channels may have been established

In the *CELL_FACH* substate, the UE performs the following actions:

- Listens to all FACHs in the cell
- Listens to the BCH transport channel of the serving cell for the decoding of system information messages
- Initiates a cell update procedure on cell change of another UTRA cell
- Uses C-RNTI assigned in the current cell as the UE identity on Common Transport Channels except for when a new cell is selected
- Transmits uplink control signals and small data packets on the RACH
- In FDD mode, transmits uplink control signals and larger data packets on CPCH when resources are allocated to the cell and UE is assigned use of those CPCH resources
- In TDD mode, transmits signaling messages or user data in the uplink and/or the downlink using USCH and/or DSCH when resources are allocated to the cell and the UE is assigned use of those USCH/DSCH resources
- In TDD mode, transmits measurement reports in the uplink using USCH when resources are allocated to it in order to trigger a handover procedure in the UTRAN

The *CELL_PCH* state is characterized by the following:

- No dedicated physical channel is allocated to the UE
- UE selects a PCH with an algorithm and uses DRX for monitoring the selected PCH via an associated PICH
- DCCHs/DTCHs are configured but cannot be used
- No Soft or Hard HO might be initiated
- No uplink activity is possible (state change to Cell_FACH is needed)
- The UE performs Cell Updates, but no URA updates
- Position of the UE is known by UTRAN on the cell level according to the cell where the UE last made a cell update in the CELL_FACH state
- The UE sends Measurement Reports to RNC according to the RNC setup

In the *CELL_PCH* state the UE performs the following actions:

- Monitor the paging occasions according to the DRX cycle and receive paging information on the PCH
- Listens to the BCH transport channel of the serving cell for the decoding of system information messages
- Initiates a cell update procedure on cell change

The DCCH logical channel cannot be used in this state. If the network wants to initiate any activity, it needs to make a paging request on the PCCH logical channel in the known cell to initiate any downlink activity.

The *URA_PCH* state is characterized by the following:

- No dedicated channel is allocated to the UE
- UE selects a PCH with an algorithm and uses DRX for monitoring the selected PCH via an associated PICH
- UE monitors Downlink PICH/PCH (paging must be detected to change into RRC connected mode)
- No uplink activity is possible
- DCCHs/DTCHs are configured but cannot be used
- No uplink activity is possible (state change to CEll_FACH is needed)
- Location of the UE is known on the URA level according to the URA assigned to the UE during the last URA update in CELL_FACH state

In the *URA_PCH* state the UE performs the following actions:

- Monitors the paging occasions according to the DRX cycle and receive paging information on the PCH
- Listens to the BCH transport channel of the serving cell for the decoding of system information messages
- Initiates a URA updating procedure on URA change

The DCCH logical channel cannot be used in this state. If the network wants to initiate any activity, it needs to make a paging request on the PCCH logical channel within the URA where the location of the UE is known. If the UE needs to transmit anything to the network, it goes to the CELL_FACH state. The transition to URA_PCH state can be controlled with an inactivity timer, and, optionally, with a counter, which counts the number of cell updates. When the number of cell updates has exceeded certain limits (a network parameter), the UE changes to the URA_PCH state.

URA updating is initiated by the UE, which, upon the detection of the Registration area, sends the network the Registration area update information on the RACH of the new cell.

Note: A UE supporting CBS should be capable of receiving BMC messages in the CELL_PCH or URA_PCH state. If PCH and the FACH carrying CTCHs are not mapped onto the same S-CCPCH, UEs with basic service capabilities may not be able to monitor Cell Broadcast messages continuously in CELL_PCH state. In this case, UEs with basic service capabilities are capable of changing from the S-CCPCH that carries the PCH selected for paging to another S-CCPCH that carries Cell Broadcast messages (for example the CTCH mapped to a FACH) and receives BMC messages during time intervals which do not conflict with the UE-specific paging occasions.

1.20.2 System Information Blocks (SIBs)

The system information elements are broadcast in System Information Blocks (SIBs) that can be monitored, for example in System Information Update messages during Node B setup/restart. A SIB groups together system information elements of the same nature. Different SIBs may have different characteristics, for example, regarding their repetition rate and the requirements on UEs to re-read the SIBs (Figure 1.99).

The system information is organized as a tree. A Master Information Block (MIB) gives references to a number of SIBs in a cell, including scheduling information for those SIBs. The SIBs contain the actual system information and optionally references to other SIBs having scheduling information for those SIBs. The referenced SIBs must have the same area scope and use the same update mechanism as the parent SIB.

Some SIBs may occur more than once with different content. In this case, scheduling information is provided for each occurrence of the SIB. This option is allowed only for SIB type 16.

All SIBs, except SIB 15.2, 15.3, and 16, use a random ID called a *value tag*. As long as the value tag contains the same value, the contents of the SIBs are unchanged. This means that a UE receives SIB and stores the value tag. The next occurrence of that SIB and UE will compare value tags. If the value is equal to the stored value, then the contents of the SIB can be discarded. If the value is different, then UE will read the SIB contents and store the new value tag.

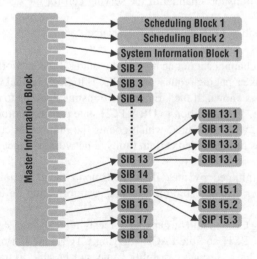

Figure 1.99   SIB overview.

Table 1.9 SIB content

Value	Meaning
SIB 1	NAS System Information, UE timer, and counter for RRC Idle and Connected Mode
SIB 2	URA Identity
SIB 3	Parameter for Cell Selection and Reselection
SIB 4	Parameter for Cell Selection and Reselection in RRC Connected Mode
SIB 5	Parameter for configuration of CPCH of actual cell
SIB 6	Parameter for configuration of Common and Shared Physical Channel of actual cell
SIB 7	Fast changing parameter for uplink Interference and Dynamic Persistence Level
SIB 8	Static CPCH Information of actual cell (FDD only)
SIB 9	CPCH Information of actual cell (FDD only)
SIB 10	Information for UE, which DCH is controlled by Dynamic Resource Allocation Control Procedure
SIB 11	Measurement Control Information of actual cell
SIB 12	Measurement Control Information of actual cell in RRC Connected Mode
SIB 13	ANSI-41 System Information
SIB 13.1	ANSI-41 RAND Information
SIB 13.2	ANSI-41 User Zone Identification
SIB 13.3	ANSI-41 Private Neighbor List
SIB 13.4	ANSI-41 Global Service Redirection
SIB 14	UL outer loop power control information for common and dedicated physical channels in RRC Idle or Connected Mode
SIB 15	Information for UE positioning method
SIB 15.1	Information for UE GPS positioning method with Differential Global Positioning System (DGPS) correction
SIB 15.2	Information for GPS Navigation Model
SIB 15.3	Information for GPS Almanac, ionospheric, and UTC Model
SIB 15.4	UE-assisted information for OTDOA UE Positioning method
SIB 15.5	UE-based information for OTDOA UE positioning method
SIB 16	Information of Radio Bearer, transport, and physical channels for UE in RRC Idle or Connected Mode in case of HO to UTRAN
SIB 17	Fast changing parameter for the configuration of Shared Physical Channels in RRC Connected Mode (FDD only)
SIB 18	PLMN Identities of neighbor cells

SIB 15.2, 15.3, and 16 contain a value tag, too, but their contents must always be read. Depending on SIB, the contents and the value tag are valid within a cell or within UTRAN (Table 1.9).

SIB Content
Note: As SIBs are defined in RRC, but transmitted in NBAP, special decoders are needed to monitor SIBs with a protocol tester, which has to cope with the fact that one protocol is octet-aligned encoded, and the other protocol is not.

Example – Broadcast System Information
The *system information* is continuously repeated on a regular basis in accordance with the scheduling defined for each system information block (Figure 1.100).

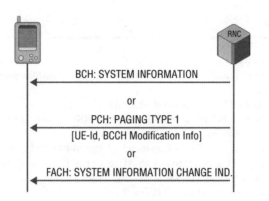

Figure 1.100   Broadcast System Information (message flow).

The UE reads **SYSTEM INFORMATION** messages broadcast on a BCH transport channel in Idle Mode as well as in states CELL_FACH, CELL_PCH, URA_PCH, and CELL_DCH (TDD only). Further, the UE reads SYSTEM INFORMATION messages broadcast on a FACH transport channel when in the CELL_FACH state. In addition, UEs that support simultaneous reception of one SCCPCH and one DPCH read system information on a FACH transport channel when in the CELL_DCH state.

Idle mode and connected mode UEs may acquire different combinations of SIBs. Before each acquisition, the UE should identify which SIBs are needed.

The UE may store SIBs (including their value tag) for different cells and different PLMNs, to be used if the UE returns to these cells. The UE considers the SIBs valid for a period of 6 hours from reception. Moreover, the UE considers all stored SIBs as invalid after the UE has been switched off.

When selecting a new cell within the currently used PLMN, the UE considers all current SIBs with area scope cell to be invalid. If the UE has stored valid SIBs for the newly selected cell, the UE may set those as current SIBs.

After selecting a new PLMN, the UE considers all current SIBs to be invalid. If the UE has previously stored valid SIBs for the selected cell of the new PLMN, the UE may set those as current SIBs. Upon selection of a new PLMN, the UE stores all information elements specified within the variable SELECTED_PLMN for the new PLMN within this variable.

For *modification of some system information* elements (for example, reconfiguration of the channels), it is important for the UE to know exactly when a change occurs. In such cases, the UTRAN should perform the following actions to indicate the change to the UEs.

Send the **PAGING TYPE 1** message on the PCCH in order to reach Idle Mode UEs as well as Connected Mode UEs in state CELL_PCH and URA_PCH. In the IE "BCCH Modification Information," UTRAN indicates the SFN when the change will occur and the new value tag that will apply for the MIB after the change has occurred. The PAGING TYPE 1 message is sent in all paging occasions.

Send the message **SYSTEM INFORMATION CHANGE INDICATION** on the BCCH mapped on FACH on all FACHs in order to reach all UEs in state CELL_FACH. In the IE "BCCH Modification Information," UTRAN indicates the SFN when the change will occur and the new value tag that will apply for the MIB after the change has occurred. UTRAN may

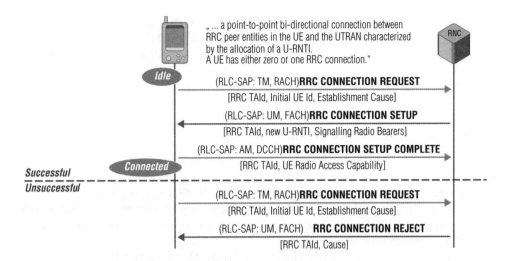

Figure 1.101   RRC Connection Establishment (message flow).

repeat the SYSTEM INFORMATION CHANGE INDICATION on all FACHs to increase the probability of proper reception in all UEs needing the information.

Example – RRC Connection Establishment

The NAS in the UE may request the establishment of only one RRC connection (Figure 1.101).
  Upon initiation of the procedure, the UE will:

- Set Connection Frame Number (CFN) in relation to System Frame Number (SFN) of current cell according to 8.5.17
- Transmit an **RRC CONNECTION REQUEST** message on the uplink CCCH, reset counter V300, and start timer T300
- Perform the mapping of the Access Class to an Access Service Class and apply the given Access Service Class when accessing the RACH
- Set the IE "Establishment cause" reflecting the cause of establishment in the higher layers
- Set the IE "Initial UE identity" to IMSI or TMSI
- Include a measurement report, as specified in the IE "Intrafrequency reporting quantity for RACH reporting" and the IE "Maximum number of reported cells on RACH" in SIB type 11

  Upon receiving an RRC CONNECTION REQUEST message, UTRAN will do one of the following:

- Transmit an **RRC CONNECTION SETUP** message on the downlink CCCH
- Transmit an **RRC CONNECTION REJECT** message on the downlink CCCH. In the RRC CONNECTION REJECT message, the UTRAN may direct the UE to another UTRA carrier or to another system. After the RRC CONNECTION REJECT message has been sent, all context information for the UE may be deleted in UTRAN

Upon receiving an RRC CONNECTION SETUP message, the UE compares the value of the IE "Initial UE identity" in the received RRC CONNECTION SETUP message with the value of the IE "Initial UE identity" in the most recent RRC CONNECTION REQUEST message sent by the UE.

If the values are different, the UE will:

- Ignore the rest of the message

If the values are identical, the UE will:

- Stop timer T300, and act upon all received information elements
- Store the value of the IE "New U-RNTI"
- Initiate the signaling link parameters according to the IE "RB mapping info"
- If neither the IE "PRACH info (for RACH)" nor the IE "Uplink DPCH info" is included, let the physical channel of type PRACH that is given in the system information be the default in uplink to which the RACH is mapped
- If neither the IE "Secondary CCPCH info" nor the IE "Downlink DPCH info" is included, start to receive the physical channel of type S-CCPCH that is given in system information to be used as the default by FACH
- Transmit an **RRC CONNECTION SETUP COMPLETE** message on the uplink DCCH after successful state transition, with the contents set as specified below:
  - Include START (*3GPP 33.102*) values to be used in ciphering and integrity protection for each CN domain
  - If requested in the IE "Capability update requirement" sent in the RRC CONNECTION SETUP message, include its UTRAN-specific capabilities in the IE "UE radio access capability"
  - If requested in the IE "Capability update requirement" sent in the RRC CONNECTION SETUP message, include its intersystem capabilities in the IE "UE system specific capability"

Example – RRC Connection Release

The purpose of the example in this procedure is to release the RRC connection including the signaling link and all radio bearers between the UE and the UTRAN. By doing so, all established signaling flows and signaling connections will be released (Figure 1.102).

When the UE is in the state CELL_DCH or CELL_FACH, the UTRAN may at anytime initiate an RRC connection release by transmitting an **RRC CONNECTION RELEASE** message using UM RLC. When UTRAN transmits an RRC CONNECTION RELEASE message as response to a received **RRC CONNECTION RE-ESTABLISHMENT REQUEST**, **CELL UPDATE**, or **URA UPDATE** message from the UE, UTRAN will use the downlink CCCH to transmit the message. In all other cases, the downlink DCCH will be used, although the downlink CCCH may be used as well.

UTRAN may transmit several RRC CONNECTION RELEASE messages to increase the probability of proper reception of the message by the UE. The number of repeated messages and the interval between the messages is a network option.

The UE will receive and act on an RRC CONNECTION RELEASE message in states CELL_DCH and CELL_FACH.

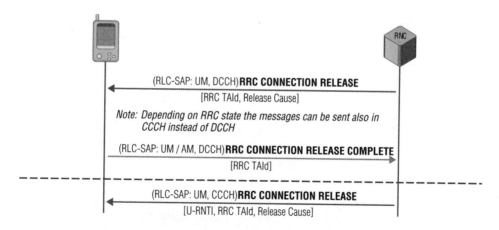

Figure 1.102 RRC Connection Release (message flow).

Furthermore, this procedure can interrupt any ongoing procedures with the UE in the above listed states.

When the UE receives the first RRC CONNECTION RELEASE message, it will:

- In state CELL_DCH:
  - Initialize the counter T308 with the value of the IE "Number of RRC Message Transmissions," which indicates the number of times the RRC CONNECTION RELEASE COMPLETE message is sent
  - Transmit an RRC CONNECTION RELEASE COMPLETE message using UM RLC on the DCCH to the UTRAN
  - Start timer T308
- In state CELL_FACH and if the RRC CONNECTION RELEASE message was received on the DCCH:
  - Transmit an RRC CONNECTION RELEASE COMPLETE message using AM RLC on the DCCH to the UTRAN

When in state CELL_FACH and if the RRC CONNECTION RELEASE message was received on the CCCH, the UE will not transmit an RRC CONNECTION RELEASE COMPLETE message.

The UE will ignore any succeeding RRC CONNECTION RELEASE messages that it receives. The UE will indicate release of all current signaling flows and radio access bearers to the NAS and pass the value of the IE "Release cause" received in the RRC CONNECTION RELEASE message to the NAS.

From the time of the indication of release to the NAS until the UE has entered idle mode, any NAS request to establish a new RRC connection will be queued. This new request may be processed only after the UE has entered idle mode. When in state CELL_FACH and if the RRC CONNECTION RELEASE message was received on the CCCH, the UE will release all its radio resources, enter idle mode, and end the procedure on the UE side.

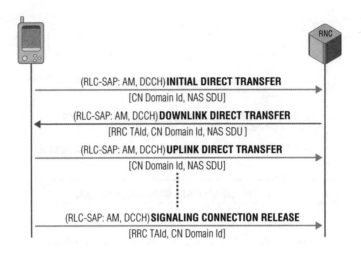

Figure 1.103   RRC Signaling Connection (message flow).

Example – RRC Signaling Connection
The INITIAL DIRECT TRANSFER procedure is used in the uplink to establish signaling connections and signaling flows. It is also used to carry the initial higher layer (NAS) messages over the radio interface.

A signaling connection comprises one or several signaling flows. This procedure requests the establishment of a new flow, and triggers, depending on the routing and if no signaling connection exists for the chosen route for the flow, the establishment of a signaling connection (Figure 1.103).

The DOWNLINK DIRECT TRANSFER procedure is used in the downlink direction to carry higher layer (NAS) messages over the radio interface.

The UPLINK DIRECT TRANSFER procedure is used in the uplink direction to carry all subsequent higher layer (NAS) messages over the radio interface belonging to a signaling flow.

The SIGNALING CONNECTION RELEASE request procedure is used by the UE to request from the UTRAN that one of its signaling connections should be released. The procedure may, in turn, initiate the signaling flow release or RRC connection release procedure.

1.21 Node B Application Part (NBAP)

NBAP (*3GPP 25.433*) is the communication protocol used on the Iub interface, between RNC and Node B, and is specified using ASN.1 PER.

1.21.1 NBAP Functions

NBAP covers a large range of different functions as described in Table 1.10.

Table 1.10 NBAP function overview

Function	Description
Cell Configuration Management	Manages the cell configuration information in a Node B
Common Transport Channel Management	Manages the configuration of Common Transport Channels in a Node B
System Information Management	Manages the scheduling of system information to be broadcast in a cell
Resource Event Management	Informs the CRNC about the status of Node B resource
Configuration Alignment	Verifies and enforces that both nodes have the same information on the configuration of the radio resources
Measurements on Common Resources	Initiate measurements in the Node B Report the result of the measurements
Radio Link Management	Manages radio links using dedicated resources in a Node B
Radio Link Supervision	Reports failures and restorations of a radio link
Compressed Mode Control [FDD]	Control the usage of compressed mode in a Node B
Measurements on Dedicated Resources	Initiate measurements in the Node B and report the result of the measurements
DL Power Drifting Correction [FDD]	Adjusts the DL power level of one or more radio links in order to avoid DL power drifting between the radio links
Reporting of General Error Situations	Reports general error situations for which function-specific error messages have not been defined
Physical Shared Channel Management [TDD]	Manages physical resources in the Node B belonging to shared channels (USCH/DSCH)
DL Power Time Slot Correction [TDD]	Enables the Node B to apply an individual offset to the transmission power in each time slot according to the downlink interference level at the UE

1.21.2 NBAP Elementary Procedures (EPs)

The NBAP protocol is used between RNC and Node B to configure and manage the Node B and setup channels on Iub and Uu interfaces. It consists of EPs. An EP is a unit of interaction between the CRNC and the Node B; the NBAP **INITIATING MESSAGE** is transporting the procedure request. The EP is identified by the parameter **Procedure Identification Code**. The CRNC Communication Context contains all info for the CRNC to communicate with a specific UE.

The Context is identified by the parameter **CRNC Communication Context Identifier**.

An EP consists of an initiating message and possibly a response message. Two kinds of EPs are used:

- *Class 1*: EPs with response (success or failure)
- *Class 2*: EPs without response

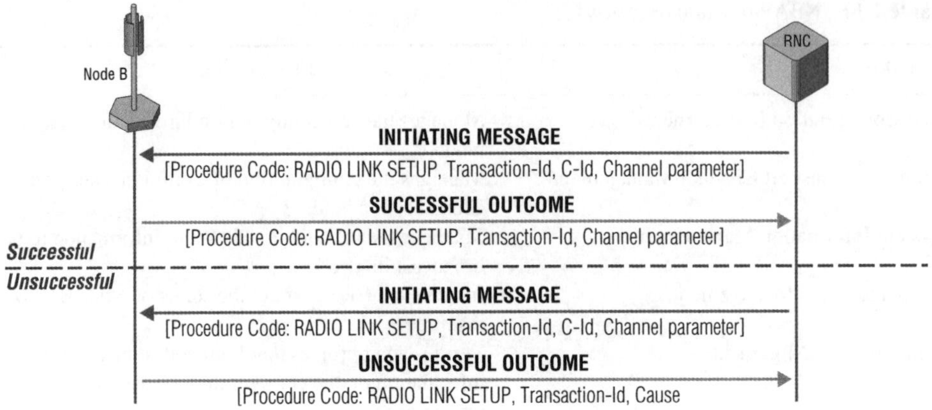

Figure 1.104   NBAP (message flow).

For *Class 1* EPs, the types of responses can be as follows:

Successful (**SUCCESSFUL OUTCOME** Message)

- A signaling message explicitly indicates that the elementary procedure has been successfully completed with the receipt of the response

Unsuccessful (**UNSUCCESSFUL OUTCOME** Message)

- A signaling message explicitly indicates that the EP failed

Class 2 EPs are considered always successful.

1.21.3 Example – NBAP

The example in Figure 1.104 shows the Radio Link Setup procedure, Class 1, both successful and unsuccessful.

1.22 Radio Network Subsystem Application Part (RNSAP)

RNSAP (*3GPP 25.423*) is the communication protocol used on the Iur interface between RNCs and is specified using ASN.1 PER.

1.22.1 RNSAP Functions

The RNSAP protocol covers different functions as described in Table 1.11.

RNSAP contains two classes of elementary procedures. The handling is the same as with NBAP.

The Iur interface RNSAP procedures are divided into four modules:

1. *RNSAP Basic Mobility procedures*
   Contain procedures used to handle the mobility within UTRAN.

Table 1.11   RNSAP function overview

Function	Description
Radio Link Management	Manages radio links using dedicated resources in a DRNS
Physical Channel Reconfiguration	Reallocates the physical channel resources for a radio link
Radio Link Supervision	Reports failures and restorations of a radio link
Compressed Mode Control [FDD]	Controls the usage of compressed mode within a DRNS
Measurements on Dedicated Resources	Initiates measurements on dedicated resources in the DRNS. The function also allows the DRNC to report the result of the measurements
DL Power Drifting Correction [FDD]	Adjusts the DL power level of one or more Radio links in order to avoid DL power drifting between the radio links
CCCH Signaling Transfer	Passes information between the UE and the SRNC on a CCCH controlled by the DRNS
Paging	Pages a UE in a URA or a cell in the DRNS
Common Transport Channel Resources Management	Utilizes Common Transport Channel Resources within the DRNS (excluding DSCH resources for FDD)
Relocation Execution	Finalizes a relocation previously prepared via other interfaces
Reporting of General Error Situations	Reports on the general error situations, for which function-specific error messages have not been defined
DL Power Time Slot Correction [TDD]	Applies an individual offset to the transmission power in each time slot according to the downlink interference level at the UE

2. *RNSAP DCH procedures*
   Contain procedures that are used to handle DCHs, DSCHs, and USCHs between two RNSs. If procedures from this module are not used in a specific Iur, then the usage of DCH, DSCH, and USCH traffic between corresponding RNSs is not possible.
3. *RNSAP Common Transport Channel procedures*
   Contain procedures that are used to control Common Transport Channel data streams (excluding the DSCH and USCH) over Iur interface.
4. *RNSAP Global procedures*
   Contain procedures that are not related to a specific UE. The procedures in this module are in contrast to the above modules involving two peer CRNCs.

1.22.2 Example – RNSAP Procedures

Example 1 in Figure 1.105 shows the transport of Layer 3 information on the Iur interface, using Class 2 elementary procedures.

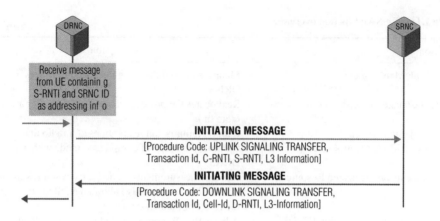

Figure 1.105   RNSAP Procedure Example 1.

Example 2 shows the Paging procedure, Class 2; Example 3 shows a successful Radio Link Setup procedure, using Class 1 (Figure 1.106).

1.23 Radio Access Network Application Part (RANAP)

RANAP (*3GPP 25.413*) provides the signaling service between UTRAN and CN which is required to fulfill the RANAP functions. RANAP services are divided into three groups on the basis of Service Access Points (SAPs):

1. *General control services*: They are related to the whole Iu interface instance between the RNC and the logical CN domain, and are accessed in CN through the General Control SAP. They utilize connectionless signaling transport provided by the Iu signaling bearer.

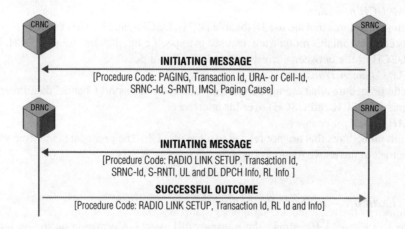

Figure 1.106   RNSAP Procedure Examples 2 and 3.

2. *Notification services*: They are related to specified UEs or all UEs in a specified area, and are accessed in CN through the Notification SAP. They utilize connectionless signaling transport provided by the Iu signaling bearer.

3. *Dedicated control services*: They are related to one UE, and are accessed in CN through the Dedicated Control SAP. RANAP functions that provide these services are associated with Iu signaling connection that is maintained for the UE in question. The Iu signaling bearer provides connection-oriented signaling transport to realize the Iu signaling connection.

The RNSAP protocol covers different functions as described in Table 1.12.

1.23.1 RANAP Elementary Procedures (EPs)

The RANAP protocol consists of EPs. An EP is a unit of interaction between the RNS and the CN; RANAP **INITIATING MESSAGE** is transporting the procedure request.

The EPs are defined separately and are intended to be used to build up complete sequences in a flexible manner. If the independence between some EPs is restricted, it is described under the relevant EP description.

Unless otherwise stated by the restrictions, the EPs may be invoked independently of each other as stand-alone procedures, which can be active in parallel.

An EP consists of an initiating message and possibly a response message. Three kinds of EPs are used:

- *Class 1*: EPs with response (success and/or failure)
- *Class 2*: EPs without response
- *Class 3*: EPs with possibility of multiple responses

For *Class 1* EPs, the types of responses can be as follows:

Successful (**SUCCESSFUL OUTCOME** Message)

- A signaling message explicitly indicates that the elementary procedure successfully completed with the receipt of the response

Unsuccessful (**UNSUCCESSFUL OUTCOME** Message)

- A signaling message explicitly indicates that the EP failed
- On time supervision expiry (for example absence of expected response)

Successful and Unsuccessful

- One signaling message reports both a successful and an unsuccessful outcome for the different included requests. The response message used is the one defined for a successful outcome

Class 2 EPs are always considered successful.

Class 3 EPs have one or several response messages reporting both successful and unsuccessful outcomes of the requests, and temporary status information about the requests. This type of EP terminates only through response(s) or the EP timer expiry; the response is transmitted as **OUTCOME** message.

Table 1.12   RANAP function overview

Function	Description
Relocating SRNC	Changes the SRNC functionality as well as the related Iu resources [RAB(s) and Signaling connection] from one RNC to another.
Overall RAB Management	Sets up, modifies, and releases RAB
Queuing the setup of RAB	Allows placing some requested RABs into a queue and indicates the peer entity about the queuing
Requesting RAB release	Requests the release of RAB (overall RAB management is a function of the CN)
Release of all Iu connection resources	Explicitly releases all resources related to one Iu connection
Requesting the release of all Iu connection resources	Requests release of all Iu connection resources from the corresponding Iu connection (Iu release is managed from the CN)
SRNS context forwarding function	Transfers SRNS context from the RNC to the CN for intersystem change in case of packet forwarding
Controlling overload in the Iu interface	Allows adjusting of the load in the Iu interface
Resetting the Iu	Resets an Iu interface
Sending the UE Common ID (permanent NAS UE identity) to the RNC	Makes the RNC aware of the UE's Common ID
Paging the user	Provides the CN for capability to page the UE
Controlling the tracing of the UE activity	Sets the trace mode for a given UE and deactivates a previously established trace
Transport of NAS information between UE and CN with two subclasses	Transport of the initial NAS signaling message from the UE to CN. This function transparently transfers the NAS information. As a consequence, the Iu signaling connection is also set up
	Transport of NAS signaling messages between UE and CN. This function transparently transfers the NAS signaling messages on the existing Iu signaling connection. It also includes a specific service to handle signaling messages differently
Controlling the security mode in the UTRAN	Sends the security keys (ciphering and integrity protection) to the UTRAN, and sets the operation mode for security functions
Controlling location reporting	Operates the mode in which the UTRAN reports the location of the UE
Location reporting	Transfers the actual location information from RNC to the CN
Data volume reporting function	Reports unsuccessfully transmitted DL data volume over UTRAN for specific RABs
Reporting general error situations	Reports general error situations, for which function-specific error messages have not been defined

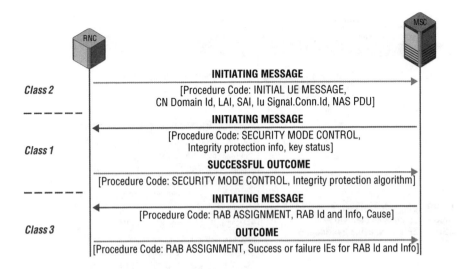

Figure 1.107 RANAP Procedure (message flow).

1.23.2 Example – RANAP Procedure

The example in Figure 1.107 shows all types of EP classes of RANAP signaling.

1. EP: INITIAL UE MESSAGE, Class 2
2. EP: SECURITY MODE CONTROL, Class 1, successful
3. EP: RAB ASSIGNMENT, Class 3, with response

1.24 ATM Adaptation Layer Type 2 – Layer 3 (AAL2L3/ALCAP)

On UMTS Iu interfaces, AAL2L3 (*ITU-T Q.2630*) represents the ALCAP function. ALCAP is a generic name for the transport signaling protocol used to set up and tear down transport bearers.

The AAL2 signaling protocol provides the signaling capability to establish, release, and maintain AAL2 point-to-point connections across a series of ATM VCCs that carry AAL2 links.

The AAL2 signaling protocol also provides maintenance functions associated with the AAL2 signaling.

In the UTRAN the RNC always starts setup and release of AAL2 SVCs using AAL2L3 signaling procedures.

1.24.1 AAL2L3 Message Format

An AAL2L3 connection is identified by a pair of Destination and Originating Signaling Association IDs (Figure 1.108).

The Binding ID provided by the radio network layer is copied into the Served User Generated Reference (SUGR) parameter of ESTABLISH.request primitive. User Plane Transport bearers for Iur interface are established and released by the AAL2L3 in the SRNC. The binding

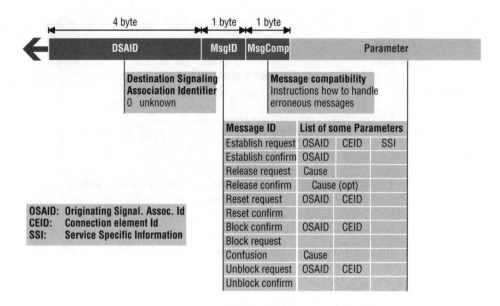

Figure 1.108   AAL2L3 message format.

identity will already have been assigned and tied to a radio application procedure when the first AAL2L3 message was received over the Iur interface in the DRNC.

User Plane Transport bearers for Iub interface are established and released by the AAL2L3 in the CRNC.

AAL2 transport layer addressing is based on embedded E.164 or AESA variants of the NSAP addressing format (E.191). Native E.164 addressing will not be used.

1.24.2 Example – AAL2L3 Procedure

Signaling Association Identifiers (SAIDs) are treated in the following way (Figure 1.109):

1. Whenever a new signaling association is created, a new protocol entity instance is created and an OSAID is allocated to it; this ID is then transported in the first message in the OSAID parameter. The DSAID in this message contains the value "unknown," meaning that all octets are set to "0." (In the figures, this is indicated by "DSAID=0.")
2. Upon receipt of a message that has a DSAID field set to "unknown," a new protocol entity instance is created and an OSAID is allocated to it.
3. In the first message returned to the originator of the association, the OSAID of the sending protocol entity instance is transported in the OSAID parameter. The DSAID field carries the previously received OSAID of the originator of the association.
4. In all subsequent messages, the DSAID field carries the previously received OSAID of the destination entity.
5. The first message returned to the originator of the association is also the last one for this signaling association (Release Confirm); no OSAID parameter is carried in the message. The SAID field carries the previously received OSAID of the originator of the association.

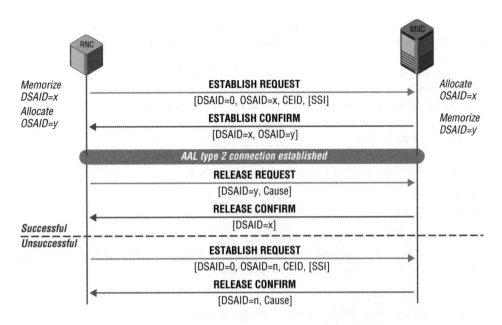

Figure 1.109   AAL2L3 establish and release example.

In order to minimize the likelihood of CEID collision, the following CEID allocation mechanism is used:

- If the AAL2 node owns the AAL2 path that carries the new connection, it allocates CEID values from CEID value 8 upwards.
- If the AAL2 node does not own the AAL2 path that carries the new connection, it allocates CEID values from CEID value 255 downwards.

Each AAL2 connection request (regardless of whether it comes directly from an AAL2 served user or from an adjacent AAL2 node) will contain an AAL2 service endpoint address, which indicates the destination of the intended AAL2 connection instance. This information is used to route the AAL2 connections via the AAL2 network to its destination endpoint. In capability set 1, the supported address formats are NSAP and E.164.

It is up to the application area or the operator of a particular network to decide what addressing plan is used in the AAL2 network. The addressing plan in the AAL2 network can be a reuse of the addressing plan in the underlying ATM network, but it can also be an independent addressing plan defined exclusively for the AAL2 network.

1.25 IU User Plane Protocol

The Iu UP protocol (*3GPP 25.415*) is located in the user plane of the radio network layer over the Iu interface, the Iu UP protocol layer. It is used to convey user data associated to RABs to meet the needs of CS and PS domain user data traffic.

One Iu UP protocol instance is associated to one and only one RAB. If several RABs are established towards one given UE, then these RABs make use of several Iu UP protocol instances.

These Iu UP protocol instances are established, relocated, and released together with the associated RAB.

The Iu UP protocol operates in modes. Modes of operation of the protocol are defined:

1. *Transparent mode (TM)*
   The transparent mode is intended for those RABs that do not require any particular feature from the Iu UP protocol other than transfer of user data.

   - Null protocol
   - Non-real-time data in plain GTP-U format

2. *Support mode for predefined SDU size (SMpSDU)*
   The support modes are intended for those RABs that do require particular features from the Iu UP protocol in addition to transfer of user data.

   - Rate control, time alignment
   - Procedure control function, such as AMR speech data

When operating in a support mode, the peer Iu UP protocol instances exchange Iu UP frames, whereas in transparent mode, no Iu UP frames are generated.

Determination of the Iu UP protocol instance mode of operation is a CN decision taken at RAB establishment based on, for example, the RAB characteristics. It is signaled in the radio network layer control plane at RAB assignment and at relocation for each RAB. It is internally indicated to the Iu UP protocol layer at user plane establishment. The choice of a mode is bound to the nature of the associated RAB and cannot be changed unless the RAB is changed.

1.25.1 Iu UP Transparent Mode

In this mode, the Iu UP protocol instance does not perform any Iu UP protocol information exchange with its peer over the Iu interface: no Iu frame is sent (Null protocol). The Iu UP protocol layer is crossed through by PDUs being exchanged between higher layers and the transport network layer (Figure 1.110).

For instance, the transfer of GTP-U PDUs could utilize the transparent mode of the Iu UP protocol.

Note that the data is transmitted on user plane channels, which have to be established earlier on by the RANAP RAB Assignment procedure. At the end of the connection, RANAP needs to release the user plane channel again.

1.25.2 Iu UP Support Mode Data Frames

Support Mode data frames represent the Iu UP NAS Data Streams specific function (Figure 1.111).

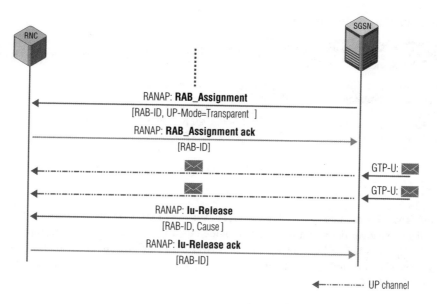

Figure 1.110 Iu UP transparent mode.

These functions are responsible for a limited manipulation of the payload and the consistency check of the frame number. If a frame loss is detected because of a gap in the sequence of the received frame numbers (for a RAB where frame numbers does not relate to time), then this gap is reported to the Procedure Control function. These functions are responsible for the CRC check and calculation of the Iu UP frame payload part. These functions are also responsible for the Frame Quality Classification handling as described below.

Transmission of data

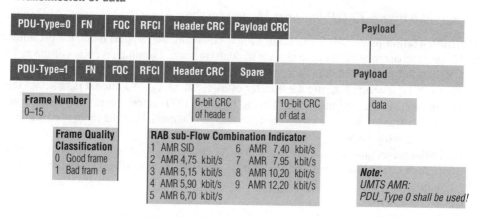

Figure 1.111 Iu UP support mode data frames.

Transmission of control information

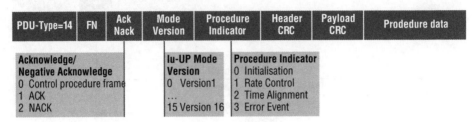

Figure 1.112   Iu UP Support Mode Control Frames.

PDU-Type 0 is defined to transfer user data over the Iu UP in support mode for predefined SDU sizes. An error detection scheme is provided over the Iu UP for the payload part.

PDU-Type 1 is defined to transfer user data over the Iu UP in support mode for predefined SDU sizes when no payload error detection scheme is necessary over the Iu UP, meaning there is no payload CRC.

1.25.3 Iu UP Support Mode Control Frames

A *Frame Number* handles the Iu UP frame numbering. The frame numbering can be based on either time or sent Iu UP PDU (Figure 1.112).

Frame Quality Classification is used to classify the Iu UP frames depending on whether errors have occurred in the frame or not. Frame Quality Classification is dependent on the RAB attribute *Delivery of Erroneous SDU* IE.

RAB Subflow Combination Indicator identifies the structure of the payload. This can be used to specify the sizes of the subflows. Subflows are AMR classes, meaning the maximum number of subflows is three and they correspond with AMR Class A, Class B, and Class C bits.

1.25.4 Example – Iu UP Support Mode Message Flow

The **Initialization** procedure is mandatory for RABs using the support mode for predefined SDU size. The purpose of the procedure is to configure both termination points of the Iu UP with RAB Subflow Combination Indicators (RFCIs) and associated RAB Subflows SDU sizes necessary to be supported during the transfer of user data phase. Additional parameters may also be passed, such as the Inter PDU Timing Interval (IPTI) information. The Initialization procedure is always controlled by the entity in charge of establishing the radio network layer user plane, meaning SRNC.

The Initialization procedure is invoked whenever indicated by the Iu UP Procedure Control function, for example, as a result of a relocation of SRNS or at RAB establishment over Iu. The Initialization procedure will not be re-invoked for the RAB without a RAB modification requested via RANAP.

The Iu user plane data could be speech, using AMR. The data packets will be transmitted with Iu UP PDU type 0 frames.

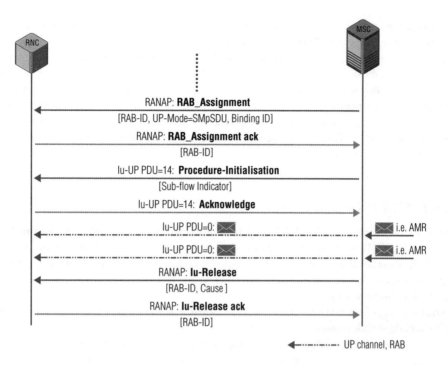

Figure 1.113   Iu UP support mode simple message flow.

There is no Iu UP release control frame. Instead the RANAP will release the resource (Figure 1.113).

1.26 Adaptive Multirate (AMR) Codec

The AMR codec (*3GPP 26.101*) offers a wide range of data rates and is used to lower codec rates as interference increases on the air interface. It is also used to harmonize codec standards amongst different cellular systems. AMR consists of the multirate speech coder, a source controlled rate scheme including a voice activity detector, a comfort noise generation system, and an error concealment mechanism to combat the effects of transmission errors and lost packets.

The multirate speech coder is a single integrated speech codec with eight source rates from 4.75 to 12.2 Kbps, and a low rate background noise-encoding mode. The speech coder is capable of switching its bitrate every 20-ms speech frame upon command.

There are two formats of AMR frames. AMR Interface Format 1 (AMR IF1) is the generic frame format for both the speech and comfort noise frames of the AMR speech codec. AMR Interface Format 2 (AMR IF2) is useful, for example, when the AMR codec is used in connection with applicable ITU-T H-series of recommendations.

The mapping of the AMR Speech Codec parameters to the Iu interface specifies the frame structure of the speech data exchanged between the RNC and the Transcoder (TC) during normal operation. This mapping is independent from the radio interface in the sense that it has the same structure for both FDD and TDD modes of the UTRAN.

The RAB parameters are defined during the RANAP **RAB Assignment** procedure initiated by the CN to establish the RAB for AMR. The AMR RAB is established with one or more RAB coordinated subflows with predefined sizes and QoS parameters. In this way, each RAB subflow combination corresponds to one AMR frame type. On the Iu interface, these RAB parameters define the corresponding parameters regarding the transport of AMR frames. Some of the QoS parameters in the RAB assignment procedure are determined from the Bearer Capability Information Element used at call setup.

1.26.1 AMR IF1 Frame Architecture

The AMR IF1 frame is used in UMTS for transmission of speech information (Figure 1.114).

Frame Type will indicate the type and size of the core frame contents. Mode Indication and Mode Type are also used to specify the AMR codec mode. The Frame Quality Indicator indicates whether the data in the frame contains errors. Note that the parameter is also used in the Iu UP protocol with inverted value meaning.

The mapping of the bits between the generic AMR frames and the PDU is the same for both uplink and downlink frames.

The number of RAB subflows, their corresponding sizes, and their attributes, such as "Delivery of erroneous SDUs," are defined at the RAB establishment and are signaled in the RANAP RAB establishment request. The number of RAB subflows is corresponding to the desired bit

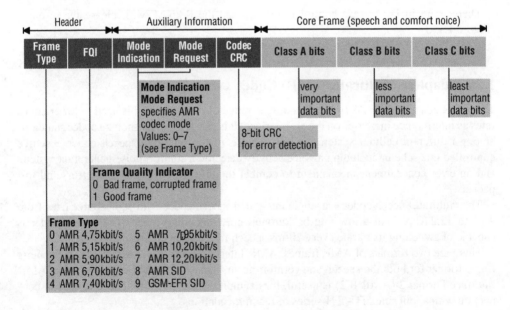

Figure 1.114   AMR IF1 frame architecture.

protection classes. The total number of bits in all subflows for one RFC has to correspond to the total number of a generic AMR frame format IF1, for the corresponding Codec Mode and Frame Type.

The RFCI definition is given in sequence of increasing SDU sizes. The definition describes Codec Type UMTS_AMR, with all eight codec modes, the Active Codec Set (ACS), and provision for Source Controlled Rate (SCR) operation.

1.27 Terminal Adaptation Function (TAF)

TAF (*3GPP 27.001, 3GPP 27.002, 3GPP 27.003*) is based on the principles of terminal adaptor functions presented in the ITU-T I-series of recommendations (I.460 to I.463).

The PLMN supports a wide range of voice and nonvoice services in the same network. To enable nonvoice traffic in the PLMN, there is a need to connect various kinds of terminal equipment to the MT.

The main functions of the MT to support data services are:

- Ensures conformity of terminal service requests to network capability
- Physically connects the reference points R and S
- Controls flow of signaling and mapping of user signaling to/from GSM PLMN access signaling
- Adapts rate of user data (see GSM 04.21) and data formatting for the transmission SAP
- Controls flow of nontransparent user data and mapping of flow control for asynchronous data services
- Supports data integrity between the MS and the IWF in the GSM PLMN
- Provides end-to-end synchronization between terminals
- Filters status information
- Supports nontransparent bearer services, for example, termination of the RLP and the Layer 2 Relay (L2R) function including optional data compression function (where applicable)
- Checks terminal compatibility
- Optionally supports local test loops

1.28 Radio Link Protocol (RLP)

The RLP (*3GPP 24.022*) utilizes reliability mechanisms of the underlying protocols in order to deliver data and terminates at the MS and IWF (typically at the MSC). It has been specified for circuit-switched data transmission within the GSM and UMTS PLMN. RLP covers the Layer 2 functionality of the ISO/OSI Reference Model. RLP has been tailored to the special needs of digital radio transmission. RLP is intended for use with nontransparent data transfer. Protocol conversion may be provided for a variety of protocol configurations. Some more features of RLP:

- Nearly identical to LAPD (Link Access Procedures on the D-Channel)
- Intended for use with nontransparent data transfer

- Foreseen data applications
  - Character-mode protocols using start–stop transmission (IA5)
  - X.25 LAP-B (Link Access Procedures on the Bearer Channel)
- Located in MT and IWF of the PLMN
- In UMTS RLP support the 576-bit frame length, in GSM 240-bit
- Two modes of operation:
  - ADM   Asynchronous Disconnected Mode
  - ABM   Asynchronous Balanced Mode
- Three RLP versions:
  - Version 0: single-link basic version
  - Version 1: single-link extended (data compression) version
  - Version 2: multilink version (1–4 physical links)

In UMTS, the RLP frame has a fixed length of 576-bits.

A frame consists of a header, an information field, and an FCS (frame check sequence) field. The size of the components depends on the radio channel type, on the RLP version, and on the RLP frame. As a benefit of using strict alignment with underlying radio transmission, there is no need for frame delimiters (such as flags) in RLP. In consequence, there is no "bit-stuffing" necessary in order to achieve code transparency.

1.29 Packet Data Convergence Protocol (PDCP)

PDCP (*3GPP 25.323*) is used to format data into a suitable structure prior to transfer over the air interface and provides its services to the NAS at the UE or the relay at the RNC.

PDCP uses the services provided by the RLC sublayer.

PDCP performs the following functions:

- Header compression and decompression of IP data streams (for example, TCP/IP and RTP/UDP/IP headers) at the transmitting and receiving entity, respectively. The header compression method is specific to the particular network layer, transport layer, or higher layer protocol combinations (for example, TCP/IP and RTP/UDP/IP)
- Transfer of user data (transmission of user data means that PDCP receives PDCP SDU from the NAS and forwards it to the RLC layer and vice versa)
- Maintenance of PDCP sequence numbers for radio bearers that are configured to support lossless SRNS relocation
- Multiplexing of different RBs onto the same RLC entity.

1.29.1 PDCP PDU Format

Data using the Transparent SAP (Tr-SAP) will use the *PDCP-No-Header-PDU*.

Data using the UM-SAP will use the *PDCP-Data-PDU* (Figure 1.115).

Data using the AM-SAP will use the *PDCP-SeqNum-PDU*. The sequence number will allow the detection of frame loss.

The *Packet Identifier* (*PID*) identifies the type of compression.

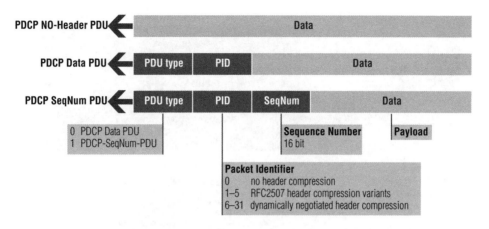

Figure 1.115 PDCP PDU Format.

1.30 Broadcast/Multicast Control (BMC)

BMC (*3GPP 25.324*) adapts broadcast and multicast services on the radio interface and is a sublayer of Layer 2 that exists in the UP only. It is located above RLC. The L2/BMC sublayer is assumed to be transparent for all services except broadcast/multicast.

BMC Functions

- Storage of Cell Broadcast Messages (CBMs)
- Traffic volume monitoring and radio resource request for CBS
- Scheduling of BMC messages
- Transmission of BMC messages to UE
- Delivery of CBMs to higher layer (NAS)
- Only one procedure: BMC Message Broadcast

At the UTRAN side, the BMC sublayer consists of one BMC protocol entity per cell. Each BMC entity requires a single CTCH, which is provided by the MAC sublayer, through the RLC sublayer. The BMC requests the Unacknowledged Mode service of the RLC.

The BMC entity on the network side predicts periodically the expected amount of CBS traffic volume (CTCH transmission rate in kbps), which is needed for the transmission of CBMs and indicated to the RRC. The algorithms used for traffic volume prediction are implementation-dependent and thus do not need to be specified. Some parameters may be set by the O&M system.

The algorithms depend on the chosen algorithms for CBM scheduling. This procedure calculates the CBS schedule periods and assigns BMC messages (for example CBS Messages and Schedule Messages) to the CBS schedule periods. The procedure then gives an indication of which of the CTCH Block Sets containing part of or the complete BMC messages has the status "new."

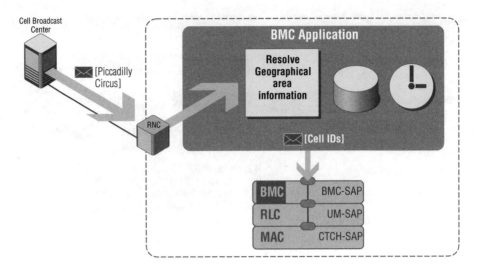

Figure 1.116 BMC architecture.

1.30.1 BMC Architecture

It is assumed that there is a function in the RNC above BMC that resolves the geographical area information of the CB message (or, if applicable, performs evaluation of a cell list) received from the Cell Broadcast Center (CBC). A BMC protocol entity serves only those messages at BMC-SAP that are to be broadcast into a specified cell (Figure 1.116).

1.31 Circuit-Switched Mobility Management (MM)

Mobility Management is a generic term for the specific mobility functions provided by a PLMN. Such functions include, e.g., tracking a mobile as it moves around a network and ensuring that communication is maintained.

The CS-specific MM part is well known from GSM and is used quite unchanged for UMTS Rel.99 (*3GPP 24.008*).

MM Functions
- MM procedures to establish and release connections
- Transfer of CM sublayer messages
- MM common procedures for security functions, for example, the Authentication procedure
- MM-specific procedures for location functions such as Periodic location updating or IMSI attach procedure
- UE identified by IMSI or TMSI

Mobility Management procedures are used to set up the connection between UE and the CS CN. Procedures like Authentication and Location Update are also part of CS MM.

A CS CN will recognize a UE by the IMSI or by a previously assigned TMSI.

1.32 Circuit-Switched Call Control (CC)

The Circuit-Switched Call Control (CC) protocol includes some basic procedures for mobile CC (no transport control!):

- Call establishment procedures
- Call clearing procedures
- Call information phase procedures
- Miscellaneous procedures

CC entities are described as communicating finite state machines that exchange messages across Radio interfaces and communicate internally with other protocol (sub)layers.

The Circuit-Switched Call Control protocol part has only slightly changed from GSM to UMTS Rel.99 (*3GPP 24.008*).

Parameters for QoS (for example, the RAB specification) have been added to the signaling protocol.

CC Functions

- Procedures similar to GSM
- CC establishes and releases CC connections between UE and CN
- Activation of voice/multimedia codec
  - Based on 3G-324M, variant of H.324 (see *3GPP 26.111*)
- Interworking with RANAP for establishment of a RAB
  - CC Setup QoS will be mapped onto RANAP RAB assignment

1.33 Example – Mobile Originated Call (Circuit Switched)

As shown in Figure 1.117, the procedure is identical to GSM from the MM and CC point of view.

However, the ciphering is not performed by the CN in the same way as known from GSM. Instead, the other main protocol of the IuCS interface, the RANAP, is in charge of all types of RAB signaling.

The initial UE message, in this example the **CM Service Request**, will transport the UE Identity, whereas the CC **Setup** message will contain the dialed telephone number.

All given messages will run on top of RANAP, which will run on top of SCCP protocol. The SCCP is responsible for defining the UE procedure connection.

1.34 Packet-Switched Mobility Management (GMM)

The **GPRS Mobility Management** protocol (*3GPP 24.008*) is used to make a UE known to the packet-switched CN and to follow its mobility. The procedures have changed only by a message, which is used as the initial UE message when connecting UE with the packet network, Session Management Activate PDP Context. This new message is the Service Request message and is used to set up a secure connection with the ability to define a QoS for the signaling information between UE and SGSN.

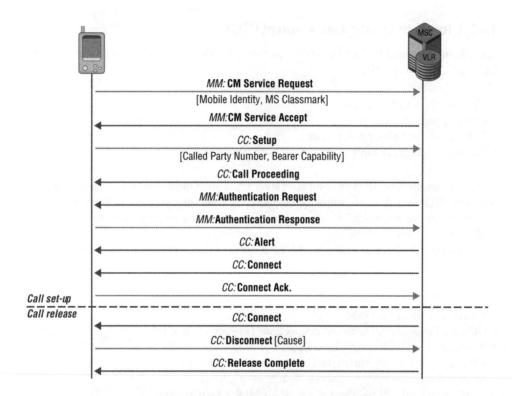

Figure 1.117   Mobile Originated Call (message flow).

GMM Functions

- Procedures similar to GPRS (GMM)
- GMM protocol makes use of a signaling connection between UE and SGSN
- GMM establishes and releases GMM contexts, for example, GPRS Attach
- GMM-specific procedures for location functions like periodic RA updating
- New message implemented to provide service to CM sublayer on top of GMM:
  – Service Request message
  – Initiated by UE, used to establish a secure connection to the network and to request the bearer establishment for sending data
- UE identified by IMSI or P-TMSI

1.35 Packet-Switched Session Management (SM)

The GPRS Session Management protocol (*3GPP 24.008*) is similar to the CS CC and is used to define the connection of a UE to a packet network.

SM exists in the UE and in the SGSN and handles PDP Context Activation, Modification, Deactivation, and Preservation Functions.

The GPRS SM protocol is used between UE and SGSN whereas the SGSN acts as relay function toward the GGSN.

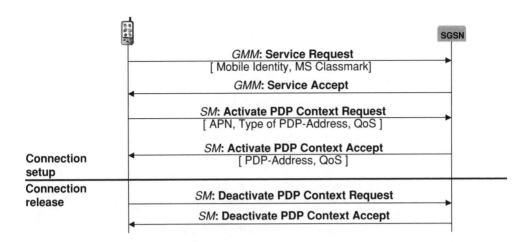

Figure 1.118   Activate PD Context (message flow).

SM Functions

- Procedures similar to GPRS (SM)
- Counterpart to CS CC protocol, meaning SM protocol is used to establish and release packet data sessions
- SM procedures to set up and release one or more PDP Contexts
- PDP Contexts are handled in UE and GGSN
- SGSN represents IWF

1.36 Example – Activate PDP Context (Packet Switched)

The example in Figure 1.118 shows the "new" signaling flow for activating and deactivating the PDP Context on the IuPS interface.

As mentioned earlier, the **GMM Service Request** and **Service Accept** are new to PS CN. The Service Request will contain the UE identity and the MS Classmark to define a QoS and an RB for the signaling. The Activate PDP Context message will contain the QoS parameter for the UP RB.

2

Short Introduction to Network Monitoring, Troubleshooting, and Network Optimization

This chapter shall give some *practical tips* and tricks regarding network monitoring, troubleshooting, and network optimization. The emphasis is on *general ideas* that help to operate and optimize the network. It must always be kept in mind that configurations and resource planning differs from manufacturer to manufacturer and from operator to operator. Also customer-specific information must be treated as confidential and cannot be published. For this reason not every gap analysis can be drilled down as deep as possible.

2.1 Iub Monitoring

Most Node Bs are not connected directly to CRNC using an STM-1 line. As a rule the STM-1 lines from CRNC lead to one or more ATM routers that also act as interface converters. From ATM routers/interface converters often E1 lines lead to the Node B. The reason for this kind of connection is that these E1 lines already exist – so they just need to be configured to fit to UTRAN configuration needs (Figure 2.1). In a minimum configuration, two E1 lines lead to one Node B, where the second line mirrors the configuration of the first one for redundancy and loadsharing reasons. However, an E1 line has total data transmission capacity of 2 Mbps, but a single user in an UTRAN FDD cell shall already be able to set up a connection with 384 kpbs. So it is clear that for high-speed data transmission services, one or two E1 lines are not enough. For this reason Inverse Multiplex Access (IMA) was introduced.

2.1.1 IMA

IMA provides the inverse multiplexing of an ATM cell stream over up to 32 physical links (E1 lines) and to retrieve the original stream at the far end from these physical links (Figure 2.2).

UMTS Signaling   Ralf Kreher and Torsten Rüdebusch
© 2005 Tektronix, Inc.   ISBN: 0-470-01351-6.

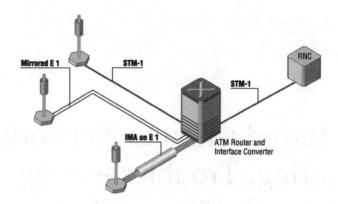

Figure 2.1   Possible transport network configurations on Iub.

The multiplexing of the ATM cell stream is performed on a cell-by-cell basis across these physical links.

The ATM inverse multiplexing technique involves inverse multiplexing and demultiplexing of ATM cells in a cyclical fashion among links grouped to form a higher bandwidth logical link whose rate is approximately the sum of the link rates. This is referred to as an IMA Group. A measurement unit like Tektronix K1297-G20 must be able to monitor all E1 lines belonging to an IMA group and to multiplex/demultiplex ATM cells in the same way as sending/receiving entities of the network.

2.1.2 Fractional ATM

Another technology that is becoming more and more important is fractional ATM. Fractional ATM allows network operators to minimize their infrastructure costs, especially during the UMTS deployment phase when the network load is low. The UMTS UTRAN and the GSM BSS share the same physical medium and exchange user and control information over this

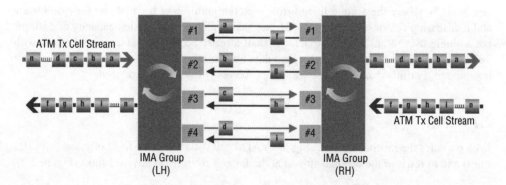

Figure 2.2   IMA: Monitoring of multiplexed ATM cells on E1 lines.

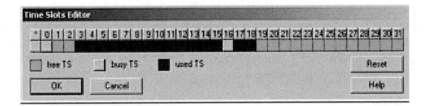

Figure 2.3   Time slot assignment for fractional ATM.

medium with the core network. The K1297-G20 time slot editor allows the assignment of an ATM fraction in any combination and is a good example to explain the fractional ATM principle. In Figure 2.3, the ATM section that forms the UMTS Iub interface is shown in dark grey. The remaining time slots can be used for GPRS Gb or GSM A interface.

2.1.3 Loadsharing and Addressing on Iub

There are several concepts for loadsharing on Iub, which means there are often several NBAP and ALCAP links between one RNC and one Node B. Loadsharing does not only increase the available transport capacity between two protocol peer entities, but it also brings redundancy to the network. In case one link crashes for any reason, there will always be alternative ways for message exchange and the connection between RNC and Node B will not be broken.

In addition, some manufacturers have divided their NBAP links into such used for common procedures and such used for dedicated procedures. A typical addressing and configuration case of a Node B with three cells is shown in Table 2.1.

The DCHs may also run in the same VPI/VCI as common transport channels. Then of course only those CID values can be used for DCH that are not occupied by common transport channels (in the example, CID 20-254 on VPI/VCI = **A/g**).

Table 2.1   Typical Node B configuration with three cells 1

Signaling link/channel	VPI/VCI	Allocated or reserved AAL 2 CID
NBAP Common Procedures 1	**A/a**	
NBAP Common Procedures 2	**A/b**	
NBAP Dedicated Procedures 1	**A/c**	
NBAP Dedicated Procedures 2	**A/d**	
ALCAP 1	**A/e**	
ALCAP 2	**A/f**	
RACH (1 per cell)	**A/g**	8; 12; 16
PCH (1 per cell)	**A/g**	9; 13; 17
FACH 1 (for control plane – 1 per cell)	**A/g**	10; 14; 18
FACH 2 (for user plane IP payload – 1 per cell)	**A/g**	11; 15; 19
Reserved for DCHs (AAL2 SVC)	**A/h**	8 - 254

To monitor the common transport channels RACH, FACH, and PCH, it is necessary to know not only the VPI/VCI, but also the correct transport format set, because here it is defined, e.g., how big RACH, FACH, or PCH RLC blocks are and how often they are sent (time transmission interval). Transport format set parameter values can follow 3GPP recommendations or can be defined by network operators/manufacturers!

2.1.4 Troubleshooting Iub Monitoring Scenarios

Three common problems when monitoring Iub links (without Autoconfiguration) are:

1. *There is no data monitored on common transport channels RACH, FACH, and PCH.*

   Solution: Remember that ATM lines are *unidirectional*! Ensure that measurement configuration is looking for RACH on uplink ATM line, while FACH and PCH can be found on downlink ATM line only!

2. *In case of NBAP or ALCAP only uplink or only downlink messages are captured, e.g., only ALCAP ECF, but ERQ messages are missed.*

   Solution: It may happen that loadsharing of NBAP is not organized following common and dedicated procedures, but following uplink and downlink traffic. In a similar way ALCAP uplink traffic may be sent on a different VPI/VCI than downlink traffic.

   Example:
   ALCAP$_1$ DL (ERQ) on VPI/VCI = **A/e** − ALCAP$_1$ UL (ECF) on VPI/VCI = **A/f**
   ALCAP$_2$ DL (ERQ) on VPI/VCI = **A/g** − ALCAP$_2$ UL (ECF) on VPI/VCI = **A/h**

3. *A monitoring configuration that worked some hours or days ago does not work anymore despite that no configuration parameter was changed.*

   Solution: Most likely a Node B reset procedure was performed. The Node B reset is performed in the same way as Node B setup (described in Section 3.1 of this book), but it may happen that ATM addressing parameters are assigned dynamically during the setup procedure. This means after successful restart the same links will have been established as before the reset, but especially the common transport channels will have assigned different CID values than before.

   Table 2.2 gives an example (based on previous configuration example).

Table 2.2 Typical Node B configuration with three cells 2

Channel name	VPI/VCI	CID before reset	CID after reset
RACH (1 per cell)	**A/g**	8; 12; 16	20; 24; 28
PCH (1 per cell)	**A/g**	9; 13; 17	21; 25; 29
FACH 1 (for control plane – 1 per cell)	**A/g**	10; 14; 18	22; 26; 30
FACH 2 (for user plane IP payload – 1 per cell)	**A/g**	11; 15; 19	23; 27; 31

```
TS 29.331 DCCH-UL (2002-03) (RRC_DCCH_UL)  rrcConnectionSetupComplete (= rrcConnectionSetupComplete)
uL-DCCH-Message
1 message
1.1 rrcConnectionSetupComplete
-00-----  1.1.1 rrc-TransactionIdentifier         0
1.1.2 startList
1.1.2.1 sTARTSingle
-----0--  1.1.2.1.1 cn-DomainIdentity             cs-domain
**b20***  1.1.2.1.2 start-Value                  '00000000000000010010'B
1.1.2.2 sTARTSingle
--1-----  1.1.2.2.1 cn-DomainIdentity             ps-domain
**b20***  1.1.2.2.2 start-Value                  '00000000000000000010'B
```

Figure 2.4   Start values for ciphering in RRC Connection Complete message.

4. *There are decoding errors in RRC messages on recently opened DCHs.*

Solution: The frames that cannot be decoded may be ciphered. The necessary input parameters for deciphering are taken from RANAP Security Mode Control procedure, and also from RRC Connection Setup Complete message on Iub interface (contains, e.g., start values for ciphering sent from UE to each domain – see Figure 2.4).

Hence, in case of soft handover scenarios a successful deciphering is only possible if the first Iub interface (UE in position 1) is monitored during call setup (see Figure 2.5).

An indicator of successful deciphering are proper decoded RLC Acknowledged Data PDUs in the VPI/VCI/CID that carries the DCH for DCCH after RRC Security Mode Complete message with rb-UL-CiphActivationTimeInfo was received from UE (Figure 2.6).

In the unsuccessful case the Iub interface that is used for call establishment is not monitored. Hence, RRC Connection Setup Complete message is not monitored and start values necessary for deciphering cannot be extracted by measurement software. If UE moves later into position

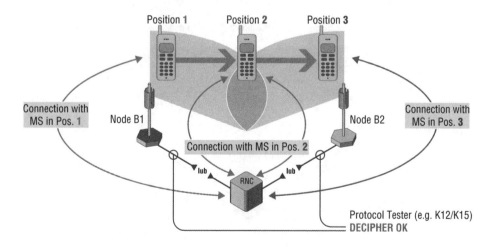

Figure 2.5   Iub deciphering.

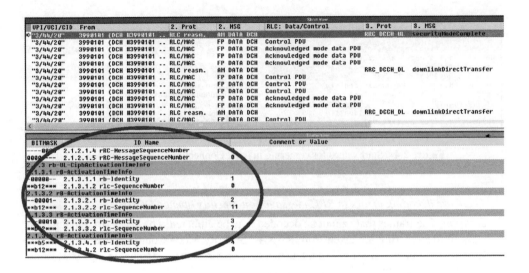

Figure 2.6   Correct decode of RLC Acknowledged Data PDU.

2 it is impossible to decipher the current connection, because correct start values are missed. This case is shown in Figure 2.7.

An indicator that the deciphering is not executed successfully is RLC Acknowledged Data PDUs, which show invalid length field information (Figure 2.8). This is because the length info of RLC frames is also ciphered and so the value is changed and becomes incorrect.

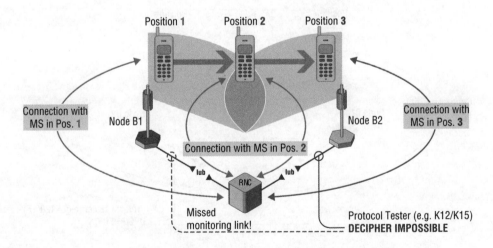

Figure 2.7   Iub deciphering problem.

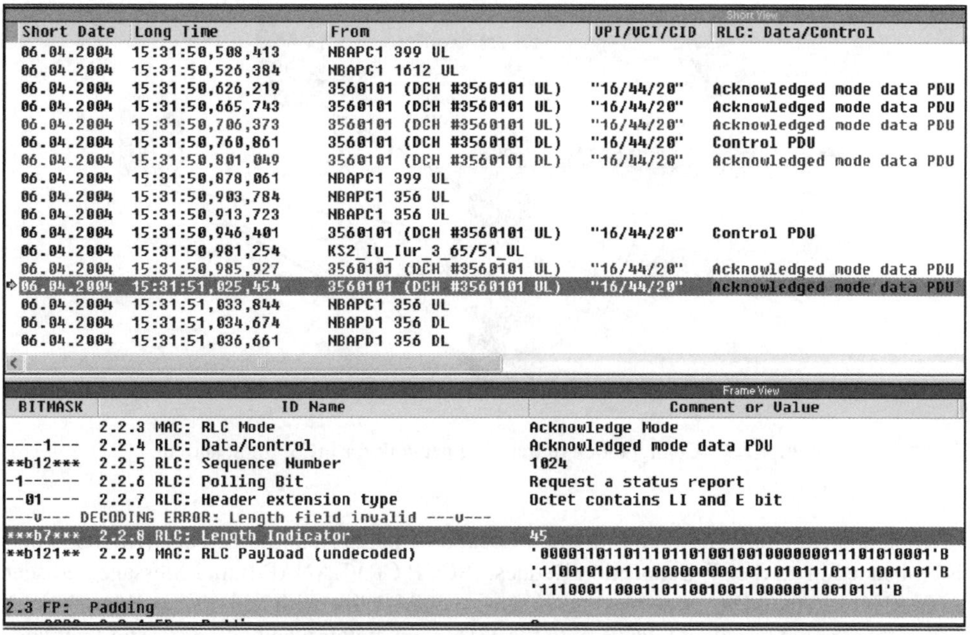

Short Date	Long Time	From	VPI/VCI/CID	RLC: Data/Control
06.04.2004	15:31:50,508,413	NBAPC1 399 UL		
06.04.2004	15:31:50,526,384	NBAPC1 1612 UL		
06.04.2004	15:31:50,626,219	3560101 (DCH #3560101 UL)	"16/44/20"	Acknowledged mode data PDU
06.04.2004	15:31:50,665,743	3560101 (DCH #3560101 UL)	"16/44/20"	Acknowledged mode data PDU
06.04.2004	15:31:50,706,373	3560101 (DCH #3560101 UL)	"16/44/20"	Acknowledged mode data PDU
06.04.2004	15:31:50,760,861	3560101 (DCH #3560101 DL)	"16/44/20"	Control PDU
06.04.2004	15:31:50,801,049	3560101 (DCH #3560101 DL)	"16/44/20"	Acknowledged mode data PDU
06.04.2004	15:31:50,878,061	NBAPC1 399 UL		
06.04.2004	15:31:50,903,784	NBAPC1 356 UL		
06.04.2004	15:31:50,913,723	NBAPC1 356 UL		
06.04.2004	15:31:50,946,401	3560101 (DCH #3560101 UL)	"16/44/20"	Control PDU
06.04.2004	15:31:50,981,254	KS2_Iu_Iur_3_65/51_UL		
06.04.2004	15:31:50,985,927	3560101 (DCH #3560101 UL)	"16/44/20"	Acknowledged mode data PDU
06.04.2004	15:31:51,025,454	3560101 (DCH #3560101 UL)	"16/44/20"	Acknowledged mode data PDU
06.04.2004	15:31:51,033,844	NBAPC1 356 UL		
06.04.2004	15:31:51,034,674	NBAPD1 356 DL		
06.04.2004	15:31:51,036,661	NBAPD1 356 DL		

BITMASK	ID Name	Comment or Value
	2.2.3 MAC: RLC Mode	Acknowledge Mode
----1---	2.2.4 RLC: Data/Control	Acknowledged mode data PDU
b12*	2.2.5 RLC: Sequence Number	1024
-1------	2.2.6 RLC: Polling Bit	Request a status report
--01----	2.2.7 RLC: Header extension type	Octet contains LI and E bit
---v----	DECODING ERROR: Length field invalid ---v---	
b7	2.2.8 RLC: Length Indicator	45
b121	2.2.9 MAC: RLC Payload (undecoded)	'00001101101110110100100100000001101010001'B '11001010111000000001011010111011110011011'B '11100011000110110010011000001100010111'B
2.3 FP: Padding		

Figure 2.8   Iub deciphering protocol trace.

2.2 Iu Monitoring

In the context of this chapter, the term "Iu monitoring" includes monitoring of IuCS, IuPS, and Iur interfaces. Looking forward to Release 5 features also Iurg interface between UTRAN RNC and GERAN BSC can be included, because information exchange on Iurg will be done using a set of RNSAP (Release 5) messages.

Also on Iu interfaces loadsharing is used for capacity and redundancy reasons. However, it may also be possible that data from several interfaces is exchanged using the same link.

Figure 2.9 shows a possible configuration scenario.

Between RNCs and core network elements there are transit exchanges (TEX 1 and TEX 2). These transit exchanges must be seen as multifunctional switches. They have an all-in-one functionality and work simultaneously as ATM router, SS#7 Signaling Transfer Point and interface/protocol converter. If CS core network domain is structured as described in Release 4 specifications, the transit exchanges may also include the Media Gateway (MGW) function.

There are 4 STM-1 fiber lines (4 fibers uplink, 4 fibers downlink) that lead from each RNC to two different transit exchanges (TEX 1 and TEX 2). In the transit exchange the VPI/VCI from the RNC (e.g., VPI/VCI = B/b) is terminated and higher layer messages like SCCP/RANAP are routed depending on Destination Point Code (DPC) of MTP Routing Label (MTP RL).

In other words, all messages on IuCS, IuPS, and Iur interfaces belonging to a single RNC are transported on the 4 STM-1 with two different VPI/VCI values between the RNC and a TEX and there is no distinguished STM-1 line for any interface like IuCS, IuPS, or Iur. The

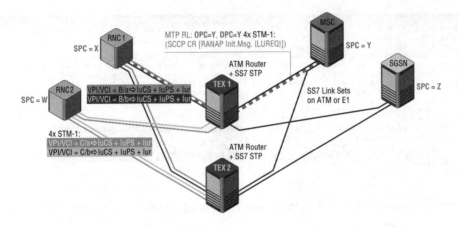

Figure 2.9   Configuration of transport network for IuCS, IuPS, and Iur.

example in the picture shows the way of a Location Update Request message (LUREQ) that
is embedded in an SCCP Connection Request (SCCP CR)/RANAP Initial Message and sent
from RNC 1 to the MSC. This message is sent on STM-1 line with VPI/VCI = B/b to TEX 1
and is then routed based on DPC = Y to the MSC – no matter whether the transport network
between TEX 1 and MSC is ATM as well or just a set of SS7 links on E1 line(s). Whether
the LUREQ message is sent on VPI/VCI = B/a or B/b is decided by loadsharing function of
RNC 1, which also does not depend on any interface characteristics.

2.2.1 Troubleshooting Iu Monitoring

Usually it is not difficult to find the ATM links on Iu interfaces. The only problem when using a
protocol tester for message analysis is to use the correct decoder when the links are monitored
on the STM-1 between RNC and TEX 1/2.

For those readers who are not familiar with protocol testers it should be mentioned that in
a protocol tester the decoder layers are arranged in quite the same way as the layers of the
protocol stack on the monitored link.

The problem on the combined Iu link (with IuCS, IuPS, and Iur) is that there must be a
dynamical decision on which messages are RANAP and which messages are RNSAP. Both
RANAP and RNSAP are users of SCCP identified by different Subsystem Numbers (RANAP:
SSN = 142; RNSAP: SSN = 143). But subsystem numbers are only exchanged during SCCP
connection setup using the SCCP Connection Request (CR) and SCCP Connection Confirmed
(CC) messages. SCCP DT1 messages used to transport NAS PDUs do not have an SSN in
header, but protocol tester decoder layers need to decide for every single DT1 message to
which higher layer decoder the DT1 contents shall be send.

A typical decoder problem looks like the example shown in Figure 2.10.

In the Iu signaling scenarios it was already explained that there is a single SCCP Class 2
connection for each RANAP or RNSAP transaction. Different SCCP Class 2 connections
are distinguished on behalf of their source local reference and destination local reference
numbers (SLR/DLR). So it is necessary to have SLR/DLR context-related protocol decoder

3. Prot	3. MSG	Procedure Code	4. Prot	4. MSG	5. Prot	5. MSG
RL	RL	id-radioLinkSetup	SCCP	CR	RNSAP370	initiatingMessage
RL	RL		SCCP	CC		
RL	RL		SCCP	DT1		
RL	RL		SCCP	DT1		
RL	RL	id-RelocationResourceAll	SCCP	DT1	RANAP err	initiatingMessage
RL	RL	8	SCCP	DT1	RANAP err	initiatingMessage
RL	RL	id-LocationReport	SCCP	DT1	RANAP err	initiatingMessage
RL	RL	8	SCCP	DT1	RANAP err	successfulOutcome
RL	RL	id-CommonID	SCCP	DT1	RANAP err	initiatingMessage
RL	RL	15	SCCP	DT1	RANAP err	successfulOutcome
RL	RL		SCCP	RLSD		
RL	RL		SCCP	RLC		

Figure 2.10 RNSAP decoding errors in DT1 messages on combined Iu link.

as implemented in Tektronix K12/K15 protocol testers to ensure correct decoding of all RANAP/RNSAP messages on the combined link (Figure 2.11). Here the SLR/DLR combination of the active SCCP connection is stored in relation to the higher layer decoder indicated by subsystem number. Using this intelligent feature the decoding errors on Iur disappear.

Another problem that might appear is that single RNSAP messages on Iur interface, especially RNSAP Radio Link Setup messages, are not shown in protocol tester monitor window. This happens because of SCCP segmentation (described in Iur handover scenarios). Figure 2.12 shows a message flow example with RNSAP frames successfully reassembled by protocol tester.

2.3 Network Optimization and Network Troubleshooting

Especially in Europe, UMTS network deployment after successful field trials and service launches entered a new critical stage: the phase of network optimization and network troubleshooting. Despite the fact that users can already use 3G services, there are still many problems in the networks, and the quality of services does not always meet expectations. The objective of network optimization is to evaluate and improve the quality of services. Network troubleshooting means to detect problems, and then find and eliminate the root causes of these problems. The fewer problems one finds, the higher quality of services can be guaranteed.

3. Prot	3. MSG	Procedure Code	4. Prot	4. MSG	5. Prot	5. MSG
RL	RL		SCCP	CR		
RL	RL		SCCP	CC		
RL	RL	id-radioLinkSetup	SCCP	DT1	RNSAP370	initiatingMessage
RL	RL		SCCP	DT1		
RL	RL		SCCP	DT1		
RL	RL	id-downlinkPowerControl	SCCP	DT1	RNSAP370	initiatingMessage
RL	RL	id-dedicatedMeasurementI	SCCP	DT1	RNSAP370	initiatingMessage
RL	RL	id-dedicatedMeasurementI	SCCP	DT1	RNSAP370	successfulOutcome
RL	RL	id-radioLinkRestoration	SCCP	DT1	RNSAP370	initiatingMessage
RL	RL	id-radioLinkDeletion	SCCP	DT1	RNSAP370	initiatingMessage
RL	RL	id-radioLinkDeletion	SCCP	DT1	RNSAP370	successfulOutcome
RL	RL		SCCP	RLSD		
RL	RL		SCCP	RLC		

Figure 2.11 Correct decoding of RNSAP messages on combined Iu link.

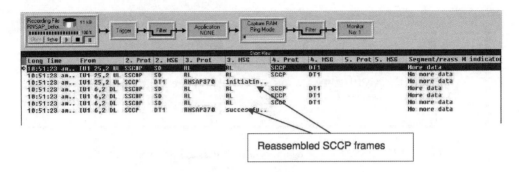

Figure 2.12   Segmented and reassembled RNSAP messages an Iur interface.

To evaluate problems (find out which problems appear and how often they appear in a network), special indicators are defined that are based on measurement results. These indicators are called key performance indicators (KPI).

In general, nowadays, the term KPI has become more and more a marketing phrase, "because it sounds good." The result is that not everybody using the term KPI really means a KPI following the correct definition. Often this abbreviation is used to cover a wide field of measurement results that includes, e.g., counters of protocol events as described in 3GPP 42.403 as well as various measurement settings and measurement reports extracted directly from signaling messages or measurements derived from analysis of data streams.

A real KPI is mostly a mathematical formula used to define a metrics ratio that describes network quality and behavior for network optimization purposes. Comparison of KPI values shall point out in a simple and understandable way whether actions that have been made to improve network and service quality have been successful or not.

All other measurements are input for KPI formulas and it is possible that also additional data is added coming, e.g., from equipment manufacturers and network operators, as shown in Figure 2.13.

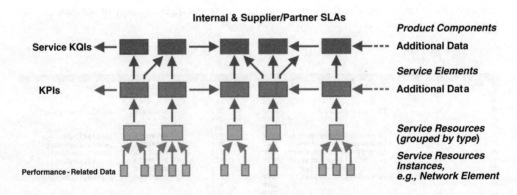

Figure 2.13   KPI as key element of KQI.

A good example of performance-related data are event counters used to count protocol messages that indicate successful or unsuccessful procedures.

A simple KPI defined based on such counters could be a success or failure rate.

Example:

Counter 1 = Σ of all GPRS Attach Request messages captured within a defined time period
Counter 2 = Σ of all GPRS Attach Reject messages captured within a defined time period

KPI: GPRS Attach Failure Ratio [in %] = $\frac{\text{Counter2}}{\text{Counter1}} \times 100$

There is a long list of similar success and failure ratios that are relatively easy to be defined using performance measurement definitions found in 3GPP 32.403. All these values are useful because they give a first overview of network quality and behavior and they may also be helpful to identify possible problems in defined areas of the network. However, simple counters and simple ratio formulas are often not enough.

For instance, if the already defined GRPS Attach Failure Ratio is calculated per SGSN, it can be used to indicate whether there is an extremely high rate of rejected GPRS Attach Requests in a defined SGSN area. However, such a high Attach Failure Ratio does not need to indicate a network problem by itself. Always a further analysis is necessary to find the root cause of network behavior. Based on the root cause analysis it can be determined whether there are problems or not. This procedure is also called drill-down analysis.

In case of rejected GPRS attach, the first step of analysis will always be to check the reject cause value of the Attach Reject message. A value that is often seen here is the cause "network failure." From 3GPP 24.008 (Mobility Management, Call Control, Session Management) it is known that the cause value "network failure" is used "if the MSC or SGSN cannot service an MS generated request because of PLMN failures, e.g. problems in MAP."

A problem in MAP may be caused by transmission problems on Gr interface between SGSN and HLR. The address of a subscriber's HLR is derived from IMSI as explained in Section 4.4 and the best way to analyze the procedure is to follow up the MAP signaling on Gr interface after GRPS Attach Request arrives at SGSN.

On Gr it can be seen whether there is a response from HLR or not and how long does it take until the response is received.

Common reasons why GPRS attach attempts are rejected with cause "network failure" are

- expiry of timers while waiting for answer from HLR, because of too much delay on signaling route between SGSN and HLR
- abortion of MAP transactions because of problems with different software versions (application contexts) in SGSN and HLR (see Section 4.4.2)
- invalid IMSI (e.g., if a service provider does not exist anymore, but its USIM cards are still out in the field)
- routing of MAP messages from foreign SGSN to home HLR of subscriber impossible, because there is no roaming contract between foreign and home network operators

The first two reasons indicate network problems that shall be solved to improve the general quality of network service. The latter two reasons show a correct behavior of the network that prevents misusage of network resources by unauthorized subscribers.

This example shows how difficult it is to distinguish between "good cases" (no problem) and "bad cases" (problem) in case of a single reject cause value. In general, four main features can be identified as main requirements of good KPI analysis:

- Intelligent multi-interface call filtering
- Provision of useful event counters
- Flexible presentation of measurement results from different points of view (sometimes called dimensions), e.g., show first Attach Rejects messages by cause values and then show IMSI of rejected subscribers related to one single cause value (to find out if they are roaming subscribers or not)
- Latency measurement to calculate time differences, e.g., between request and response messages

Another example that demonstrates these needs is shown in Figure 2.14.

The call flow diagram in the figure shows that MSC rejects a location update request belonging to a combined Location/Routing Area Update procedure, because RNC is obviously not able to execute Security Mode functions required by CS core network domain within an acceptable time frame. Once again the reject cause value in this case is "network failure," but this time the root cause of the problem is not in core network. A classical location update failure rate would show the problem related to MSC only, but using multi-interface call trace

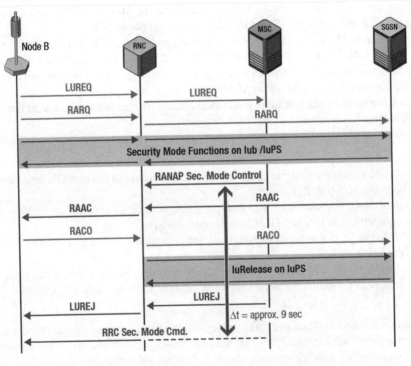

Figure 2.14 Drill-down analysis of rejected Location Update procedure.

function and call-related latency measurement it becomes possible to identify the RNC as the problem child of the network in this case.

In addition to root cause analysis, latency measurement is also useful to calculate call setup times as well as delivery times for short messages and data frames. Further important network performance parameters are throughput (data transmission rate on single interfaces or for single applications, e.g., file transfer).

Quality of services on radio interface can be determined on behalf of so-called radio link performance indicators to which belong the following:

- *BLER*: This is long-time average block error rate calculated from transport blocks. A transport block is considered to be erroneous if a CRC error is indicated by appropriate information element of Framing Protocol for uplink data. Unfortunately there is not good downlink BLER report specified yet that could be sent by user equipment. Only RRC measurement report with event-ID *e5a* indicates that downlink BLER exceeded a defined threshold.
- *BER*: Bit error rate (BER) can either be measured as Transport Channel BER or Physical Channel BER. Reports are sent by Node B to RNC for uplink data. The uplink BER is encoded in Framing Protocol Quality Estimate value.
- *SIR Error*: This shows the gap between the assigned SIR target and measured SIR. Analysis of SIR error per connection shows how good the SRNC is able to adjust uplink transmission power of UE, which means accuracy of Open Loop Power Control.
- *Transmitted Code Power*: This is the power allocated per connection. Based on this measurement, DL load of any user per connection can be estimated. The purpose of this procedure in RNC is to avoid the blockage of all available DL power resources by single UEs.

  In order to establish the radio link with the defined bit rate, the appropriate transmission power is also needed: the higher the required bit rate, the higher the output power per connection. Admission controller and packet scheduler should allow UE to use as much power as needed to reach the Node B at a predefined quality level. The allowed UE transmitted power should be on as low a level as possible, to save the short radio resources.

  Quality level per connection is defined in the planning phase and should be kept constant as long as customer satisfaction is ensured.
- *RTWP*: Received Total Wideband Power (RTWP) reflects the total noise level within the UMTS frequency band of one single cell. Call admission control and packet scheduler functions in RNC may use RTWP for calculation of necessary dedicated resources (load-based admission control function). A high RTWP level indicates increasing interferences in cell. To prevent excessive cell breathing, RNC may reconfigure all existing radio bearers used in this cell. As a result, short-time peaks of intra-cell handover rate KPIs can be measured.

How event counters and performance parameters depend on each other is still pretty unknown. Hence, one of the main challenges of UMTS network optimization is to define so-called co-related KPIs that link protocol events with QoS parameters. The general difficulty is to define useful KPIs on one hand and to give a correct interpretation of measured KPIs on the other hand. Indeed, a high level of expertise is necessary to work on this task and just to list all possible known problems would exceed the contents of this book.

However, finally another nice example shall be given that shows how protocol analysis can be used to optimize the network.

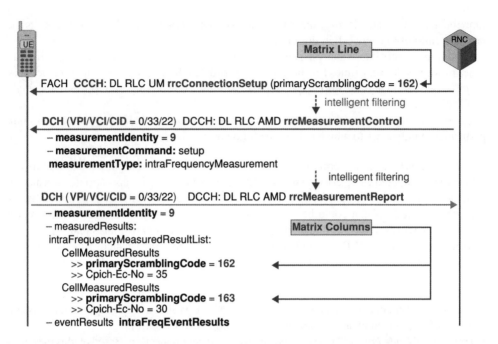

Figure 2.15 RRC messages used to calculate cell neighbor matrix.

As the reader will learn in Chapter 3, softer and soft handovers play an important role to guarantee a stable radio link quality. The prerequisite to perform handovers is that cells overlap, but overlapping shall only happen in border areas. If one UMTS cell overlaps another one too much, the interference level of the overlapped cells will rise, which is indicated by increased RTWP level and this will in turn lower the QoS of radio links. However, if overlapping areas are too small there is a quite big risk that UE can lose contact with network, which leads to a dropped call.

At any time of radio network planning and optimization there is the question of how much neighbor cells overlap and whether the expected overlapping factor is reached or not. The tool that helps radio network planners to optimize their settings is called "cell overlapping matrix" or "cell neighbor matrix."

Using intelligent call filter functions and statistic functions this matrix can easily be calculated based on data found in different RRC messages.

As shown in Figure 2.15, primary scrambling code included in RRC Connection Setup message helps to identify the cell in which a UE is located during call setup. Then, after call setup, this UE will send RRC Measurement Report messages to SRNC (see Section 3.7). Those RRC Measurement Reports include measured signal strength of primary CPICH of all cells the UE is able to measure on radio interface: the cell in which the call was set up as well as neighbor cells of this cell. All cells are identified by their primary scrambling codes and now these primary scrambling codes are used to name columns and lines of cell neighbor matrix.

The cell in which the call was originally set up and where the measurement reports come from is used to define a line of the matrix, e.g., SC = 9 in the first line of Figure 2.16. Neighbor cell primary scrambling codes are found in the column names of the matrix.

cell neighbor matrix						
SC	220	380	444	355	345	152	84
9	214	202	106	92	90	36	32
	24.0991 %	22.747747 %	11.936937 %	10.36036 %	10.135135 %	4.054054 %	3.6036036%
SC	220	177	150	113			
18	2	2	2	2			
	25.0 %	25.0 %	25.0 %	25.0 %			
SC	417	414	220				
37	4	4	4				
	33.333336 %	33.333336 %	33.333336 %				
SC	220	355	380	72	444	479	152
60	602	92	66	32	28	22	16
	65.72052 %	10.043668 %	7.2052402 %	3.49345 %	3.0567684 %	2.4017467 %	1.746725 %
SC	220	479	380	355	444	345	60
72	10	6	6	6	4	4	4
	20.833332 %	12.5 %	12.5 %	12.5 %	8.333334 %	8.333334 %	8.333334 %
SC	380	220	355	444	9	417	399
84	12	8	6	4	4	2	2
	28.57143 %	19.047619 %	14.285715 %	9.523809 %	9.523809 %	4.7619047 %	4.7619047 %

Figure 2.16　Cell neighbor matrix calculated by analysis tool.

The final cell neighbor matrix (Figure 2.16) shows how often a single neighbor cell was reported by an UE with an active radio link in the cell that stands in front of a matrix line. If the cell was reported in 2 out of 10 total measurement reports the overlapping factor is 20%. The figure shows neighbor cells for cells with primary scrambling codes (SC) = 9, 18, 37, 60, 72, and 84.

3

UMTS UTRAN
Signaling Procedures

After the comprehensive UMTS refresher the focus now changes toward details of multiple examples of common signaling procedures on UMTS UTRAN line interfaces.

To achieve a better understanding of how the protocols of the different interfaces interact, the first focus is on Iub procedures. Thereafter it is shown how messages from Iub are forwarded to the core network domains. Finally, the handover procedures that include Iur signaling are explained.

All signaling procedures are based on real trace files from real network operation, field trials, or testbeds. However, it should be noticed that 3GPP standards offer a wide range of possibilities of how procedures could be designed. So the focus here is not to show what is possible following the 3GPP standards but what is implemented in present UMTS equipment. It also must be mentioned that especially on Iub interface a wide range of manufacturer-specific solutions can be found and there are also many options for network operators. Not all of these options and specific solutions can be explained in text and graphics, and hence only a few are highlighted.

As often as possible we have added examples of signaling messages and parameters so that not only the variables in the call flow diagrams are presented, but also shown are examples of how real values of these variables look like in a protocol tester's environment. Unfortunately, many UMTS signaling messages are pretty voluminous. Hence, we often can only show the message header and some selected information elements related to the call flow procedures. All parameter values that would allow identification of subscribers, network operators, or equipment manufacturers have been changed. Our intention in showing message examples is to give a better understanding of call parameters. The given examples shall not be taken as recommendation for network configuration and settings. They have been extracted from different trace file sources and it cannot be guaranteed that two consecutive message examples (also if they are presented in the same procedure description) have the same origin.

Procedures, messages, and parameters are based on 3GPP standards as described in Release 99 and Release 4 specifications. The reader should keep in mind that these specifications are improved constantly. As a rule every three months a new protocol version is released.

UMTS Signaling   Ralf Kreher and Torsten Rüdebusch
© 2005 Tektronix, Inc.   ISBN: 0-470-01351-6.

Note: In the context of this book the term "DCH" is often used to describe a dedicated AAL2 SVC on Iub interface that is actually only a transport bearer for dedicated signaling or user traffic. Data transmitted in the same AAL2 SVC is sent on more than just one radio interface DCH. In a similar way the term "DCCH" or "DTCH" is used in some cases, but of course there is more than just one DCCH or DTCH assigned to a single connection. Indeed, when Iub interface is monitored, the differences between used radio DCHs are not significant, but control plane and user plane traffic is always running on different transport bearers. Within a single connection these transport bearers can be easily distinguished on behalf of different VPI/VCI/CID values. To indicate the purpose of each AAL2 SVC we named these transport bearers DCH, DTCH, or DCCH, which will hopefully increase the understandability of the text despite that it is incorrect from the point of view of 3GPP standards.

3.1 Iub – Node B Setup

A Node B Setup needs to be performed if a new Node B has been installed, changes in configuration have been made, or after a system reset (e.g. for installation of a new software version). To "announce" these changes to the network the Node B initiates the Setup scenario.

3.1.1 Overview

If a Node B is set up against a Radio Network Controller (RNC) this setup will happen in three steps (Figure 3.1).

Step 1: The Node B requests to be audited by the RNC. During the audit, the RNC is informed of how many (just one or more) cells belong to the Node B and which local cell identifiers they have.

Step 2: For each cell a Cell Setup is performed by the RNC. During the Cell Setup, the physical (radio interface) channels are parameterized. These channels are mandatory in case of a

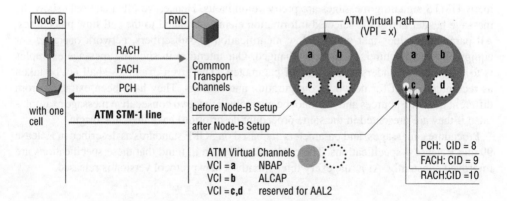

Figure 3.1 Node B Setup overview.

User Equipment (UE) initial access. In other words, if they are not available it is impossible for the UE after it is switched on to get access to the network via the radio interface.

Step 3: The common transport channels Paging Channel (PCH), Forward Access Channel (FACH), and Random Access Channel (RACH) are set up and optionally parameterized in each cell of the new Node B. On Iub interface these common transport channels are carried by AAL2 connections on ATM lines. ATM/AAL2 header values (VPI/VCI/CID) are important because without knowing them it is impossible to monitor signaling and data transport on PCH, RACH, and FACH. If these channels are not monitored some of the most important messages for call setup and mobility management procedures like Paging messages and RRC Connection Setup will be missed in call traces. Once the AAL2 connection for a common transport channel is installed during Node B Setup, it will not be released until this Node B is taken out of service or reset.

3.1.2 Message Flow

The Node B Setup scenario is executed when a new Node B is taken into service or after reset (Figure 3.2).

With an auditRequired message, the Node B requests an audit sequence by RNC. One audit sequence consists of one or more audit procedures (one in our example). The auditRequired procedure code is transmitted in an NBAP Class 2 Elementary procedure without response (connectionless). Hence, longTransactionID has no meaning (the value is 0).

NBAP UL initiatingMessage **Id-auditRequired**(longTransActionID=**a**)

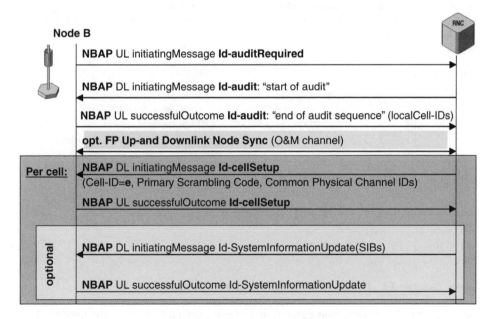

Figure 3.2 Node B Setup call flow 1/4.

The audit procedure belongs to NBAP Class 1 Elementary procedures with response (connection-oriented). Both messages, Initiating Message and Successful Outcome, are linked with the same longTransactionID value **b**.

> **NBAP** DL initiatingMessage **Id-audit** (longTransActionID=**b**)
> **NBAP** UL successfulOutcome **Id-audit** (longTransAction ID=**b**, id-local Cell IDs= {**0,1,2,...**})

With SuccessfulOutcome response of the audit procedure, the RNC is informed of how many cells belong to the Node B, which is audited. A local Cell-ID is assigned by the Node B to each of its cells. In addition, power consumption law values for common and dedicated channels for all cells are reported to the RNC, so that from then on the RNC is able to control the power resources of each cell, which is one of the most critical parameters for UMTS air interface operation.

```
+---------+----------------------------+-------------------------+
|BITMASK  |ID Name                     |Comment or Value         |
+---------+----------------------------+-------------------------+
|16:12:24,151,397  frm NodeB UL  SSCOP  SD  NBAP  succesfulOutcome |
|id-audit                                                         |
|UNI SSCOP (SSCOP)  SD (= Seq. Conn.mode Data) [Layer Name Only]  |
|TS 25.433 V3.6.0 (2001-06) (NBAP)  succesfulOutcome             |
|(= succesfulOutcome)                                            |
|nbapPDU                                                         |
|1 succesfulOutcome                                             |
|1.1 procedureID                                               |
|00000000 |1.1.1 procedureCode         |id-audit                 |
/\/\/\/\/\/\/\/\/\/\/\/\/\/\/\/\/\/\/\/\/\/\/\/\/\/\/\/\/\/\/\/\/\/\/\/\/\/\/\/\/\/\/\/\/\
|1.4 transactionID                                              |
|***B2*** |1.4.1 longTransActionId     |43                       |
/\/\/\/\/\/\/\/\/\/\/\/\/\/\/\/\/\/\/\/\/\/\/\/\/\/\/\/\/\/\/\/\/\/\/\/\/\/\/\/\/\/\/\/\/\
|1.5.1.1 sequence                                              |
|***B2*** |1.5.1.1.1 id                |id-End-Of-Audit-         |
|                                      Sequence-Indicator        |
|01------ |1.5.1.1.2 criticality       |ignore                   |
|0------- |1.5.1.1.3 value             |end-of-audit-sequence    |
/\/\/\/\/\/\/\/\/\/\/\/\/\/\/\/\/\/\/\/\/\/\/\/\/\/\/\/\/\/\/\/\/\/\/\/\/\/\/\/\/\/\/\/\/\
|1.5.1.3.3.1 sequenceOf                                        |
|***B2*** |1.5.1.3.3.1.1 id            |id-Local-Cell-           |
|                                      InformationItem-AuditR    |
|01------ |1.5.1.3.3.1.2 criticality   |ignore                   |
|1.5.1.3.3.1.3 value                                           |
|00000000 |1.5.1.3.3.1.3.1 local-Cell-ID |0                      |
-----------------------------------------------------------------
```

Message Example 3.1   Extract NBAP SuccessfulOutcome (Id-audit).

Note: The criticality information element indicates for each parameter in the message how a peer entity that receives this message shall react if the parameter is not known on the receiver side.

In the next step it is possible that framing protocol (**FP**) Uplink and Downlink Node Synchronization frames can be monitored on a Switched Virtual AAL2 Connection (AAL2 SVC) if manufacturer-specific node operation and maintenance protocol is running on such an AAL2 channel. Synchronization in case of framing protocol means alignment of frame numbers and timers on the RNC and the Node B side.

The following procedure is executed "per cell." In the example call flow we will have a look at the cell with id-C-ID = **e**.

With the CellSetup message, the RNC assigns a Cell-ID (id-C-ID) to each single local Cell ID. Other important parameters inside the CellSetup message are:

- Primary Scrambling Code
- Common Physical Channel IDs of:
  - Primary Synchronization Channel (P-SCH)
  - Secondary Synchronization Channel (S-SCH)
  - Primary Common Pilot Channel (CPICH)
  - Common Control Physical Channel (CCPCH) that carries the Broadcast Channel (BCH)

The common physical channels are necessary to ensure successful initial UE access. In addition, the message also contains information about UMTS absolute radio frequency code number (UARFCN) and maximum transmission power of the cell's antenna as well as further antenna parameters.

> **NBAP** DL initiatingMessage **Id-CellSetup** (longTransActionID=**c**, Id-local Cell ID={**0**}, id-C-ID=**e**, Primary Scrambling Code, Common Physical Channel Info, Common Transport Channel ID of BCH)

```
+---------+------------------------------------+----------------+
|BITMASK  |ID Name                             |Comment or Value |
+---------+------------------------------------+----------------+
|16:12:27,166,380   frm RNC DL  SSCOP  SD  NBAP  initiatingMessage  |
|id-cellSetup                                                       |
|UNI SSCOP (SSCOP)  SD (= Seq. Conn.mode Data) [Layer Name Only]    |
|TS 25.433 V3.6.0 (2001-06) (NBAP)  initiatingMessage              |
|(= initiatingMessage)                                             |
|nbapPDU                                                           |
|1 initiatingMessage                                              |
|1.1 procedureID                                                  |
|00000101 |1.1.1 procedureCode                 |id-cellSetup     |
```
〰〰
```
|1.4 transactionID                                                |
|***B2*** |1.4.1 longTransActionId             |8                |
```
〰〰
```
|1.5.1.1 sequence                                                 |
|***B2*** |1.5.1.1.1 id                        |id-Local-Cell-ID |
|00------ |1.5.1.1.2 criticality               |reject           |
|00000000 |1.5.1.1.3 value                     |0                |
|1.5.1.2 sequence                                                 |
|***B2*** |1.5.1.2.1 id                        |id-C-ID          |
```

```
|00------ |1.5.1.2.2 criticality              |reject       |
|***B2*** |1.5.1.2.3 value                    |0            |
```
~~~~~~
```
|1.5.1.8 sequence                             |
|***B2*** |1.5.1.8.1 id          |id-PrimaryScramblingCode |
|00------ |1.5.1.8.2 criticality              |reject       |
|***B2*** |1.5.1.8.3 value                    |1            |
```
~~~~~~
```
|1.5.1.11 sequence                            |
|***B2*** |1.5.1.11.1 id         |id-PrimarySCH-           |
|                                 Information-Cell-Setup   |
|00------ |1.5.1.11.2 criticality             |reject       |
|1.5.1.11.3 value                             |
|00000001 |1.5.1.11.3.1 commonPhysicalChannelID |1         |
|***B2*** |1.5.1.11.3.2 primarySCH-Power      |0            |
|1------- |1.5.1.11.3.3 tSTD-Indicator        |inactive     |
|1.5.1.12 sequence                            |
|***B2*** |1.5.1.12.1 id         |id-SecondarySCH-         |
|                                 Information-Cell-Set     |
|00------ |1.5.1.12.2 criticality             |reject       |
|1.5.1.12.3 value                             |
|00000010 |1.5.1.12.3.1 commonPhysicalChannelID |2         |
|***B2*** |1.5.1.12.3.2 secondarySCH-Power    |0            |
|1------- |1.5.1.12.3.3 tSTD-Indicator        |inactive     |
|1.5.1.13 sequence                            |
|***B2*** |1.5.1.13.1 id         |id-PrimaryCPICH-         |
|                                 Information-Cell-Set     |
|00------ |1.5.1.13.2 criticality             |reject       |
|1.5.1.13.3 value                             |
|00000000 |1.5.1.13.3.1 commonPhysicalChannelID |0         |
|***B2*** |1.5.1.13.3.2 primaryCPICH-Power    |300          |
|1------- |1.5.1.13.3.3 transmitDiversityIndicator |inactive  |
|1.5.1.14 sequence                            |
|***B2*** |1.5.1.14.1 id         |id-PrimaryCCPCH-         |
|                                 Information-Cell-Set     |
|00------ |1.5.1.14.2 criticality             |reject       |
|1.5.1.14.3 value                             |
|00000011 |1.5.1.14.3.1 commonPhysicalChannelID |3         |
|1.5.1.14.3.2 bCH-information                 |
|00000100 |1.5.1.14.3.2.1 commonTransportChannelID |4       |
--------------------------------------------------------------------
```

Message Example 3.2 Extract NBAP Initiating Message (Id-CellSetup).

Node B confirms the transmission of parameters with

   NBAP UL successfulOutcome **Id-CellSetup** (longTransActionID=**c**)

```
+---------+------------------------------------------+-------------------+
|BITMASK  |ID Name                                   |Comment or Value   |
+---------+------------------------------------------+-------------------+
|16:13:07,436,710   frm NodeB UL  SSCOP  SD  NBAP  succesfulOutcome |
|id-cellSetup                                                       |
|UNI SSCOP (SSCOP)  SD (= Seq. Conn.mode Data) [Layer Name Only]    |
|TS 25.433 V3.6.0 (2001-06) (NBAP)  succesfulOutcome               |
|(= succesfulOutcome)                                              |
|nbapPDU                                                           |
|1 succesfulOutcome                                               |
|1.1 procedureID                                                  |
|00000101 |1.1.1 procedureCode              |id-cellSetup      |
|-01----- |1.1.2 ddMode                     |fdd               |
|---00--- |1.2 criticality                  |reject            |
|-----0-- |1.3 messageDiscriminator         |common            |
|1.4 transactionID                                                |
|***B2*** |1.4.1 longTransActionId          |8                 |
+-------------------------------------------------------------------+
```

Message Example 3.3   Extract NBAP SuccessfulOutcome (Id-CellSetup).

Optional procedure

In the System Information Update that follows optionally the Cell Setup, a number of System Information Blocks (SIBs) are transmitted. These SIBs contain parameters like timers and counters for changing RRC states and UMTS Registration Area (URA) Identity. A Master Information Block (MIB) contains information about which of the many different SIBs are provided for a cell that is defined by its C-ID. System Information Update can also be executed at the end of the whole Node B Setup procedure. In this case, all necessary SIBs are transmitted to the Node B at once.

> **NBAP** DL initiatingMessage **Id-SystemInformationUpdate**
> (longTransActionID=**d**, id-C-ID=**e**, MIB + SIBs)
> **NBAP** UL successfulOutcome **Id-SystemInformationUpdate**
> (longTransActionID=**d**)

```
+---------+---------------------------+------------------------+
|BITMASK  |ID Name                    |Comment or Value        |
+---------+---------------------------+------------------------+
|16:13:07,467,723   frm RNC DL  NBAP-SIB  masterInfo..        |
|NBAP SIB from TS 25.433 V3.6.0 (2001-06) (NBAP-SIB)         |
|masterInfoBlock (= masterInfoBlock)                         |
|sib_description                                             |
|1 sib_choice                                                |
|1.1 masterInfoBlock                                         |
|-000---- |1.1.1 mib-ValueTag         |1                      |
|                                                            |
|1.1.3 sibSb-ReferenceList                                   |
```

```
∿∿∿∿∿∿∿∿∿∿∿∿∿∿∿∿∿∿∿∿∿∿∿∿∿∿∿∿∿∿∿∿∿∿∿∿∿∿∿∿∿∿∿∿∿∿∿∿∿∿∿∿∿∿∿∿∿∿∿
|1.1.3.1.1 sibSb-Type                                               |
|***b8*** |1.1.3.1.1.1 sysInfoType1      |1                         |
∿∿∿∿∿∿∿∿∿∿∿∿∿∿∿∿∿∿∿∿∿∿∿∿∿∿∿∿∿∿∿∿∿∿∿∿∿∿∿∿∿∿∿∿∿∿∿∿∿∿∿∿∿∿∿∿∿∿∿
|1.1.3.2 schedulingInformationSIBSb                                 |
|1.1.3.2.1 sibSb-Type                                               |
|------00 |1.1.3.2.1.1 sysInfoType2      |1                         |
∿∿∿∿∿∿∿∿∿∿∿∿∿∿∿∿∿∿∿∿∿∿∿∿∿∿∿∿∿∿∿∿∿∿∿∿∿∿∿∿∿∿∿∿∿∿∿∿∿∿∿∿∿∿∿∿∿∿∿
|1.1.3.3.1 sibSb-Type                                               |
|-00----- |1.1.3.3.1.1 sysInfoType3      |1                         |
∿∿∿∿∿∿∿∿∿∿∿∿∿∿∿∿∿∿∿∿∿∿∿∿∿∿∿∿∿∿∿∿∿∿∿∿∿∿∿∿∿∿∿∿∿∿∿∿∿∿∿∿∿∿∿∿∿∿∿
|1.1.3.4 schedulingInformationSIBSb                                 |
|1.1.3.4.1 sibSb-Type                                               |
|---00--- |1.1.3.4.1.1 sysInfoType11     |1                         |
-------------------------------------------------------------------
```

Message Example 3.4   Extract of Master Information Block (MIB) from NBAP System Information Update.

```
+---------+-------------------------------------+-----------------+
|BITMASK  |ID Name                              |Comment or Value |
+---------+-------------------------------------+-----------------+
|16:13:07,467,723  frm RNC DL  NBAP-SIB  sibType11                |
|NBAP SIB from TS 25.433 V3.6.0 (2001-06) (NBAP-SIB)  sibType11   |
|(= sibType11)                                                    |
|sib_description                                                  |
|1 sib_choice                                                     |
|1.1 sibType11                                                    |
|--0----- |1.1.1 sib12indicator                |0                |
|1.1.2 measurementControlSysInfo                                  |
∿∿∿∿∿∿∿∿∿∿∿∿∿∿∿∿∿∿∿∿∿∿∿∿∿∿∿∿∿∿∿∿∿∿∿∿∿∿∿∿∿∿∿∿∿∿∿∿∿∿∿∿∿∿∿∿∿∿∿
|1.1.2.1.1.1 cellSelectQualityMeasure                             |
|1.1.2.1.1.1.1 cpich-Ec-No                                        |
|1.1.2.1.1.1.1.1 intraFreqMeasurementSysInfo                      |
|1000---- |1.1.2.1.1.1.1.1.1 intraFreqMeasurementID  |9           |
|1.1.2.1.1.1.1.1.2 intraFreqCellInfoSI-List                       |
|1.1.2.1.1.1.1.1.2.1 newIntraFreqCellList                         |
|1.1.2.1.1.1.1.1.2.1.1 newIntraFreqCellSI-ECN0                    |
|---00000 |1.1.2.1.1.1.1.1.2.1.1.1 intraFreqCellID   |0           |
|1.1.2.1.1.1.1.1.2.1.1.2 cellInfo                                 |
|***b6*** |1.1.2.1.1.1.1.1.2.1.1.2.1 cellIndividualOff.. |-20      |
|1.1.2.1.1.1.1.1.2.1.1.2.2 modeSpecificInfo                       |
|1.1.2.1.1.1.1.1.2.1.1.2.2.1 fdd                                  |
|1.1.2.1.1.1.1.1.2.1.1.2.2.1.1 primaryCPICH-Info                  |
|***b9*** |1.1.2.1.1.1.1.1.2.1.1.2.2.1.1.1 primaryScra.. |1        |
|-----1-- |1.1.2.1.1.1.1.1.2.1.1.2.2.1.2 readSFN-Indic.. |1        |
|------0- |1.1.2.1.1.1.1.1.2.1.1.2.2.1.3 tx-DiversityI.. |0        |
```

```
|1.1.2.1.1.1.1.1.2.1.1.2.3 cellSelectionReselectionInfo        |
|***b7*** |1.1.2.1.1.1.1.1.2.1.1.2.3.1 q-Offset1S-N       |0     |
|-0110010 |1.1.2.1.1.1.1.1.2.1.1.2.3.2 q-Offset2S-N       |0     |
|1010011- |1.1.2.1.1.1.1.1.2.1.1.2.3.3 maxAllowedUL-TX..  |33    |
----------------------------------------------------------------
```

Message Example 3.5 Extract of System Information Block 11 (SIB 11) that contains broadcast information for cell (re)selection and intrafrequency cell measurement.

Note: To learn more about intrafrequency cell measurement see Section 3.7.

After successful Cell Setup, RNC starts Common Transport Channel (CTrCH) Setup for each cell (Figure 3.3). A CTrCH Setup request is sent to the Node B that serves the cell, which is registered on the RNC side with its C-ID. The message contains a list of parameters for the transport channel (in this case, PCH). It includes information on which physical channel the CTrCH will be mapped onto and – besides other radio-related items – the Common Transport Channel ID (CTrCH-ID) and the Transport Format Set (TFS) of the CTrCH.

In case of CTrCH Setup for the PCH, the message also contains the parameters for the appropriate paging indicator channel (PICH).

> **NBAP** DL initiatingMessage **Id-commonTransportChannelSetup**
> (longTransActionID=**f**, id-C-ID=**e**, Common Physical Channel Type for CTrCH
> Setup, PCH-Parameters, commonTransportChannelID=**g**, PICH Parameters)

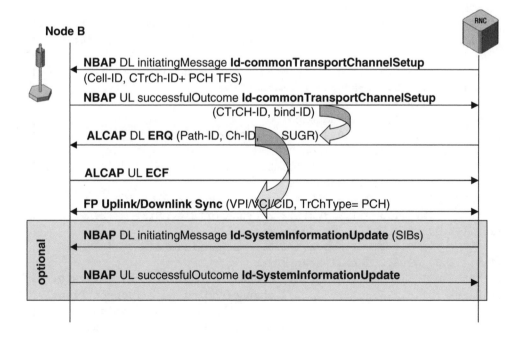

Figure 3.3 Node B Setup call flow 2/4.

```
+---------+-------------------------------------------+----------+
|16:13:09,026,000  frm RNC DL  SSCOP  SD  NBAP  initiatingMessage  |
|id-commonTransportChannelSe                                       |
|UNI SSCOP (SSCOP)  SD (= Seq. Conn.mode Data) [Layer Name Only]   |
|TS 25.433 V3.6.0 (2001-06) (NBAP)  initiatingMessage              |
|(= initiatingMessage)                                             |
|nbapPDU                                                           |
|1 initiatingMessage                                               |
|1.1 procedureID                                                   |
|00001100 |1.1.1 procedureCode    |id-commonTransportChannelSetup  |
~~~~~~~~~~~~~~~~~~~~~~~~~~~~~~~~~~~~~~~~~~~~~~~~~~~~~~~~~~~~~~~~~~~~~~~
|1.4 transactionID                                                 |
|***B2*** |1.4.1 longTransActionId                    |22         |
~~~~~~~~~~~~~~~~~~~~~~~~~~~~~~~~~~~~~~~~~~~~~~~~~~~~~~~~~~~~~~~~~~~~~~~
|***B2*** |1.5.1.1.1 id                              |id-C-ID     |
|00------ |1.5.1.1.2 criticality                     |reject      |
|***B2*** |1.5.1.1.3 value                           |1           |
~~~~~~~~~~~~~~~~~~~~~~~~~~~~~~~~~~~~~~~~~~~~~~~~~~~~~~~~~~~~~~~~~~~~~~~
|***B2*** |1.5.1.3.1 id      |id-CommonPhysicalChannelType-CTCH-Se |
|01------ |1.5.1.3.2 criticality                     |ignore      |
|1.5.1.3.3 value                                                   |
|1.5.1.3.3.1 secondary-CCPCH-parameters                            |
|00001010 |1.5.1.3.3.1.1 commonPhysicalChannelID     |10          |
|***B2*** |1.5.1.3.3.1.3 fdd-DL-ChannelisationCodeNumber|4         |
~~~~~~~~~~~~~~~~~~~~~~~~~~~~~~~~~~~~~~~~~~~~~~~~~~~~~~~~~~~~~~~~~~~~~~~
|1.5.1.3.3.1.9 pCH-Parameters                                      |
|***B2*** |1.5.1.3.3.1.9.1 id |id-PCH-ParametersItem-CTCH-SetupRqst|
|00------ |1.5.1.3.3.1.9.2 criticality               |reject      |
|1.5.1.3.3.1.9.3 value                                             |
|00001100 |1.5.1.3.3.1.9.3.1 commonTransportChannelID |12         |
|1.5.1.3.3.1.9.3.2 transportFormatSet                              |
|1.5.1.3.3.1.9.3.2.1 dynamicParts                                  |
|1.5.1.3.3.1.9.3.2.1.1 sequence                                    |
|***B2*** |1.5.1.3.3.1.9.3.2.1.1.1 nrOfTransportBlocks |0         |
|1.5.1.3.3.1.9.3.2.1.1.2 mode                                      |
|         |1.5.1.3.3.1.9.3.2.1.1.2.1 notApplicable   |0           |
|1.5.1.3.3.1.9.3.2.1.2 sequence                                    |
|***B2*** |1.5.1.3.3.1.9.3.2.1.2.1 nrOfTransportBlocks |1         |
|***B2*** |1.5.1.3.3.1.9.3.2.1.2.2 transportBlockSize |240        |
|1.5.1.3.3.1.9.3.2.1.2.3 mode                                      |
|         |1.5.1.3.3.1.9.3.2.1.2.3.1 notApplicable   |0           |
|1.5.1.3.3.1.9.3.2.2 semi-staticPart                               |
|***b3*** |1.5.1.3.3.1.9.3.2.2.1 transmissionTimeInter..|msec-10   |
|--01---- |1.5.1.3.3.1.9.3.2.2.2 channelCoding       |convolutional-|
|                                                   coding         |
|-----0-- |1.5.1.3.3.1.9.3.2.2.3 codingRate          |half        |
|11100101 |1.5.1.3.3.1.9.3.2.2.4 rateMatcingAttribute |230        |
|-011---- |1.5.1.3.3.1.9.3.2.2.5 cRC-Size            |v16         |
```

```
wwwwwwwwwwwwwwwwwwwwwwwwwwwwwwwwwwwwwwwwwwwwwwwwwwwwwwwwwwwwwwwwwwwwwwww
|1.5.1.3.3.1.9.3.6 pICH-Parameters                                    |
|00001011 |1.5.1.3.3.1.9.3.6.1 commonPhysicalChannelID     |11        |
|***B2*** |1.5.1.3.3.1.9.3.6.2 fdd-dl-ChannelisationCo..   |3         |
----------------------------------------------------------------------
```

Message Example 3.6   NBAP Initial Message (Common Transport Channel Setup) for PCH.

This message example shows an NBAP Common Transport Channel Setup Request for a PCH with common transport channel ID = 12. Transport blocks of this PCH will have a size of 240 bits and there is only one transport block sent every 10 ms. To define the first transport format with 0 transport blocks (1.5.1.3.3.1.9.3.2.1.1.1) is mandatory to ensure synchronization of the transport channel also if no information needs to be currently transmitted, a situation that is also known as "silent mode."

To provide redundancy on radio interface, a convolutional code with a coding rate of 1.5 is used. This means that for every real bit of PCH information, 2 bits of information are sent on radio interface to decrease the number of transmission errors on air. The additional CRC size of PCH frames is 16 bit. The PCH in the example will be mapped onto a Secondary Common Control Physical Channel (S-CCPCH) with physical channel ID = 10, which can be identified on radio interface using downlink channelization code = 4. Related to this PCH is a PICH that has common physical channel ID = 11 and is encoded using downlink channelization code = 3.

After the message is sent, RNC awaits the appropriate CTrCH Setup Response:

NBAP UL successfulOutcome **Id-commonTransportChannelSetup**
(longTransActionID=**f**, commonTransportChannelID=**g**, bindingID=**h**)

The Node B answers with an NBAP Successful Outcome message including the same procedure code "Common Transport Channel Setup" and the same CTrCH-ID. In addition, a bindingID is provided.

```
+---------+-----------------------------------------+-----------------+
|BITMASK  |ID Name                                  |Comment or Value |
+---------+-----------------------------------------+-----------------+
|16:13:09,043,721  frm NodeB UL  SSCOP  SD  NBAP  succesfulOutcome |
|id-commonTransportChannelSe                                       |
|                                                                  |
|UNI SSCOP (SSCOP)  SD (= Seq. Conn.mode Data) [Layer Name Only]   |
|TS 25.433 V3.6.0 (2001-06) (NBAP)  succesfulOutcome               |
|(= succesfulOutcome)                                              |
|nbapPDU                                                           |
|1 succesfulOutcome                                               |
|1.1 procedureID                                                  |
|00001100 |1.1.1 procedureCode     |id-commonTransportChannelSetup |
wwwwwwwwwwwwwwwwwwwwwwwwwwwwwwwwwwwwwwwwwwwwwwwwwwwwwwwwwwwwwwwwwwwwwwww
|1.4 transactionID                                                |
|***B2*** |1.4.1 longTransActionId              |22                |
|1.5 value                                                        |
```

```
|1.5.1 protocolIEs                                                      |
|1.5.1.1 sequence                                                       |
|***B2*** |1.5.1.1.1 id                       |id-PCH-Parameters-|
|         |                                   |CTCH-SetupRsp      |
|01------ |1.5.1.1.2 criticality             |ignore             |
|1.5.1.1.3 value                                                        |
|00001100 |1.5.1.1.3.1 commonTransportChannelID |12                   |
|***B2*** |1.5.1.1.3.2 bindingID             |01 80              |
------------------------------------------------------------------------
```

Message Example 3.7   NBAP SuccessfulOutcome (Common Transport Channel Setup).

This bindingID connects the NBAP layer with the ALCAP function, which is realized in our example message flow by an AAL2L3 signaling procedure. However, some manufacturers have integrated the ALCAP function in their (proprietary) NBAP software. The advantage of such a solution is increased efficiency regarding the usage of network resources and a most likely faster setup of AAL2 SVC. The disadvantage of any proprietary solution is that it anticipates deployment of multivendor environment for network operators, because interoperability between network nodes of different manufacturers becomes quite impossible.

If AAL2L3 is used, the value of the bind-ID can be found back as Served User Generated Reference (SUGR) in the AAL2L3 Establish Request (ERQ) message. It may happen that bindingID and SUGR are decoded in different formats since NBAP specification defines bind-ID as a 4-octet string only, while AAL2L3 says the coding of SUGR depends on implementation. Hence, for example, the NBAP bind-ID could be shown in hexadecimal format while the SUGR is decoded as a decimal number – but the value remains the same. This is also true for our message examples: **01 80** (hex) = **384** (dec)!

> **ALCAP** DL **ERQ** (Originating Signal. Ass. ID=**i**, AAL2 Path=**k**, AAL2
> ChannelId=**l**, served user gen reference=**h**)
> **ALCAP** UL **ECF** (Originating Signal. Ass. ID=**m**, Destination Sign. Assoc. ID=**i**)

```
+----------+----------------------------------+----------------------+
|BITMASK   |ID Name                           |Comment or Value      |
+----------+----------------------------------+----------------------+
|16:13:09,070,714  frm RNC DL  SSCOP  SD  AAL2L3  ERQ                 |
|UNI SSCOP (SSCOP)  SD (= Seq. Conn.mode Data) [Layer Name Only]      |
|ITU-T Q.2630.1/2 AAL2 Signalling CS1/2 (AAL2L3)  ERQ                 |
|(= Establish Request)                                               |
|Establish Request                                                   |
|***B4*** |Dest. Sign Assoc. Id.             |0                     |
```
```
|***B4*** |Originating signal. ass. Id.      |16777245              |
|***B4*** |AAL2 type 2 path id.              |1234                  |
|00001010 |channel id. (0, 8-255)            |9                     |
```
```
|***B4*** |served user gen reference         |384                   |
------------------------------------------------------------------------
```

Message Example 3.8   Extract from AAL2L3 ERQ.

As already mentioned, ALCAP/AAL2L3 is used to set up an AAL2 SVC. An AAL2 SVC is required because this will be the physical layer for the common transport channel on Iub interface, which is yet to be installed. Each AAL2 virtual connection is uniquely identified by:

- ATM Virtual Path Identifier (VPI)
- ATM Virtual Channel Identifier (VCI)
- AAL2 Connection Identifier (CID)

The AAL2L3 ERQ sent by the RNC already includes two important parameters:

- Path-ID
- Channel-ID (Ch-ID)

However, the Path-ID in the ERQ message is *not identical* with the VPI! It is a pointer to an entry in an RNC configuration table. While the Ch-ID of the ERQ message will be used as value for AAL2 CID, the VPI/VCI combination of the ATM header can be found in this configuration table. It only depends on the manufacturer-specific implementations of the switching software as to how Path-ID and VPI/VCI are linked. Table 3.1 introduces three typical examples.

Table 3.1 Configuration example

	AAL2L3		ATM AAL2	
	Path-ID	Ch-ID	VPI/VCI	CID
Example 1	65	10	0/65	10
Example 2	2002004001	12	2/39	12
Example 3 (from message examples 3.9 and 3.10)	1234	9	10/26	9

On behalf of these examples it emerges very clearly that Ch-ID and CID will always have the same value, while Path-ID and VCI may have the same value or nothing in common at all. Hence, this *Path-ID in ERQ message must be seen as a pointer* to a specific record in the ATM configuration table.

The AAL2L3 ERQ message is answered by an Establish Confirm (ECF) message. Originating/Destination Signaling Association ID (OSAID/DSAID) links both AAL2L3 messages. The OSAID value sent with ERQ comes back as DSAID value with the ECF message.

```
+---------+---------------------------------------+---------------------+
|BITMASK  |ID Name                                |Comment or Value     |
+---------+---------------------------------------+---------------------+
|16:13:09,087,725  frm NodeB UL   SSCOP   SD   AAL2L3   ECF              |
|UNI SSCOP (SSCOP)   SD (= Seq. Conn.mode Data) [Layer Name Only]        |
|ITU-T Q.2630.1/2 AAL2 Signalling CS1/2 (AAL2L3)   ECF                   |
|(= Establish Confirm)                                                   |
|Establish Confirm                                                       |
```

```
|***B4***  |Dest. Sign Assoc. Id.              |16777245                    |
/\/\/\/\/\/\/\/\/\/\/\/\/\/\/\/\/\/\/\/\/\/\/\/\/\/\/\/\/\/\/\/\/\/\/\/\/\/\/\/\/\/\/\/\/\/\/\/\/\/\
|***B4***  |Originating signal. ass. Id.   |2                           |
----------------------------------------------------------------------
```

Message Example 3.9   Extract from AAL2L3 ECF.

Finally, FP Synchronization frames are seen on the VPI/VCI/CID that carries the PCH.

FP Uplink and Downlink Synchronization (in AAL2 Path/Connection=**k′** [**VPI/VCI**] and AAL2 ConnectionID=l [**CID**])

```
+---------+----------------------------------------+----------------+
|BITMASK  |ID Name                                 |Comment or Value |
+---------+----------------------------------------+----------------+
|16:13:09,133,708  Cell 0 PCH  "10/26/9"  RLC/MAC   FP CTRL PCH    |
|TS 25.322 V3.7.0 (RLC) / 25.321 V3.8.0 (MAC) / 25.435, 25.427
V3.7.0 (FP) - (2001-06) (RLC/MAC)  FP CTRL PCH (= FP Control
Frame PCH)  |
|FP Control Frame PCH                                              |
|         |FP:  VPI/VCI/CID                   |"10/26/9"          |
|         |FP:  Radio Mode                    |FDD (Frequency     |
|         |                                   |Division Duplex)   |
|         |FP:  Direction                     |Downlink           |
|         |FP:  Transport Channel Type        |PCH(Paging Channel)|
|         |FP:  CRC Check Result              |OK                 |
|         |FP:  Control Frame Type (Iub Common)|DL Synchronisation |
|**b12*** |FP:  Connection Frame Number       |2811               |
|----0000 |FP:  Spare                         |'0000'B            |
----------------------------------------------------------------------
```

Message Example 3.10   Framing Protocol Downlink Synchronization.

RNC starts sending Downlink (DL) synchronization frames to the Node B and Node B responds with UL synch frames. Uplink (UL) synchronization frames include the Time of Arrival (ToA) value related to the Connection Frame Number (CFN) value included in both of the synch frames. Based on the ToA the RNC adjusts the sending time instant of the next DL synchronization frame. The connection is synchronized when the DL synch frames are received at Node B within the predefined timing window, i.e., the ToA values in UL synch frames are within the set boundaries. After this start up synchronization the PCH can be used to transmit paging messages.
Note: In addition to the PCH, the PICH is taken into service as well. The PICH is used to carry the paging indicators. The PICH is always associated with an S-CCPCH to which a PCH transport channel is mapped. If a paging indicator is set to "1," it is an indication that UEs associated with this paging indicator should read the corresponding frame of the associated S-CCPCH. The PICH parameters can be found in the section after the PCH parameters in the first NBAP message of PCH Common Transport Channel Setup.

Optional once again, a System Information Update procedure may follow PCH setup to transmit SIB 5 information to Node B. SIB 5 contains information about physical channels PICH, PRACH, and AICH as well as TFS definitions of PCH, RACH, and FACH that will be broadcast on radio interface. Indeed, SIB 5 would fill up the next few pages if we would give a message example, but from point of view of a signaling expert it is drop-dead gorgeous.

NBAP DL initiatingMessage **Id-SystemInformationUpdate**
(longTransActionID=**n**, id-C-ID=**e**, MIB + SIB 5)
NBAP UL successfulOutcme **Id-SystemInformationUpdate**
(longTransActionID=**n**)

The setup procedure for the FACH (Figure 3.4) deals with the same messages for CTrCH Setup in NBAP and ALCAP layer that have already been introduced in the PCH setup. However, there may be some differences if more than just one FACH is installed. Depending on manufacturer-specific software implementation, it is possible that two FACHs with different TFSs will be used, e.g., to have one FACH for RRC signaling and the other one for transmitting? IP payload frames in downlink direction when UE is in RRC CELL_FACH state (see Sections 3.8 and 3.9).

NBAP DL initiatingMessage **Id-commonTransportChannelSetup**
(longTransActionID=**q**, id-C-ID=**e**, FACH-Parameters:
commonTransportChannelID=**o** + Transport Format Set 1,
commonTransportChannelID=**p** + Transport Format Set 2)

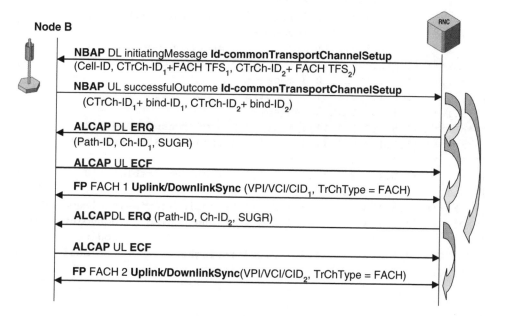

Figure 3.4   Node B Setup call flow 3/4.

```
+---------+----------------------------+------------------------+
|BITMASK  |ID Name                     |Comment or Value        |
+---------+----------------------------+------------------------+
|16:13:10,880,719  frm RNC DL  SSCOP  SD  NBAP  initiatingMessage |
|id-commonTransportChannelSe                                      |
|                                                                 |
|UNI SSCOP (SSCOP)  SD (= Seq. Conn.mode Data) [Layer Name Only]  |
|TS 25.433 V3.6.0 (2001-06) (NBAP)  initiatingMessage             |
|(= initiatingMessage)                                            |
|nbapPDU                                                          |
|1 initiatingMessage                                             |
|1.1 procedureID                                                 |
|00001100 |1.1.1 procedureCode       |id-commonTransportChannelSetup|
```
~~~~~~
```
|1.5.1.3.3.1 secondary-CCPCH-parameters                          |
|00010100 |1.5.1.3.3.1.1 commonPhysicalChannelID       |20        |
|0000---- |1.5.1.3.3.1.3 dl-ScramblingCode             |0         |
|***B2*** |1.5.1.3.3.1.4 fdd-DL-ChannelisationCodeNumber|5        |
```
~~~~~~
```
|1.5.1.3.3.1.11.3.1 fACH-ParametersItem-CTCH-SetupRqstFDD        |
|00010111 |1.5.1.3.3.1.11.3.1.1 commonTransportChannelID|23        |
|1.5.1.3.3.1.11.3.1.2 transportFormatSet                         |
|1.5.1.3.3.1.11.3.1.2.1 dynamicParts                             |
|1.5.1.3.3.1.11.3.1.2.1.1 sequence                               |
|***B2*** |1.5.1.3.3.1.11.3.1.2.1.1.1 nrOfTransportBlo..|0           |
|1.5.1.3.3.1.11.3.1.2.1.1.2 mode                                 |
|         |1.5.1.3.3.1.11.3.1.2.1.1.2.1 notApplicable  |0          |
|1.5.1.3.3.1.11.3.1.2.1.2 sequence                               |
|***B2*** |1.5.1.3.3.1.11.3.1.2.1.2.1 nrOfTransportBlo..|1           |
|***B2*** |1.5.1.3.3.1.11.3.1.2.1.2.2 transportBlockSize|168         |
|1.5.1.3.3.1.11.3.1.2.1.2.3 mode                                 |
|         |1.5.1.3.3.1.11.3.1.2.1.2.3.1 notApplicable  |0          |
|1.5.1.3.3.1.11.3.1.2.1.3 sequence                               |
|***B2*** |1.5.1.3.3.1.11.3.1.2.1.3.1 nrOfTransportBlo..|2           |
|***B2*** |1.5.1.3.3.1.11.3.1.2.1.3.2 transportBlockSize|168         |
|1.5.1.3.3.1.11.3.1.2.1.3.3 mode                                 |
|         |1.5.1.3.3.1.11.3.1.2.1.3.3.1 notApplicable  |0          |
|1.5.1.3.3.1.11.3.1.2.2 semi-staticPart                          |
|***b3*** |1.5.1.3.3.1.11.3.1.2.2.1 transmissionTimeIn..|msec-10    |
|--01---- |1.5.1.3.3.1.11.3.1.2.2.2 channelCoding  |convolutional-|
|                                                 coding        |
|-----0-- |1.5.1.3.3.1.11.3.1.2.2.3 codingRate      |half          |
|11011011 |1.5.1.3.3.1.11.3.1.2.2.4 rateMatcingAttribute|220        |
|-011---- |1.5.1.3.3.1.11.3.1.2.2.5 cRC-Size        |v16           |
|1.5.1.3.3.1.11.3.1.2.2.6 mode                                   |
|         |1.5.1.3.3.1.11.3.1.2.2.6.1 notApplicable  |0           |
|***B2*** |1.5.1.3.3.1.11.3.1.3 toAWS               |55            |
```

```
|***B2*** |1.5.1.3.3.1.11.3.1.4 toAWE                       |1          |
|***B2*** |1.5.1.3.3.1.11.3.1.5 maxFACH-Power               |60         |
|1.5.1.3.3.1.11.3.2 fACH-ParametersItem-CTCH-SetupRqstFDD             |
|00011000 |1.5.1.3.3.1.11.3.2.1 commonTransportChannelID|24         |
|1.5.1.3.3.1.11.3.2.2 transportFormatSet                              |
|1.5.1.3.3.1.11.3.2.2.1 dynamicParts                                  |
|1.5.1.3.3.1.11.3.2.2.1.1 sequence                                    |
|***B2*** |1.5.1.3.3.1.11.3.2.2.1.1.1 nrOfTransportBlo..|0         |
|1.5.1.3.3.1.11.3.2.2.1.1.2 mode                                      |
|        |1.5.1.3.3.1.11.3.2.2.1.1.2.1 notApplicable  |0         |
|1.5.1.3.3.1.11.3.2.2.1.2 sequence                                    |
|***B2*** |1.5.1.3.3.1.11.3.2.2.1.2.1 nrOfTransportBlo..|1         |
|***B2*** |1.5.1.3.3.1.11.3.2.2.1.2.2 transportBlockSize|360       |
|1.5.1.3.3.1.11.3.2.2.1.2.3 mode                                      |
|        |1.5.1.3.3.1.11.3.2.2.1.2.3.1 notApplicable  |0         |
|1.5.1.3.3.1.11.3.2.2.2 semi-staticPart                               |
|***b3*** |1.5.1.3.3.1.11.3.2.2.2.1 transmissionTimeIn..|msec-10  |
|--10---- |1.5.1.3.3.1.11.3.2.2.2.2 channelCoding   |turbo-coding |
|-----1-- |1.5.1.3.3.1.11.3.2.2.2.3 codingRate      |third        |
|10000001 |1.5.1.3.3.1.11.3.2.2.2.4 rateMatcingAttribute|130      |
|-011---- |1.5.1.3.3.1.11.3.2.2.2.5 cRC-Size        |v16          |
```

Message Example 3.11 NBAP Initiating Message (Common Transport Channel Setup) for two FACHs.

Compared to the CTrCH Setup for PCH, it is evident if one looks at both message examples that PCHs and FACHs are mapped onto separate secondary CCPCHs (which is an option because 3GPP defines that, in general, one S-CCPCH can serve all common transport channels). The downlink scrambling code is the same because the FACHs are established in the same cell as PCH before, but common physical channel ID and DL channelization code number are different.

If the TFSs of the two FACHs are compared to each other, it emerges that both sets transmit blocks in the same time interval of 10 ms, but block size is different. The faster channel sends larger transport blocks (360 bit compared to 168 bit) within the same time interval of 10 ms. The larger blocks are sent with more redundancy using a turbo coding with coding rate 1/3 (3 radio bits contain 1 bit of FACH information), while for smaller blocks (168 bit) convolutional coding with coding rate 1/2 is enough. Most likely, the FACH with larger blocks will be used for downlink transmission of IP payload if necessary and the other one for carrying RRC signaling of the connection that usually does not require highest data transmission rates.

NBAP UL successfulOutcome **Id-commonTransportChannelSetup**
(longTransActionID=**q**, commonTransportChannelID=**o**, bindingID=**r**,
commonTransportChannelID=**p**, bindingID=**v**)

Already with the NBAP CTrCH Setup request message, two different FACHs with their CTrCH-IDs and TFSs are defined. The TFS parameters indicate differences in transmission quality and transmission speed.

For each FACH a new bind-ID is assigned, which leads to two independent Establish procedures in ALCAP/AAL2L3. The result is that both FACHs will have their own physical transport layer in the form of an AAL2 SVC.

ALCAP DL ERQ (Originating Signal. Ass. ID=**s**, AAL2 Path=**k**, AAL2 ChannelId=**u**, served user gen reference=**r**)
ALCAP UL **ECF** (Originating Signal. Ass. ID=**t**, Destination Sign. Assoc. ID=**s**)

FP FACH1 **Downlink Synchronization** (in AAL2 Path=**k**′ and AAL2 ChannelId=**u**)

ALCAP DL **ERQ** (Originating Signal. Ass. ID=**w**, AAL2 Path=**k**, AAL2 ChannelId=**x**, served user gen reference=**v**)
ALCAP UL **ECF** (Originating Signal. Ass. ID=**x**, Destination Sign. Assoc. ID=**w**)

FP FACH2 **Downlink Synchronization** (in AAL2 Path=**k**′ and AAL2 ChannelId=**x**)

Optional possible

NBAP DL initiatingMessage **Id-SystemInformationUpdate**
(longTransActionID=**y**, id-C-ID=**e**, MIB + SIBs)

NBAP UL successfulOutcome **Id-SystemInformationUpdate**
(longTransActionID=**y**)

The RACH (Figure 3.5) is set up in the same way as the PCH. The only difference from PCH and FACH setup is that after AAL2L3 ECF no FP Synchronization frames are sent.

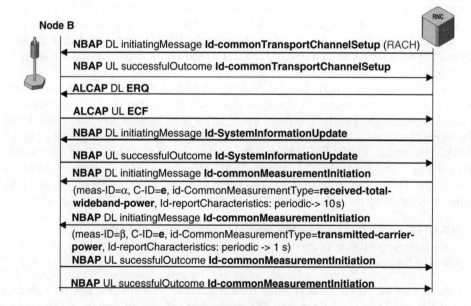

Figure 3.5   Node B Setup call flow 4/4.

```
+---------+---------------------------+---------------------------+
|BITMASK  |ID Name                    |Comment or Value           |
+---------+---------------------------+------ -------------------+
|16:13:14,680,711  frm RNC DL  SSCOP  SD  NBAP  initiatingMessage |
|id-commonTransportChannelSe                                      |

|UNI SSCOP (SSCOP)  SD (= Seq. Conn.mode Data) [Layer Name Only]  |
|TS 25.433 V3.6.0 (2001-06) (NBAP)  initiatingMessage             |
|(= initiatingMessage)                                            |
|nbapPDU                                                          |
|1 initiatingMessage                                              |
|1.1 procedureID                                                  |
|00001100 |1.1.1 procedureCode                          |id-       |
commonTransportChannelSetup                                       |
```
〰〰
```
|1.5.1.3.3.1 pRACH-parameters                                     |
|00110010 |1.5.1.3.3.1.1 commonPhysicalChannelID       |50        |
|0000---- |1.5.1.3.3.1.2 scramblingCodeNumber          |0         |
```
〰〰
```
|1.5.1.3.3.1.9 rACH-Parameters                                    |
|***B2*** |1.5.1.3.3.1.9.1 id                          |id-RACH-  |
ParametersItem-CTCH-SetupRqs                                      |
|00------ |1.5.1.3.3.1.9.2 criticality                 |reject    |
|1.5.1.3.3.1.9.3 value                                            |
|00110100 |1.5.1.3.3.1.9.3.1 commonTransportChannelID  |52        |
|1.5.1.3.3.1.9.3.2 transportFormatSet                             |
|1.5.1.3.3.1.9.3.2.1 dynamicParts                                 |
|1.5.1.3.3.1.9.3.2.1.1 sequence                                   |
|***B2*** |1.5.1.3.3.1.9.3.2.1.1.1 nrOfTransportBlocks |1         |
|***B2*** |1.5.1.3.3.1.9.3.2.1.1.2 transportBlockSize  |168       |
|1.5.1.3.3.1.9.3.2.1.1.3 mode                                     |
|         |1.5.1.3.3.1.9.3.2.1.1.3.1 notApplicable    |0         |
|1.5.1.3.3.1.9.3.2.1.2 sequence                                   |
|***B2*** |1.5.1.3.3.1.9.3.2.1.2.1 nrOfTransportBlocks |1         |
|***B2*** |1.5.1.3.3.1.9.3.2.1.2.2 transportBlockSize  |360       |
|1.5.1.3.3.1.9.3.2.1.2.3 mode                                     |
|         |1.5.1.3.3.1.9.3.2.1.2.3.1 notApplicable    |0         |
|1.5.1.3.3.1.9.3.2.2 semi-staticPart                              |
|***b3*** |1.5.1.3.3.1.9.3.2.2.1 transmissionTimeInter..|msec-20  |
|--01---- |1.5.1.3.3.1.9.3.2.2.2 channelCoding         |convolutional-|
|                                                        coding    |
|-----0-- |1.5.1.3.3.1.9.3.2.2.3 codingRate            |half      |
|10010101 |1.5.1.3.3.1.9.3.2.2.4 rateMatcingAttribute  |150       |
|-011---- |1.5.1.3.3.1.9.3.2.2.5 cRC-Size              |v16       |
|1.5.1.3.3.1.9.3.2.2.6 mode                                       |
|         |1.5.1.3.3.1.9.3.2.2.6.1 notApplicable      |0         |
|1.5.1.3.3.1.10 aICH-Parameters                                   |
```

```
|00110011 |1.5.1.3.3.1.10.1 commonPhysicalChannelID        |51       |
|0------- |1.5.1.3.3.1.10.2 aICH-TransmissionTiming         |v0       |
|***B2*** |1.5.1.3.3.1.10.3 fdd-dl-ChannelisationCodeN..|2       |
|10110--- |1.5.1.3.3.1.10.4 aICH-Power                      |0        |
|------1- |1.5.1.3.3.1.10.5 sTTD-Indicator                  |inactive |
--------------------------------------------------------------------------
```

Message Example 3.12   NBAP Initiating Message (Common Transport Channel Setup) for RACH.

RACH is mapped onto physical random access channel (PRACH) that has its own uplink scrambling code. The TFSs of RACH correspond to those defined for FACH(s) earlier (transport block size either 168 or 360 bit). The message also contains acquisition indication channel (AICH) parameters.

The setup of the CTrCHs is always completed with at least one System Information Update procedure that is used to transmit to Node B all SIBs that have not been sent yet. It is also possible that some SIBs will be retransmitted, especially SIB 5, to update the TFS settings of RACH that have not been available in an earlier stage. All SIBs will be sent if they have not been sent before.

As a rule, Common Measurement Initiation procedure for each single cell follows. Since there are mostly at least two different common measurement types (received total wideband power and transmitted carrier power), there will be two different Common Measurement Initiation procedures per cell. In addition to received total wideband power and transmitted carrier power, Common Measurement Initiation message may also include information about used method and accuracy of location measurement.

The first initialization is related to received total wideband power, a parameter that indicates the overall level of all received signals in the UMTS frequency band of the cell. The measurement results of received total wideband power are used by admission control and packet scheduler function of RNC to (re)calculate the allocated radio bearers of all UEs in the cell.

In the example, received total wideband power measurement is initiated for cell with Cell-ID = **e** and an appropriate NBAP Common Measurement Report message with same measurement-ID = α is expected every 10 s.

> **NBAP** DL initiatingMessage **Id-CommonMeasurementInitiation**
> (longTransActionID=**z**, id-Measurement-ID=α,
> id-CommonMeasurementObjectType-CM-Rqst -> c-ID=**e**,
> id-CommonMeasurementType= **received-total-wideband-power**,
> id-ReportCharacteristics: periodic – 10 s)

A second measurement initiation for the same cell contains settings for transmitted carrier power measurement. This is the TX power of the cell's antenna reported in a range from 0 to 100 %. An appropriate NBAP Common Measurement Report message is expected every 1 s.

> **NBAP** DL initiatingMessage **Id-CommonMeasurementInitiation**
> (longTransActionID=**aa**, id-Measurement-ID=β
> (id-CommonMeasurementObjectType-CM-Rqst -> c-ID=**e**,

id-CommonMeasurementType = **transmitted-carrier-power**,
id-ReportCharacteristics: periodic – 1 s)

Note: 3GPP 25.133 recommends for all NBAP common measurement types a reporting interval of 100 ms, but as long as not many subscribers are seen in the present UMTS networks, the reporting intervals are longer (in some cases up to 2 min), because as long as only 1, 2, or 3 users are served by one cell, there will be no significant changes in measurement results.

Finally, the Node B Setup procedure is finished with the confirmation of NBAP Common Measurement Initiation:

NBAP UL SuccessfulOutcome **Id-CommonMeasurementInitiation**
(longTransActionID=**z**)

NBAP UL SuccessfulOutcome **Id-CommonMeasurementInitiation**
(longTransActionID=**aa**)

3.2 Iub – IMSI/GPRS Attach Procedure

A Location Update (IMSI Attach) and/or Attach (for GPRS) procedure (Figure 3.6) is performed if a UE is switched on in a defined area of the network. The Location Update procedure – as known from GSM standards – is also used to indicate the change of a location area after the UE is IMSI-attached to the CS core network domain.

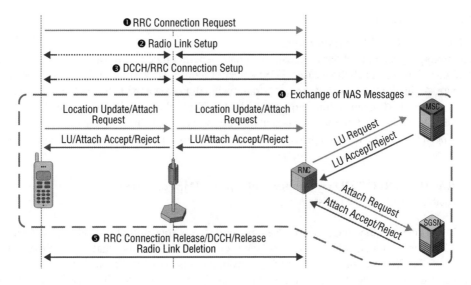

Figure 3.6 Iub IMSI/GPRS Attach procedure overview.

3.2.1 Overview

A Location Update Type (LUT) identifier in the Non-Access Stratum (NAS) message shows if the type of location update is:

- IMSI attach
- Normal location updating
- Periodic location updating

Before NAS messages can be exchanged between the USIM and the databases in the core network domains, it is necessary to build up an RRC connection between the UE and the RNC:

Step 1: An RRC Connection Request is sent from UE to RNC.

Step 2: Radio resources must be provided for the setup of a dedicated (transport) channel (DCH) that carries the logical dedicated control channels (DCCHs). The DCCHs are used for transmission of RRC and NAS messages.

Step 3: As long as DCH with DCCHs are not available, the signaling messages for RRC connection setup are transmitted using CTrCHs RACH (uplink) and FACH (downlink). RRC Connection Setup is completed after DCH setup.

Step 4: Location Update/GPRS Attach NAS messages embedded in RRC Direct Transfer messages are sent from UE to RNC and then forwarded to CS or PS Domain via Iu interfaces. The domains either accept or reject the Attach requests coming from the UE. An appropriate answer message is sent and once again included in RRC messages transported from RNC to UE.

Step 5: RRC connection is released; DCH and radio link are deleted.

3.2.2 Message Flow

Iub CS IMSI Attach/Location Update

The following example shows a Location Update general procedure toward CS core network domain as it is seen in all three cases: IMSI Attach (Figure 3.7), Normal, and Periodic Location Update. As in all other cases of Access Stratum signaling exchange between UE and RNC, it starts with an RRC Connection Request sent by UE. Since there is still no dedicated transport channel available, this first message is sent on Common Control Channel (logical channel) that is mapped onto RACH, which provides transport services for data and signaling in the uplink direction.

> **RACH**: UL RLC TMD **rrcConnectionRequest** (<u>**IMSI or TMSI**</u>,
> establishmentCause=**registration**)

On air interface the rrcConnectionRequest message is transmitted in Radio Link Control (RLC) transparent mode (TMD). The message contains a UE identifier that can be either IMSI or TMSI if a TMSI was provided by Visitor Location Register (VLR) before. This UE identifier is necessary because RACH is used by all users of the cell where the UE is located and there is still no RNTI assigned to identify this UE on common air interface channels uniquely.

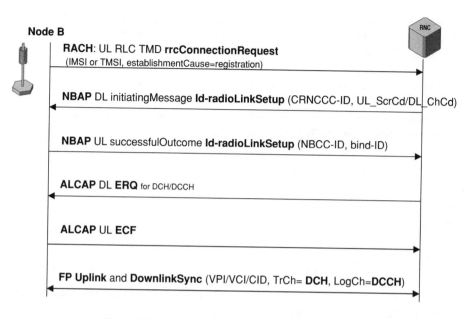

Node B

RACH: UL RLC TMD **rrcConnectionRequest**
(IMSI or TMSI, establishmentCause=registration)

NBAP DL initiatingMessage **Id-radioLinkSetup** (CRNCCC-ID, UL_ScrCd/DL_ChCd)

NBAP UL successfulOutcome **Id-radioLinkSetup** (NBCC-ID, bind-ID)

ALCAP DL **ERQ** for DCH/DCCH

ALCAP UL **ECF**

FP Uplink and **DownlinkSync** (VPI/VCI/CID, TrCh= **DCH**, LogCh=**DCCH**)

Figure 3.7   Iub IMSI Attach procedure call flow 1/4.

The next step is the Radio Link Setup procedure performed by NBAP protocol. The radio link setup is used to establish the necessary air interface resources for a DCH that is related to a Node B Communication Context in the Node B. Since this is an NBAP Class 1 Elementary Procedure, Initiating Message and Successful Outcome are linked with a TransactionID (in this case, longTransActionID). A pair of CRNC Communication Context ID (CRNCCC-ID) and Node B Communication Context ID identifies all NBAP messages regarding a single UE exchanged between this Node B and this single (C)RNC during the whole Location Update procedure. The cell-ID (C-ID) of the used cell is also applied as well as the radio link ID of this link for this specific UE. A pair of uplink Scrambling Code (ScrCd) and downlink Channelization Code (ChCd) is used to identify uniquely all messages from this UE on air interface. It describes a dedicated physical channel (DPCH) on radio interface (or to say it more exactly, a DPCH in downlink and a pair of separated dedicated physical data channel/dedicated physical control channel [DPDCH/DPCCH] in uplink direction). As shown later, this DPCH will carry DCHs. For the DCHs there need to be defined uplink and downlink TFSs in the Radio link setup request. Logical DCCHs that transport RRC messages including NAS messages will be mapped onto these DCHs. There is also a downlink scrambling code assigned in the same message, but this is just an identifier of the cell antenna that sends the signals on radio interface. From the point of view of UE this parameter is of course very important, but for signaling analysis on Iub it is not useful because it cannot help filtering out messages related to a single UE on Iub interface. That is the reason why DL ScrCd is not highlighted in our call flow examples. For the Uplink Channelization Code, only the code length is predefined by the network. The code itself is selected by UE using a random procedure. In the uplink direction the spreading code can indeed only be used for signal spreading, but identification function is limited. Only DPCCH and DPDCH of a single UE can be distinguished by different uplink spreading codes.

Uplink channels of different UEs cannot be distinguished because the orthogonality concept of the spreading code table requires synchronization of all senders for error-free code detection – and different UEs cannot be synchronized to each other.

NBAP DL initiatingMessage **Id-radioLinkSetup** (longTransActionID=**c**, id-CRNC-CommunicationContextID=**d**, ULscramblingCode/ DLChannelizationCode=**b**/β, DCH-SpecificInformationList: DCH-ID=**z**, UL Transport Format Set, DL Transport Format Set)

```
------------------------------------------------------------------
|TS 25.433 V3.6.0 (2001-06) (NBAP)  initiatingMessage            |
|(= initiatingMessage)                                           |
|nbapPDU                                                         |
|1 initiatingMessage                                            |
|1.1 procedureID                                                |
|00011011 |1.1.1 procedureCode                  |id-radioLinkSetup|
~~~~~~~~~~~~~~~~~~~~~~~~~~~~~~~~~~~~~~~~~~~~~~~~~~~~~~~~~~~~~~~~~~~~~
|***B2*** |1.5.1.1.1 id              |id-CRNC-CommunicationContextID |
|00------ |1.5.1.1.2 critic          |reject                       |
|***B3*** |1.5.1.1.3 value           |65538                        |
~~~~~~~~~~~~~~~~~~~~~~~~~~~~~~~~~~~~~~~~~~~~~~~~~~~~~~~~~~~~~~~~~~~~~

|1.5.1.2.3.1 ul-ScramblingCode                                   |
|***B3*** |1.5.1.2.3.1.1 uL-ScramblingCodeNumber       |1068459  |
|1------- |1.5.1.2.3.1.2 uL-ScramblingCodeLength       |long     |
|--110--- |1.5.1.2.3.2 minUL-ChannelizationCodeLength  |v256     |
~~~~~~~~~~~~~~~~~~~~~~~~~~~~~~~~~~~~~~~~~~~~~~~~~~~~~~~~~~~~~~~~~~~~~
|1.5.1.4.3.1.5 dCH-SpecificInformationList                       |
|1.5.1.4.3.1.5.1 dCH-Specific-FDD-Item                           |
|00011111 |1.5.1.4.3.1.5.1.1 dCH-ID                     |31       |
|1.5.1.4.3.1.5.1.2 ul-TransportFormatSet                         |
|1.5.1.4.3.1.5.1.2.1 dynamicParts                                |
|1.5.1.4.3.1.5.1.2.1.1 sequence                                  |
|***B2*** |1.5.1.4.3.1.5.1.2.1.1.1 nrOfTransportBlocks  |0        |
|1.5.1.4.3.1.5.1.2.1.1.2 mode                                    |
|         |1.5.1.4.3.1.5.1.2.1.1.2.1 notApplicable      |0        |
|1.5.1.4.3.1.5.1.2.1.2 sequence                                  |
|***B2*** |1.5.1.4.3.1.5.1.2.1.2.1 nrOfTransportBlocks  |1        |
|***B2*** |1.5.1.4.3.1.5.1.2.1.2.2 transportBlockSize   |148      |
|1.5.1.4.3.1.5.1.2.1.2.3 mode                                    |
|         |1.5.1.4.3.1.5.1.2.1.2.3.1 notApplicable      |0        |
|1.5.1.4.3.1.5.1.2.2 semi-staticPart                             |
|***b3*** |1.5.1.4.3.1.5.1.2.2.1 transmissionTimeInter..|msec-40  |
|--01---- |1.5.1.4.3.1.5.1.2.2.2 channelCoding       |convolutional-|
|                                                   coding       |
|-----1-- |1.5.1.4.3.1.5.1.2.2.3 codingRate             |third    |
|10011111 |1.5.1.4.3.1.5.1.2.2.4 rateMatcingAttribute   |160      |
```

```
|-011---- |1.5.1.4.3.1.5.1.2.2.5 cRC-Size                    |v16        |
|1.5.1.4.3.1.5.1.2.2.6 mode                                             |
|         |1.5.1.4.3.1.5.1.2.2.6.1 notApplicable            |0         |
|1.5.1.4.3.1.5.1.3 dl-TransportFormatSet                                |
|1.5.1.4.3.1.5.1.3.1 dynamicParts                                       |
|1.5.1.4.3.1.5.1.3.1.1 sequence                                         |
|***B2*** |1.5.1.4.3.1.5.1.3.1.1.1 nrOfTransportBlocks   |0         |
|1.5.1.4.3.1.5.1.3.1.1.2 mode                                           |
|         |1.5.1.4.3.1.5.1.3.1.1.2.1 notApplicable        |0         |
|1.5.1.4.3.1.5.1.3.1.2 sequence                                         |
|***B2*** |1.5.1.4.3.1.5.1.3.1.2.1 nrOfTransportBlocks   |1         |
|***B2*** |1.5.1.4.3.1.5.1.3.1.2.2 transportBlockSize    |148       |
|1.5.1.4.3.1.5.1.3.1.2.3 mode                                           |
|         |1.5.1.4.3.1.5.1.3.1.2.3.1 notApplicable        |0         |
|1.5.1.4.3.1.5.1.3.2 semi-staticPart                                    |
|***b3*** |1.5.1.4.3.1.5.1.3.2.1 transmissionTimeInter.. |msec-40   |
|--01---- |1.5.1.4.3.1.5.1.3.2.2 channelCoding           |convolutional-|
|                                               coding    |
|-----1-- |1.5.1.4.3.1.5.1.3.2.3 codingRate              |third     |
|10011111 |1.5.1.4.3.1.5.1.3.2.4 rateMatcingAttribute    |160       |
|-011---- |1.5.1.4.3.1.5.1.3.2.5 cRC-Size                |v16       |
/\/\/\/\/\/\/\/\/\/\/\/\/\/\/\/\/\/\/\/\/\/\/\/\/\/\/\/\/\/\/\/\/\/\/\/\/\/\/\
|***b5*** |1.5.1.5.3.1.3.1 rL-ID                         |0         |
|***B2*** |1.5.1.5.3.1.3.2 c-ID                          |13466     |
/\/\/\/\/\/\/\/\/\/\/\/\/\/\/\/\/\/\/\/\/\/\/\/\/\/\/\/\/\/\/\/\/\/\/\/\/\/\/\
|1.5.1.5.3.1.3.7 dl-CodeInformation                                     |
|1.5.1.5.3.1.3.7.1 fDD-DL-CodeInformationItem                           |
|***b4*** |1.5.1.5.3.1.3.7.1.1 dl-ScramblingCode            |0      |
|***B2*** |1.5.1.5.3.1.3.7.1.2 fdd-DL-ChannelizationCodeNumber|4    |
------------------------------------------------------------------------
```

Message Example 3.13   NBAP Initiating Message (Radio Link Setup).

The answer to NBAP Initiating Message is the successful outcome:

> **NBAP** UL successfulOutcome **Id-radioLinkSetup** (longTransAction ID=**c**,
> id-CRNC-CommunicationContextID=**d**, DCH-ID=**z**, bindingID=**e**,
> NodeBCommunicationsContext-ID=**p**)

As already seen in the setup of CTrCHs, the bindingID links the NBAP procedure with the appropriate ALCAP procedure. In addition, the Node B Communication Context ID is sent to the RNC.

```
|TS 25.433 V3.6.0 (2001-06) (NBAP)  succesfulOutcome
(= succesfulOutcome)                                                    |
|nbapPDU                                                                 |
|1 succesfulOutcome                                                      |
|1.1 procedureID                                                         |
|00011011 |1.1.1 procedureCode  |id-radioLinkSetup                       |
```

```
/\/\/\/\/\/\/\/\/\/\/\/\/\/\/\/\/\/\/\/\/\/\/\/\/\/\/\/\/\/\/\/\/\/\/\/\/\/\/\/\/\/\/\/\/\/\/\/\/\/\/\/\/\/\
|***B2*** |1.5.1.1.1 id          |id-CRNC-CommunicationContextID   |
|01------ |1.5.1.1.2 criticality                 |ig2ore         |
|***B3*** |1.5.1.1.3 value                       |65538          |
|1.5.1.2 sequence                                               |
|***B2*** |1.5.1.2.1 id          |id-NodeB-CommunicationContextID  |
|01------ |1.5.1.2.2 criticality                 |ignore         |
|00000000 |1.5.1.2.3 value                       |0              |
/\/\/\/\/\/\/\/\/\/\/\/\/\/\/\/\/\/\/\/\/\/\/\/\/\/\/\/\/\/\/\/\/\/\/\/\/\/\/\/\/\/\/\/\/\/\/\/\/\/\/\/\/\/\
|---00000 |1.5.1.4.3.1.3.1 rL-ID                 |0              |
|00000--- |1.5.1.4.3.1.3.2 rL-Set-ID             |0              |
|***B2*** |1.5.1.4.3.1.3.3 received-total-wide-band-power|70         |
/\/\/\/\/\/\/\/\/\/\/\/\/\/\/\/\/\/\/\/\/\/\/\/\/\/\/\/\/\/\/\/\/\/\/\/\/\/\/\/\/\/\/\/\/\/\/\/\/\/\/\/\/\/\
|00011111 |1.5.1.4.3.1.3.4.1.1.1.1 dCH-ID        |31             |
|***B2*** |1.5.1.4.3.1.3.4.1.1.1.2 bindingID     |02 10          |
|1------- |1.5.1.4.3.1.3.5 sSDT-SupportIndicator |sSDT-not       |
                                                 -supported|
------------------------------------------------------------------
```

Message Example 3.14   NBAP Successful Outcome (Radio Link Setup).

Further parameters in the successful outcome of Radio Link Setup are the radio link ID (rL-ID) and radio link set ID (rL-Set-ID), both of which are important for signaling analysis of soft handovers. In addition, the Node B reports an actual received-total-wide-band-power measurement result to the RNC. A specific indicator shows whether the Site Selection Diversity Transmission (SSDT) feature is supported by Node B or not.

The ALCAP Establish procedure works in the same way as seen during CTrCH Setup. The served user generated reference has the same value as the bindingID in Radio Link Setup response.

> **ALCAP** DL **ERQ** (Originating Signal. Ass. ID=**f**, AAL2 Path=**g**, AAL2 Channel id=**h**, served user gen reference=**e**)
> **ALCAP** UL **ECF** (Originating Signal. Ass. ID=**i**, Destination Sign. Assoc. ID=**f**)

After ALCAP Establish procedure the AAL 2 SVC for the DCH between UE and RNC on Iub interface is taken into service with a synchronization procedure. The VPI/VCI/CID values correspond to Path-ID and Ch-ID in ALCAP ERQ message as described for CTrCH Setup before.

> **DCH** in AAL2 Path=**g**′ and Connection=**h**: downlink and uplink synchronization FP frames

Now the resources for the DCH are established, but still the UE is in CELL_FACH state. To change from CELL_FACH to CELL_DCH state it waits for an incoming message that completes the DCH assignment by telling the UE (identified by its IMSI or TMSI) which physical and logical resources – that means which dedicated physical channels identified by a pair of UL scrambling code/DL channelization code and which Signaling Radio Bearers (logical control

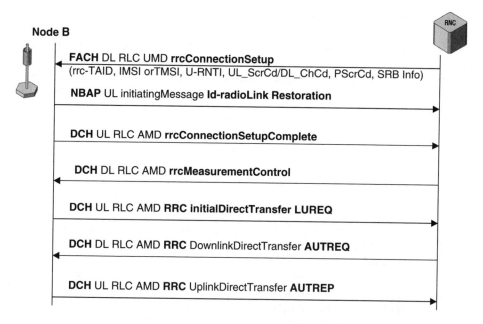

Node B

FACH DL RLC UMD **rrcConnectionSetup**
(rrc-TAID, IMSI orTMSI, U-RNTI, UL_ScrCd/DL_ChCd, PScrCd, SRB Info)

NBAP UL initiatingMessage **Id-radioLink Restoration**

DCH UL RLC AMD **rrcConnectionSetupComplete**

DCH DL RLC AMD **rrcMeasurementControl**

DCH UL RLC AMD **RRC initialDirectTransfer LUREQ**

DCH DL RLC AMD **RRC** DownlinkDirectTransfer **AUTREQ**

DCH UL RLC AMD **RRC** UplinkDirectTransfer **AUTREP**

Figure 3.8   Iub IMSI Attach procedure call flow 2/4.

channels) mapped onto which dedicated transport channel (DCH) – have been provided for the requested RRC connection. In addition to the DL channelization code the primary scrambling code (PScrCd) of the cell is included. The message called RRC Connection Setup is sent on FACH in downlink direction:

FACH: DL RLC UMD **rrcConnectionSetup** (rrc-Transaction Identifier
[rrc-TAID]=**a**, **IMSI or TMSI**, U-RNTI=**r**,
ULscramblingCode/DLchannelizationCode=**b**/β, Primary Scrambling Code $= \delta$,
MappingInfo for Signaling Radio Bearers)

The rrcConnectionSetup message (Figure 3.8) contains a rrcTransactionID that will be used to identify all following RRC messages of this RRC Setup procedure. rrcTransactionID value is only valid for a single RRC procedure of a single UE. That is the reason why most RRC messages of the same connection have the same rrcTransactionID value.

ULscramblingCode/DLChannelizationCode in this RRC connection setup are the same as in NBAP Radio Link Setup procedure before. The difference is that NBAP is the "language" in the dialog between Node B and (C)RNC, while RRC is "spoken" between UE and (S)RNC. So we have two different communications with different partners, but they are "talking" about the same topic!

After receiving RRC Connection Setup message, the UE knows which physical resources (DPCH) have been provided for it to be used on radio interface. IMSI or TMSI is used to ensure correct UE identification and in addition a U-RNTI is assigned. U-RNTI consists of Serving RNC identity (SRNC-ID) and SRNC Radio Network Temporary Identity (S-RNTI), which is a 20-bit random number that identifies a UE with an RRC connection within the UTRAN uniquely.

Furthermore, rrcConnectionSetup message contains the channel mapping info for the Signaling Radio Bearers (SRBs). Each SRB stands for a logical channel (DCCH) that carries specific signaling messages, e.g. there are different channels used for RRC control messages and RRC frames that contain NAS messages. The used channel is later indicated by C/T field value of MAC Data PDU.

```
|TS 29.331 CCCH-DL (2001-06) (RRC_CCCH_DL)  rrcConnectionSetup
(= rrcConnectionSetup)                                                  |
|dL-CCCH-Message                                                        |
|1 message                                                             |
|1.1 rrcConnectionSetup                                                |
/\/\/\/\/\/\/\/\/\/\/\/\/\/\/\/\/\/\/\/\/\/\/\/\/\/\/\/\/\/\/\/\/\/\/\/\/\/\
|1.1.1.1.1 initialUE-Identity                                          |
|1.1.1.1.1.1 imsi                                                      |
|***b4*** |1.1.1.1.1.1.1 digit                      |2                 |
|---1001- |1.1.1.1.1.1.2 digit                      |9                 |
|***b4*** |1.1.1.1.1.1.3 digit                      |9                 |
/\/\/\/\/\/\/\/\/\/\/\/\/\/\/\/\/\/\/\/\/\/\/\/\/\/\/\/\/\/\/\/\/\/\/\/\/\/\
|---00--- |1.1.1.1.2 rrc-TransactionIdentifier      |0                 |
|1.1.1.1.3 new-U-RNTI                                                  |
|**b12*** |1.1.1.1.3.1 srnc-IDentity                |44                |
|**b20*** |1.1.1.1.3.2 S-RNTI                       |17980             |
|-----00- |1.1.1.1.4 rrc-StateIndicator             |cell-DCH          |
/\/\/\/\/\/\/\/\/\/\/\/\/\/\/\/\/\/\/\/\/\/\/\/\/\/\/\/\/\/\/\/\/\/\/\/\/\/\
|1.1.1.1.6.1.3 rb-MappingInfo                                          |
|1.1.1.1.6.1.3.1.1.1.1 ul-TransportChannelType                         |
|--11111- |1.1.1.1.6.1.3.1.1.1.1.1 dch              |32                |
|***b4*** |1.1.1.1.6.1.3.1.1.1.2 logicalChannelIdentity |1             |
/\/\/\/\/\/\/\/\/\/\/\/\/\/\/\/\/\/\/\/\/\/\/\/\/\/\/\/\/\/\/\/\/\/\/\/\/\/\
|1.1.1.1.6.1.3.1.2.1.1 dl-TransportChannelType                         |
|***b5*** |1.1.1.1.6.1.3.1.2.1.1.1 dch              |32                |
|-0001--- |1.1.1.1.6.1.3.1.2.1.2 logicalChannelIdentity |1             |
|1.1.1.1.6.2 sRB-InformationSetup                                      |
|***b5*** |1.1.1.1.6.2.1 rb-Identity                |1                 |
/\/\/\/\/\/\/\/\/\/\/\/\/\/\/\/\/\/\/\/\/\/\/\/\/\/\/\/\/\/\/\/\/\/\/\/\/\/\
|1.1.1.1.6.2.3 rb-MappingInfo                                          |
|1.1.1.1.6.2.3.1.1.1.1 ul-TransportChannelType                         |
|***b5*** |1.1.1.1.6.2.3.1.1.1.1.1 dch              |32                |
|----0010 |1.1.1.1.6.2.3.1.1.1.2 logicalChannelIdentity |2             |
|1.1.1.1.6.2.3.1.2.1.1 dl-TransportChannelType                         |
|-11111-- |1.1.1.1.6.2.3.1.2.1.1.1 dch              |32                |
|***b4*** |1.1.1.1.6.2.3.1.2.1.2 logicalChannelIdentity |2             |
|1.1.1.1.6.3 sRB-InformationSetup                                      |
|---00010 |1.1.1.1.6.3.1 rb-Identity                |2                 |
/\/\/\/\/\/\/\/\/\/\/\/\/\/\/\/\/\/\/\/\/\/\/\/\/\/\/\/\/\/\/\/\/\/\/\/\/\/\
```

```
|1.1.1.1.11 ul-ChannelRequirement                                              |
|-------1 |1.1.1.1.11.1.2.1.1 scramblingCodeType       |longSC    |
|***B3*** |1.1.1.1.11.1.2.1.2 scramblingCode           |1068459   |
|         |1.1.1.1.11.1.2.1.3 numberOfDPDCH            |1         |
|110----- |1.1.1.1.11.1.2.1.4 spreadingFactor          |sf256     |
wwwwwwwwwwwwwwwwwwwwwwwwwwwwwwwwwwwwwwwwwwwwwwwwwwwwwwwwwwwwwwwwwwwwwwwwwwwwwwwww
|1.1.1.1.13 dl-InformationPerRL-List                                           |
|1.1.1.1.13.1.2.1.3.1 dL-ChannelizationCode                                    |
|1.1.1.1.13.1.2.1.3.1.1 sf-AndCodeNumber                                       |
|***b8*** |1.1.1.1.13.1.2.1.3.1.1.1 sf256              |4         |
-------------------------------------------------------------------------------
```

Message Example 3.15   Extract from RRC Connection Setup.

The message example shows a very comprehensive extract of RRC connection setup message contents. As seen, initial UE identifier is IMSI, which is embedded in the message digit by digit (in the example only the first three digits [MCC] are shown). Then we see rrcTransactionID and U-RNTI. The RRC state indicator orders the UE to change its state into CELL_DCH after the dedicated resource information was received. Then follows radio bearer mapping info: all logical channels (DCCHs) are mapped onto a dedicated (transport) channel (DCH) identified by RRC transport channel identity = 32. However, if looking back at NBAP Radio Link Setup procedure, it emerges that CRNC set up a dedicated channel with DCH-ID = 31. Why this difference?

As written in 3GPP 25.433 (NBAP), "The DCH ID is the identifier of an active dedicated transport channel. It is unique for each active DCH among the active DCHs simultaneously allocated for the same UE."

DCH-ID in NBAP is described as INTEGER (0 . . . 255).

RRC (3GPP 25.331) defines the parameter Transport Channel Identity as "the ID of a DCH . . . that . . . Radio Bearer could be mapped onto."

Transport Channel ID in RRC is described as INTEGER (1 . . . 32).

Indeed, there is a correlation between both identities despite that the number range is different. As a rule it can be monitored that a DCH that was set up with DCH-ID = 31, e.g. during NBAP Radio Link Setup procedure, will be identified in RRC Radio Bearer Setup by Transport Channel ID = 32, this means RRC Transport Channel Identity = NBAP DCH-ID + 1. However, not all equipment manufacturers adhere to this unwritten rule. There have been also cases monitored in the field where values of RRC and NBAP are equal. That could become another reason for interoperability problems in a multivendor environment.

SRB ID in RRC Connection Setup example is the same as logical channel ID. In total there are four SRBs defined (only two of them are shown in the message example). Each SRB represents a DCCH for signaling messages transported in RLC unacknowledged mode (SRB 1), for RLC acknowledged mode (SRB 2), for NAS signaling with high priority (SRB 3), and for NAS signaling with lower priority (SRB 4). Unfortunately, the numbering scheme for SRBs sometimes depends on the vendor – there are cases where radio bearer ID 2, 3, 4, and 5 are used instead of 1, 2, 3, and 4.

Finally, one can find uplink scrambling code and uplink channelization code length of the dedicated physical channels in the message as well as downlink spreading factor (equal to spreading code length) and downlink spreading code number of downlink DPCH. Using this information the UE is able to find and use the already provided dedicated physical channel resources on radio interface.

In the next step, NBAP Radio Link Restoration message indicates that UE and Node B are now synchronized on air interface. In other words, the UE has found the provided dedicated physical channels. CRNC Communication Context ID identifies once again the NBAP signaling connection regarding this single UE.

> **NBAP** UL initiatingMessage **id-radioLinkRestoration** (shortTransActionID=**j**, id-CRNC-CommunicationContextID=**d**)

Now, after synchronization on air interface the DCH is available and RRC messages (RLC AMD) are carried on Iub interface in AAL2 Path=**g′** and Connection=**h**:

> DL RLC AMD **rrcMeasurementControl**

One or more RRC Measurement Control messages are sent to initialize RRC measurement functions on the UE side. To learn more about the specific options of this message see Section 3.7.

UL RLC AMD **rrcConnectionSetupComplete** (rrc-TransactionIdentifier=**a**) confirms the RRC connection establishment. The rrc-TransactionIdentifier links this message to the previous RRC Connection Setup that was sent on FACH. In case that ciphering is switched on, the message contains the start values for CS and PS core network domains that are necessary to initialize encryption functions.

Then the transport of NAS messages starts. They are sent on DCH embedded in RRC messages. The function and parameters of NAS messages are well known from GSM standards, so only a general description is given.

*UL initialDirectTransfer **LUREQ*** – Location Update Request is sent to VLR to indicate change of Location or IMSI attach of UE. The message contains UE identity (IMSI or TMSI) and Location Area Information (LAI). Besides this, a number of MS Classmark elements are included that inform the network about capabilities like supported algorithms for ciphering and integrity protection.

DownlinkDirectTransfer AUTREQ – Authentication Request is sent by the network to check UE identity.

UplinkDirectTransfer AUTREP – Authentication Response contains Signed Response (SRES) Information Element constructed by UE which is compared with Expected Response (XRES) in VLR. If SRES = XRES, then authentication UE was authenticated successfully.

If ciphering is requested by the network, the **RRC SecurityModeCommand** message is used to start/stop ciphering and to start or modify integrity protection (Figure 3.9).

RRC SecurityModeComplete message confirms the configuration of ciphering and/or integrity protection. Here a list can be found that shows for each SRB the RLC sequence number of the first ciphered RLC frame.

Now the transfer of NAS messages can be continued.

> **RRC** DownlinkDirectTransfer **LUACC** (opt. TMSI) *or* **LUREJ**: Location Update Accept *or* Location Update Reject

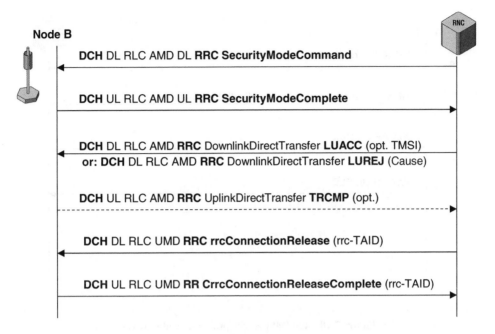

Node B

DCH DL RLC AMD DL **RRC SecurityModeCommand**

DCH UL RLC AMD UL **RRC SecurityModeComplete**

DCH DL RLC AMD **RRC** DownlinkDirectTransfer **LUACC** (opt. TMSI)
or: DCH DL RLC AMD **RRC** DownlinkDirectTransfer **LUREJ** (Cause)

DCH UL RLC AMD **RRC** UplinkDirectTransfer **TRCMP** (opt.)

DCH DL RLC UMD **RRC rrcConnectionRelease** (rrc-TAID)

DCH UL RLC UMD **RR CrrcConnectionReleaseComplete** (rrc-TAID)

Figure 3.9 Iub IMSI Attach procedure call flow 3/4.

These messages show whether the Location Update was accepted by the core network domain or rejected. If the request was accepted, a new TMSI may be allocated to the UE. This always happens in case of IMSI attach, but in case of normal or periodic location update it is just an option. In case the location update is rejected, the LUREJ message contains a cause value that indicates the reason for rejection.

Note: In general, all requested actions can be either accepted or rejected by the core network. In the same way as described for location update request and also GPRS attach request, connection management service request that starts a voice call or SMS or activate PDP context request can be either accepted or rejected. A reject message in such cases often indicates a gap in the network that requires troubleshooting. For this reason such procedures are often called a "bad case" instead of a "good case" when the request is accepted.

If a new TMSI is assigned with LUACC message, an **RRC** UplinkDirectTransfer TMSI Reallocation Complete (TRCMP) message is sent by the UE to complete the message exchange with the core network domain. Now the mobile uses the (new) TMSI.

Then it is no longer necessary to keep the RRC connection active because the location update procedure is finished. So RRC connection is released with:

(RLC UMD) DL **RRC rrcConnectionRelease** (rrc-Transaction Identifier
[rrc-TAID]=**a**)
(RLC UMD) UL **RRC rrcConnectionReleaseComplete** (rrc-Transaction Identifier
[rrc-TAID]=**a**)

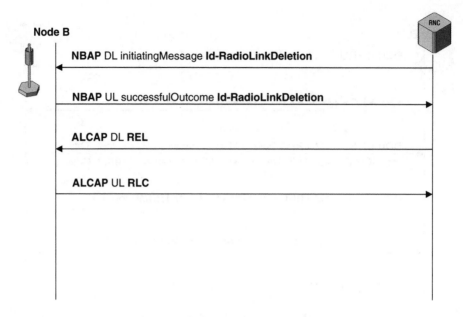

Figure 3.10   Iub IMSI Attach procedure call flow 4/4.

On behalf of the next two messages (see Figure 3.10),

> **NBAP** DL initiatingMessage **Id-RadioLinkDeletion** (shortTransAction ID=**j**,
> id-CRNC-CommunicationContextID=**d**, NodeBCommunicationsContext-ID=**p**)

and

> **NBAP** UL successfulOutcome **Id-RadioLinkDeletion** (shortTransActionID=**j**,
> id-CRNC-CommunicationContextID=**d**)

the radio resources for the DCH/DCCH are released. And with

> **ALCAP** DL **REL**ease Request (Dest. Sign. Assoc. ID=**i**)
> **ALCAP** UL **Re**Lease Confirm (Dest. Sign. Assoc. ID=**f**)

the physical layer (AAL2 SVC) for this DCH on Iub interface is released too.

Iub PS Attach
In case of GPRS Attach (Figure 3.11) the RRC connection setup is the same as for CS location update. Depending on the type and features of the UE, it is possible to have a combined IMSI/GPRS attach. In this case the same DCHs/DCCHs as in the Location Update procedure example would be used for the transport of NAS messages. NAS messages will be forwarded by SRNC to Serving GPRS Support Node (SGSN) that has its own Location Register function (sometimes called SGSN Location Register [SLR]).

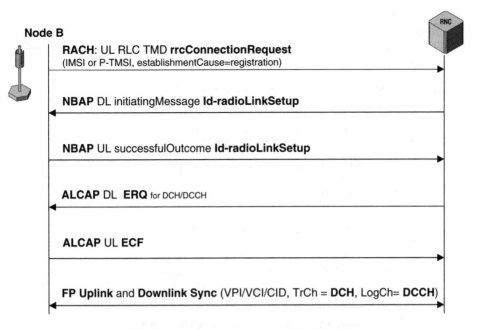

Figure 3.11   Iub IMSI Attach procedure call flow 1/3.

Messages in detail

RACH: UL RLC TMD rrcConnectionRequest (**IMSI or P-TMSI**, establishmentCause=**registration**)

NBAP DL initiatingMessage **Id-radioLinkSetup** (longTransActionID=**c**, id-CRNC-CommunicationContextID=**d**, ULscramblingCode/ DLChannelizationCode=**b**/β, DCH-SpecificInformationList: DCH-ID=**z**, UL Transport Format Set, DL Transport Format Set)

NBAP UL successfulOutcome **Id-radioLinkSetup** (longTransAction ID=**c**, id-CRNC-CommunicationContextID=**d**, DCH-ID=**z**, bindingID=**e**, NodeBCommunicationsContext-ID=**p**)

ALCAP DL **ERQ** (Originating Signal. Ass. ID=**f**, AAL2 Path=**g**, AAL2 Channel id=**h**, served user gen reference=**e**)

ALCAP UL **ECF** (Originating Signal. Ass. ID=**i**, Destination Sign. Assoc. ID=**f**)

DCH in AAL2 Path=**g**$'$ and Channel=**h** will be initialized by sending downlink and uplink synchronization FP frames (see Figure 3.12).

Since the RRC connection will be used for a GPRS Attach, the P-TMSI (packet TMSI) will be used as temporary identity of the UE in the rrcConnectionSetup message if any P-TMSI was assigned by SGSN location register function before.

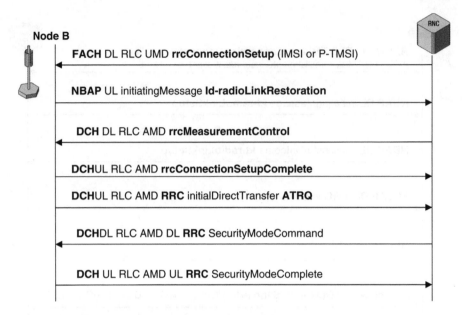

Figure 3.12   Iub GPRS Attach procedure call flow 2/3.

FACH: DL RLC UMD **rrcConnectionSetup** (rrc-Transaction Identifier
[rrc-TAID]=**a**, **IMSI or P-TMSI**, U-RNTI=**r**,
ULscramblingCode/DLChannelizationCode=**b**/β, Primary Scrambling Code $= \delta$,
MappingInfo for Signaling Radio Bearers)

NBAP UL initiatingMessage **id-radioLinkRestoration** (shortTransAction ID=**j**,
id-CRNC-communicationContextID=**d**)

The following **RRC messages** (RLC AMD) are carried on Iub interface in AAL2 Path=**g'**
and Connection=**h**:

DL RLC AMD **rrcMeasurementControl** (rrc-TransactionIdentifier=**a**)
UL RLC AMD **rrcConnectionSetupComplete** (rrc-TransactionIdentifier=**a**).

UL initialDirectTransfer ATRQ — GPRS Attach Request message contains UE
identifier IMSI or P-TMSI, Location and Routing Area Identity (LAI, RAI)

The network may or may not perform the authentication procedure. Ciphering and/or integrity protection are activated with:

DL **RRC SecurityModeCommand**
UL **RRC SecurityModeComplete**

RRC DownlinkDirectTransfer ATAC — GPRS Attach Accept message (Figure 3.13) confirms from SGSN/SLR that subscriber is GPRS attached now, an optional (new) P-TMSI is
assigned

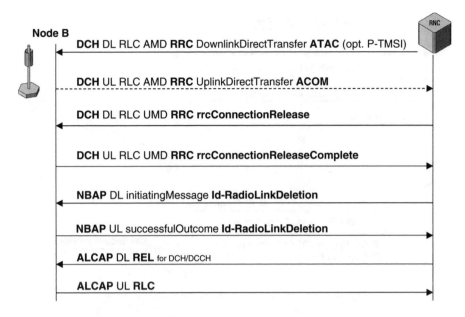

Figure 3.13   Iub GPRS Attach procedure call flow 3/3.

RRC UplinkDirectTransfer ACOM — GPRS Attach Complete message is optionally sent in case of (new) P-TMSI assignment using ATAC message.
Release of RRC Connection:

> (RLC UMD) DL **rrcConnectionRelease**
> (RLC UMD) UL **rrcConnectionReleaseComplete**

Release of radio resources:

> **NBAP** DL initiatingMessage **Id-RadioLinkDeletion** (shortTransActionID=**j**,
> id-CRNC-communicationContextID=**d**, NodeBCommunicationsContext-ID=**p**)
> **NBAP** UL successfulOutcme **Id-RadioLinkDeletion** (shortTransActionID=**j**,
> id-CRNC-communicationContextID=**d**)

Release of AAL2 SVC for DCH/DCCH on Iub:

> **ALCAP** DL **REL** (Dest. Sign. Assoc. ID=**i**)
> **ALCAP** UL **RLC** (Dest. Sign. Assoc. ID=**f**)

3.3 Iub CS – Mobile Originated Call

This scenario describes the message flow for a user-initiated voice call, which includes the allocation and release of radio (access) bearers.

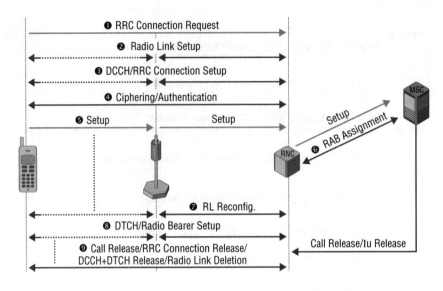

Figure 3.14   Iub mobile-originated voice call overview.

3.3.1 Overview

Steps 1 to 3 for the mobile-originated voice call (MOC) are the same as described for IMSI/GPRS Attach procedure (see Figure 3.14).

Step 4: The optional ciphering/authentication procedure requested by the network is used to double-check UE identity and to switch on ciphering between RNC and UE if necessary.

Step 5: The voice call setup starts with a SETUP message in MM/SM/CC layer. The SETUP message includes the dialed called party number and is forwarded by the RNC to the CS core network domain.

Step 6: The CS core network domain defines a Quality of Service (QoS) for the voice call. QoS values are key parameters of the Radio Access Bearer (RAB). The RAB Assignment procedure can be compared with the setup of a bearer channel in CCS#7-based networks. The RAB Service provides a "channel" for user data (voice packets) between the Mobile Termination (MT) of the UE and the serving MSC in the CS core network domain.

Step 7: The Radio Link Reconfiguration provides radio resources for the establishment of the Radio Bearer in the next step.

Step 8: On behalf of the parameter values negotiated in RAB Assignment procedure, a new radio bearer is set up to carry the (logical) dedicated traffic channels (DTCHs). If AMR codec is used to encode the voice information, three DTCHs are set up, one for each class of AMR bits: class A, class B, and class C bits.

Step 9: The release of the voice call follows the release of the RRC connection if no other services are active. Then both the dedicated control channel and the dedicated traffic channel are released as well. Finally, the RNC releases the radio resources that have been blocked for both channels. The AAL2 SVCs are also deleted.

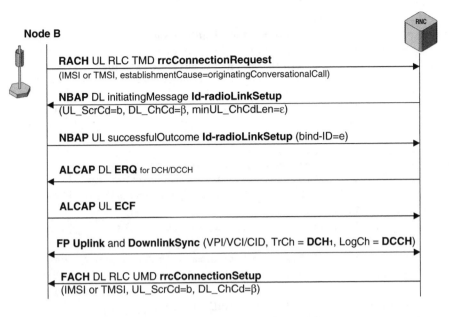

Figure 3.15 Iub MOC call flow 1/6.

3.3.2 Message Flow

Already the rrcConnectionRequest message contains the call establishment cause that indicates that an MOC is started (Figure 3.15). Since in the call flow example there is still no dedicated control channel available, this first message is sent via RACH:

RACH: UL RLC TMD **rrcConnectionRequest** (<u>IMSI or TMSI</u>, establishmentCause=**originatingConversationalCall**)

As in the procedures described before, the NBAP is responsible for the Radio link setup. However, this time it will be interesting to watch a little bit closer the downlink channelization code and minimum uplink channelization code length as well as DCH-IDs:

NBAP DL initiatingMessage Id-radioLinkSetup (longTransActionID=**c**, id-CRNC-communicationContextID=**d**, ULscramblingCode/ DLChannelizationCode=**b**/β, minULChCdLen=ε, DCH-SpecificInformationList: DCH-ID=**z**, UL Transport Format Set, DL Transport Format Set)

```
|TS 25.433 V3.6.0 (2001-06) (NBAP)  initiatingMessage        |
|(= initiatingMessage)                                        |
|nbapPDU                                                      |
|1 initiatingMessage                                          |
|1.1 procedureID                                              |
|00011011 |1.1.1 procedureCode              |id-radioLinkSetup|
```

```
|1.5.1.2.3.1 ul-ScramblingCode                                             |
|***B3*** |1.5.1.2.3.1.1 uL-ScramblingCodeNumber          |1068457   |
|1------- |1.5.1.2.3.1.2 uL-ScramblingCodeLength          |long      |
|--110--- |1.5.1.2.3.2 minUL-ChannelizationCodeLength     |v256      |
\wwwwwwwwwwwwwwwwwwwwwwwwwwwwwwwwwwwwwwwwwwwwwwwwwwwwwwwwwwwwwwwwwwwwwwwwwww\
|1.5.1.4.3.1.5.1 dCH-Specific-FDD-Item                                     |
|00011111 |1.5.1.4.3.1.5.1.1 dCH-ID                        |31        |
\wwwwwwwwwwwwwwwwwwwwwwwwwwwwwwwwwwwwwwwwwwwwwwwwwwwwwwwwwwwwwwwwwwwwwwwwwww\
|1.5.1.4.3.1.5.1.2 ul-TransportFormatSet                                   |
|1.5.1.4.3.1.5.1.2.1 dynamicParts                                          |
\wwwwwwwwwwwwwwwwwwwwwwwwwwwwwwwwwwwwwwwwwwwwwwwwwwwwwwwwwwwwwwwwwwwwwwwwwww\
|***B2*** |1.5.1.4.3.1.5.1.2.1.2.1 nrOfTransportBlocks    |1         |
|***B2*** |1.5.1.4.3.1.5.1.2.1.2.2 transportBlockSize     |148       |
|1.5.1.4.3.1.5.1.2.1.2.3 mode                                              |
|         |1.5.1.4.3.1.5.1.2.1.2.3.1 notApplicable        |0         |
|1.5.1.4.3.1.5.1.2.2 semi-staticPart                                       |
|***b3*** |1.5.1.4.3.1.5.1.2.2.1 transmissionTimeInter..|msec-40   |
|--01---- |1.5.1.4.3.1.5.1.2.2.2 channelCoding            |convolutional-|
|                                                             coding   |
|-----1-- |1.5.1.4.3.1.5.1.2.2.3 codingRate               |third     |
|10011111 |1.5.1.4.3.1.5.1.2.2.4 rateMatcingAttribute     |160       |
|-011---- |1.5.1.4.3.1.5.1.2.2.5 cRC-Size                 |v16       |
|1.5.1.4.3.1.5.1.3 dl-TransportFormatSet                                   |
\wwwwwwwwwwwwwwwwwwwwwwwwwwwwwwwwwwwwwwwwwwwwwwwwwwwwwwwwwwwwwwwwwwwwwwwwwww\
|***B2*** |1.5.1.4.3.1.5.1.3.1.2.1 nrOfTransportBlocks    |1         |
|***B2*** |1.5.1.4.3.1.5.1.3.1.2.2 transportBlockSize     |148       |
|1.5.1.4.3.1.5.1.3.1.2.3 mode                                              |
|         |1.5.1.4.3.1.5.1.3.1.2.3.1 notApplicable        |0         |
|1.5.1.4.3.1.5.1.3.2 semi-staticPart                                       |
|***b3*** |1.5.1.4.3.1.5.1.3.2.1 transmissionTimeInter..|msec-40   |
|--01---- |1.5.1.4.3.1.5.1.3.2.2 channelCoding            |convolutional-|
|                                                             coding   |
|-----1-- |1.5.1.4.3.1.5.1.3.2.3 codingRate               |third     |
|10011111 |1.5.1.4.3.1.5.1.3.2.4 rateMatcingAttribute     |160       |
\wwwwwwwwwwwwwwwwwwwwwwwwwwwwwwwwwwwwwwwwwwwwwwwwwwwwwwwwwwwwwwwwwwwwwwwwwww\
|1.5.1.5.3.1.3.7 dl-CodeInformation                                        |
|1.5.1.5.3.1.3.7.1 fDD-DL-CodeInformationItem                              |
|***b4*** |1.5.1.5.3.1.3.7.1.1 dl-ScramblingCode          |0         |
|***B2*** |1.5.1.5.3.1.3.7.1.2 fdd-DL-ChannelizationCo..|4         |
---------------------------------------------------------------------
```

Message Example 3.16   NBAP Radio Link Setup for voice call.

NBAP UL successfulOutcome Id-radioLinkSetup (longTransActionID=**c**,
id-CRNC-CommunicationContextID=**d**, DCH-ID=**z**, bindingID=**e**,
NodeBCommunicationContext-ID=**p**)

Then ALCAP protocol sets up the AAL2 SVC for the dedicated control channel (DCH/DCCH):

ALCAP DL **ERQ** (Originating Signal. Ass. ID=**f**, AAL2 Path=**g**, AAL2
ChannelId=**h**, served user gen reference=**e**)
ALCAP UL **ECF** (Originating Signal. Ass. ID=**i**, Destination Sign. Assoc. ID=**f**)

Downlink and uplink **Framing Protocol Synchronization** messages are monitored on this
DCH in AAL2 Path=**g**′ and Channel=**h**.

However, the DCCH cannot be used for transmitting RRC messages before the UE has
not received the parameters for the assigned physical resources on radio interface (UL scrambling code, DL channelization code, etc.). The scrambling code/channelization code is sent in
downlink direction with:

FACH: DL RLC UMD **rrcConnectionSetup** (rrc-Transaction Identifier [rrc-TAID]=
a, **IMSI or TMSI**, U-RNTI=**r**, ULscramblingCode/DLChannelizationCode=**b**/β,
Primary Scrambling Code $= \delta$, MappingInfo for Signaling Radio Bearers)

As shown in the Location Update scenario, this message also contains mapping info for the
logical channels (SRB). They will be mapped onto the DCH defined already in NBAP Radio
link setup message.

That UE found the dedicated physical channels on air interface is indicated by sending:

NBAP UL initiating Message **Id-radioLinkRestoration** (shortTransActionID=**j**,
id-CRNC-communicationContextID=**d**)

All following RRC messages (RLC AMD) run in AAL2 SVC with CID=**h** (see Figure
3.16):

DL RLC AMD **rrcMeasurementControl**

UL RLC AMD **rrcConnectionSetupComplete** (rrc-TransactionIdentifier=**a**).

MM/SM/CC messages in the RRC Signaling Connection are:

UL initialDirect Transfer **CMSREQ** – Connection Management Service Request (can be
answered optionally with Connection Management Service Accept message **CMSACC** or
can be rejected by sending **CMSREJ**)
DownlinkDirectTransfer **AUTREQ** – Authentication Request, network requests double-check
of UE identity
UplinkDirectTransfer **AUTREP** – Authentication Response, UE answers the authentication
request with a signed response

In case that the signed response is identical with the expected response, ciphering/integrity
protection is activated between UE and RNC by sending **RRC SecurityModeCommand**
message (Figure 3.17).

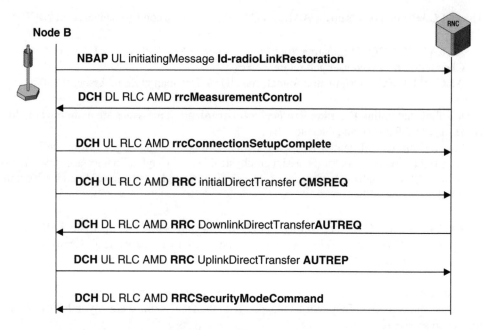

Figure 3.16   Iub MOC call flow 2/6.

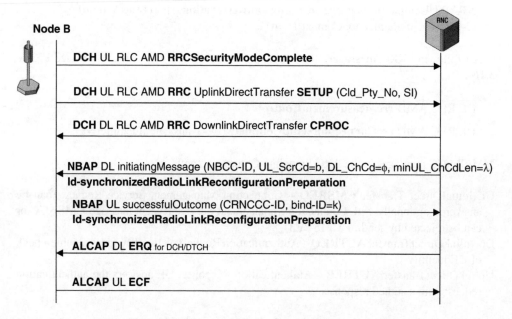

Figure 3.17   Iub MOC call flow 3/6.

RRC SecurityModeComplete message completes ciphering activation procedure.

Further MM/SM/CC messages embedded in RRC uplink/downlink direct transfer messages are as follows:

UplinkDirectTransfer SETUP – includes the called party (B-party) number that was dialed by the UMTS subscriber and as an option the stream identifier (SI) if the UE supports multicall capability. SI value will later be used as RAB-ID value by RRC protocol entities. All embedded MM/SM/CC messages belonging to this call are marked with the same transaction ID (TIO) value.

```
|TS 25.322 V3.7.0 (2001-06) reassembled (RLC reasm.)  AM DATA DCH  |
|(= Acknowledged Mode Data DCH)                                    |
|Acknowledged Mode Data DCH                                        |
|       |FP:  VPI/VCI/CID              |"10/26/183"                |
|       |FP:  Direction               |Uplink                     |
|       |FP:  Transport Channel Type  |DCH (Dedicated Channel)    |
|       |MAC: Target Channel Type  |DCCH (Dedicated Control Channel)|
|       |MAC: C/T Field               |Logical Channel 3          |
|       |MAC: RLC Mode                |Acknowledge Mode           |
|TS 29.331 DCCH-UL (2001-06) (RRC_DCCH_UL)  uplinkDirectTransfer   |
|(= uplinkDirectTransfer)                                          |
|uL-DCCH-Message                                                   |
|1 message                                                         |
|1.1 uplinkDirectTransfer                                          |
|0------- |1.1.1 cn-DomainIdentity      |cs-domain                 |
|**b136** |1.1.2 nas-Message  |c4 40 01 69 10 80 e7 19 00 44 52...|
|TS 24.008 Call Control V3.8.0 (CC-DMTAP) SETUP (= Setup)          |
|Setup                                                             |
|----0011 |Protocol Discriminator       |call control, call related|
|         |                             |SS messag                 |
|-000---- |Transaction Id value (TIO)   |TI value 0                |
/\/\/\/\/\/\/\/\/\/\/\/\/\/\/\/\/\/\/\/\/\/\/\/\/\/\/\/\/\/\/\/\/\/\/\/\
|CaLleD party BCD number                                           |
|01011110 |IE Name                      |CaLleD party BCD number   |
|00000111 |IE Length                    |7                         |
|----0000 |Number plan                  |Unknown                   |
|-000---- |Type of number               |Unknown                   |
|1------- |Extension bit                |No Extension              |
|**b44*** |Called party number          |'0800123456'              |
|1111---- |Filler                       |15                        |
|Stream Identifier                                                 |
|00101101 |IE Name                      |Stream Identifier         |
|00000001 |IE Length                    |1                         |
|00000001 |Stream Identifier            |1                         |
---------------------------------------------------------------------
```

Message Example 3.17 Call Control SETUP (MOC).

In the message example it is further shown on which channels the message is transported. Physical transport bearer on Iub interface is an AAL2 SVC with VPI/VCI/CID = 10/26/183. On this physical transport bearer a dedicated (transport) channel (DCH) is running and because SETUP is an NAS signaling message it is carried by a logical DCCH with logical channel ID = 3, which is identical with signaling radio bearer ID = 3 (NAS signaling with high priority). The CN-DomainIdentity of RRC Uplink Direct Transfer message indicates that the message will be forwarded to CS core network domain represented by MSC (Rel. 99) or MSC Server (Rel. 4 and higher). Since type of number and number plan of called party number are unknown, it is sure that the called party number string contains all the digits as they have been dialed by the subscriber.

Reception of SETUP is answered by core network with

DownlinkDirectTransfer **CPROC** – This message indicates that the call is being processed by core network entities. On UE side the call control state is changed when this message is received. With entering the new state it becomes impossible to send any additional dialing information.

Now Radio Resources for DTCHs need to be provided by (C)RNC. A dedicated physical channel in uplink and downlink direction already exists and it is identified by UL scrambling code number and DL channelization code number. Hence, this channel needs to be reconfigured because more data traffic between UE and network is expected when DTCHs are mapped on this physical channel. The higher the data transfer rate on radio interface is, the lower must be the chosen spreading factor assigned with channelization code. So a new downlink channelization code with lower spreading factor than before is found in the Synchronized Radio Link Reconfiguration Preparation message as well as a new minimum uplink channelization code length (minULChCdLen), the value of which is also smaller than in Radio Link Setup message before ($\lambda < \epsilon$).

NBAP DL: initiatingMessage
Id-synchronizedRadioLinkReconfigurationPreparation (shortTransActionID=**j**, NodeBCommunicationsContext-ID=**p**, ULscramblingCode/
DLChannelizationCode=**b**/ϕ, minULChCdLen= λ, Transport Format Sets of DCHs)

Please note in the message example that for the protocol tester decoder unit already the name of this procedure code is too long to display it correctly, which is with no doubt one of the big disadvantages of excessive ASN.1 PER usage.

```
|TS 25.433 V3.6.0 (2001-06) (NBAP)  initiatingMessage                   |
|(= initiatingMessage)                                                  |
|nbapPDU                                                                |
|1 initiatingMessage                                                    |
|1.1 procedureID                                                        |
|00011111 |1.1.1 procedureCode|id-synchronisedRadioLinkReconfigurat|
~~~~~~~~~~~~~~~~~~~~~~~~~~~~~~~~~~~~~~~~~~~~~~~~~~~~~~~~~~~~~~~~~~~~~~~~~~~~~~~~~~~~
|1.5.1.2.3.1 ul-ScramblingCode                                          |
|***B3*** |1.5.1.2.3.1.1uL-ScramblingCodeNumber        |1068457  |
|1------- |1.5.1.2.3.1.2 uL-ScramblingCodeLength        |long     |
|10011000 |1.5.1.2.3.2 ul-SIR-Target                    |70       |
|-100---- |1.5.1.2.3.3 minUL-ChannelizationCodeLength   |v64      |
```

```
/\/\/\/\/\/\/\/\/\/\/\/\/\/\/\/\/\/\/\/\/\/\/\/\/\/\/\/\/\/\/\/\/\/\/\/\/\/\/\/\/\/\/\/\/\/\
|1.5.1.4.3.1.5 dCH-SpecificInformationList                             |
|1.5.1.4.3.1.5.1 dCH-Specific-FDD-Item                                 |
|00000000 |1.5.1.4.3.1.5.1.1 dCH-ID                              |0      |
/\/\/\/\/\/\/\/\/\/\/\/\/\/\/\/\/\/\/\/\/\/\/\/\/\/\/\/\/\/\/\/\/\/\/\/\/\/\/\/\/\/\/\/\/\/\
|1.5.1.4.3.1.5.1.2 ul-TransportFormatSet                               |
|***B2*** |1.5.1.4.3.1.5.1.2.1.2.1 nrOfTransportBlocks |1             |
|***B2*** |1.5.1.4.3.1.5.1.2.1.2.2 transportBlockSize   |39            |
|***B2*** |1.5.1.4.3.1.5.1.2.1.3.1 nrOfTransportBlocks |1             |
|***B2*** |1.5.1.4.3.1.5.1.2.1.3.2 transportBlockSize   |81            |
|1.5.1.4.3.1.5.1.2.2 semi-staticPart                                   |
|***b3*** |1.5.1.4.3.1.5.1.2.2.1 transmissionTimeInter..|msec-20    |
|--01---- |1.5.1.4.3.1.5.1.2.2.2 channelCoding    |convolutional-|
|                                                 coding        |
|-----1-- |1.5.1.4.3.1.5.1.2.2.3 codingRate       |third         |
|11000111 |1.5.1.4.3.1.5.1.2.2.4 rateMatcingAttribute  |200          |
|-010---- |1.5.1.4.3.1.5.1.2.2.5 cRC-Size         |v12           |
/\/\/\/\/\/\/\/\/\/\/\/\/\/\/\/\/\/\/\/\/\/\/\/\/\/\/\/\/\/\/\/\/\/\/\/\/\/\/\/\/\/\/\/\/\/\
|1.5.1.4.3.1.5.1.3 dl-TransportFormatSet                               |
|***B2*** |1.5.1.4.3.1.5.1.3.1.1.1 nrOfTransportBlocks |1             |
|***B2*** |1.5.1.4.3.1.5.1.3.1.1.2 transportBlockSize   |0             |
|***B2*** |1.5.1.4.3.1.5.1.3.1.2.1 nrOfTransportBlocks |1             |
|***B2*** |1.5.1.4.3.1.5.1.3.1.2.2 transportBlockSize   |39            |
|***B2*** |1.5.1.4.3.1.5.1.3.1.3.1 nrOfTransportBlocks |1             |
|***B2*** |1.5.1.4.3.1.5.1.3.1.3.2 transportBlockSize   |81            |
|1.5.1.4.3.1.5.1.3.2 semi-staticPart                                   |
|***b3*** |1.5.1.4.3.1.5.1.3.2.1 transmissionTimeInter..|msec-20    |
|--01---- |1.5.1.4.3.1.5.1.3.2.2 channelCoding    |convolutional-|
|                                                 coding        |
|-----1-- |1.5.1.4.3.1.5.1.3.2.3 codingRate       |third         |
|11010001 |1.5.1.4.3.1.5.1.3.2.4 rateMatcingAttribute  |210          |
|-010---- |1.5.1.4.3.1.5.1.3.2.5 cRC-Size         |v12           |
/\/\/\/\/\/\/\/\/\/\/\/\/\/\/\/\/\/\/\/\/\/\/\/\/\/\/\/\/\/\/\/\/\/\/\/\/\/\/\/\/\/\/\/\/\/\
|00000001 |1.5.1.4.3.1.5.2.1dCH-ID                       |1            |
|***B2*** |1.5.1.4.3.1.5.2.2.1.2.1 nrOfTransportBlocks |1             |
|***B2*** |1.5.1.4.3.1.5.2.2.1.2.2transportBlockSize    |103           |
/\/\/\/\/\/\/\/\/\/\/\/\/\/\/\/\/\/\/\/\/\/\/\/\/\/\/\/\/\/\/\/\/\/\/\/\/\/\/\/\/\/\/\/\/\/\
|1.5.1.4.3.1.5.2.3 dl-TransportFormatSet                               |
|***B2*** |1.5.1.4.3.1.5.2.3.1.2.1 nrOfTransportBlocks |1             |
|***B2*** |1.5.1.4.3.1.5.2.3.1.2.2transportBlockSize    |103           |
/\/\/\/\/\/\/\/\/\/\/\/\/\/\/\/\/\/\/\/\/\/\/\/\/\/\/\/\/\/\/\/\/\/\/\/\/\/\/\/\/\/\/\/\/\/\
|00000010 |1.5.1.4.3.1.5.3.1dCH-ID                       |2            |
|1.5.1.4.3.1.5.3.2 ul-TransportFormatSet                               |
|***B2*** |1.5.1.4.3.1.5.3.2.1.2.1 nrOfTransportBlocks |1             |
|***B2*** |1.5.1.4.3.1.5.3.2.1.2.2transportBlockSize    |60            |
/\/\/\/\/\/\/\/\/\/\/\/\/\/\/\/\/\/\/\/\/\/\/\/\/\/\/\/\/\/\/\/\/\/\/\/\/\/\/\/\/\/\/\/\/\/\
|1.5.1.4.3.1.5.3.3 dl-TransportFormatSet                               |
|***B2*** |1.5.1.4.3.1.5.3.3.1.2.1 nrOfTransportBlocks |1             |
|***B2*** |1.5.1.4.3.1.5.3.3.1.2.2transportBlockSize    |60            |
```

```
∿∿∿∿∿∿∿∿∿∿∿∿∿∿∿∿∿∿∿∿∿∿∿∿∿∿∿∿∿∿∿∿∿∿∿∿∿∿∿∿∿∿∿∿∿∿∿∿∿∿∿∿∿
|1.5.1.5.3.1.3.2 dl-CodeInformation                               |
|1.5.1.5.3.1.3.2.1 fDD-DL-CodeInformationItem                     |
|----0000 |1.5.1.5.3.1.3.2.1.1 dl-ScramblingCode      |0          |
|***B2*** |1.5.1.5.3.1.3.2.1.2 fdd-DL-ChannelizationCo.. |3         |
-------------------------------------------------------------------
```

Message Example 3.18   NBAP Synchronized Radio Link Reconfiguration Preparation for voice call.

The message example shows how uplink channelization code length and downlink scrambling code values are changed and that three DCHs are set up, each with a different transport block size that is characteristic in this combination for AMR encoded voice calls. Also for all AMR encoded speech the transmission time interval is always 20 ms.

The answer from Node B is given – as already seen in case of Radio link setup – by sending a Successful Outcome message:

> **NBAP** UL: successfulOutcome **Id-synchronisedRadioLinkReconfiguration Preparation** (shortTransActionID=**j**, id-CRNC-communicationContextID=**d**, bindingID=**k**)

Now another AAL2 SVC is set up by ALCAP function. Despite that there are three DCHs, they are seen as a set of coordinated transport channels that are always set up and released in combination. Individual DCHs within such a set cannot operate individually. If the setup of one of these channels fails, the setup of all other channels in the set fails as well. And a set of coordinated DCHs is transferred over one transport bearer. Hence, only one AAL2 SVC is necessary to transport the DTCHs on Iub interface:

> **ALCAP** DL **ERQ** (Originating Signal. Ass. ID=**l**, AAL2 Path=**m**, AAL2 ChannelId=**n**, served user gen reference=**k**)
> **ALCAP** UL **ECF** (Originating Signal. Ass. ID=**q**, Destination Sign. Assoc. ID=**l**)

When this traffic channel AAL2 SVC (Path=**m′** and Connection=**n**) on Iub is opened, downlink and uplink FP synchronization frames can be monitored (see Figure 3.18). Sending an **NBAP** DL initiatingMessage **Id-synchronizedRadioLinkReconfiguration-Commit** (shortTransActionID=**j**, NodeBCommunicationsContext-ID=**p**) is necessary to order the Node B to switch to the new radio link previously prepared by the Synchronized Radio Link Reconfiguration Preparation procedure.

The messages in this DCH on AAL2 Path=**g′** and Channel=**h** are:

> DL **RRC RadioBearerSetup** (ULscramblingCode/DLChannelizationCode=**b**/ϕ, minULChCdLen=λ, Transport Format Sets of DCHs, MappingInfo for radio bearers)

This message starts setup of a radio bearer that is part of a user plane RAB. The connection is identified by its RAB-ID, and the user is identified by its uplink scrambling code. The message contains the TFSs of the DCHs, this time sent to UE, plus the appropriate info of which radio bearer (DTCH) is mapped onto which DCH.

The example message shows that radio bearer ID = 5 (remember that ID 1–4 are already occupied by SRBs) will be mapped onto the first dedicated channel for AMR info. In a similar

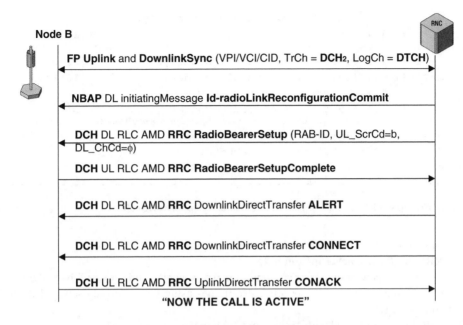

Figure 3.18 Iub MOC call flow 4/6.

way that is not shown in the message example, radio bearers 6 and 7 are mapped onto DCH 2 and DCH 3 (*remember: NBAP DCH-ID + 1!*). In addition, the information about changed uplink channelization code length and downlink channelization code number is included.

```
|TS 29.331 DCCH-DL (2001-06) (RRC_DCCH_DL)  radioBearerSetup       |
|(= radioBearerSetup)                                              |
/\/\/\/\/\/\/\/\/\/\/\/\/\/\/\/\/\/\/\/\/\/\/\/\/\/\/\/\/\/\/\/\/\/\/\/\/\/\/\/\
|1.1.1.1.4.1.1.1  rab-Identity                                     |
|***b8*** |1.1.1.1.4.1.1.1.1 gsm-MAP-RAB-Identity        |1        |
|1.1.1.1.4.1.2 rb-InformationSetupList                             |
|1.1.1.1.4.1.2.1 rB-InformationSetup                               |
|00101--- |1.1.1.1.4.1.2.1.1 rb-Identity                 |5        |
|1.1.1.1.4.1.2.1.3 rb-MappingInfo                                  |
|1.1.1.1.4.1.2.1.3.1 rB-MappingOption                              |
|1.1.1.1.4.1.2.1.3.1.1 ul-LogicalChannelMappings                   |
|1.1.1.1.4.1.2.1.3.1.1.1 oneLogicalChannel                         |
|1.1.1.1.4.1.2.1.3.1.1.1.1 ul-TransportChannelType                 |
|00000--- |1.1.1.1.4.1.2.1.3.1.1.1.1.1 dch               |1        |
|1.1.1.1.4.1.2.1.3.1.1.1.2 rlc-SizeList                            |
|         |1.1.1.1.4.1.2.1.3.1.1.1.2.1 configured       |0        |
|***b3*** |1.1.1.1.4.1.2.1.3.1.1.1.3 mac-LogicalChanne..|1        |
|1.1.1.1.4.1.2.1.3.1.2 dl-LogicalChannelMappingList                |
|1.1.1.1.4.1.2.1.3.1.2.1 dL-LogicalChannelMapping                  |
```

```
|1.1.1.1.4.1.2.1.3.1.2.1.1 dl-TransportChannelType                       |
|***b5*** |1.1.1.1.4.1.2.1.3.1.2.1.1.1 dch                    |1           |
~~~~~~~~~~~~~~~~~~~~~~~~~~~~~~~~~~~~~~~~~~~~~~~~~~~~~~~~~~~~~~~~~~~~~~~~~~~~~~~~~~
|-------1 |1.1.1.1.9.1.1.1.1 scramblingCodeType              |longSC      |
|***B3*** |1.1.1.1.9.1.1.1.2 scramblingCode                  |1068457     |
|         |1.1.1.1.9.1.1.1.3 numberOfDPDCH                   |1           |
|100----- |1.1.1.1.9.1.1.1.4 spreadingFactor                 |sf64        |
~~~~~~~~~~~~~~~~~~~~~~~~~~~~~~~~~~~~~~~~~~~~~~~~~~~~~~~~~~~~~~~~~~~~~~~~~~~~~~~~~~
|1.1.1.1.12.1.2.1.3.1 dL-ChannelizationCode                              |
|1.1.1.1.12.1.2.1.3.1.1 sf-AndCodeNumber                                 |
|***b7*** |1.1.1.1.12.1.2.1.3.1.1.1 sf128                    |3           |
----------------------------------------------------------------------------
```

Message Example 3.19   Extract of RRC Radio Bearer Setup for voice call.

The answer of UE to RRC Radio Bearer Setup request is:
UL **RRC RadioBearerSetupComplete**
It indicates that new radio bearer configuration was accepted and activated. Now the UE is ready to send and receive AMR speech codec frames using the previously set-up dedicated transport channels.

The call flow is continued with exchanging NAS messages:

DownlinkDirectTransfer **ALERT** – indicates call presentation (ringing) to B-party, B-party "receives" the call

DownlinkDirectTransfer **CONNECT** – B-party has accepted the call, they talk to each other

UplinkDirectTransfer **CONACK** – A-party confirms receiving CONNECT to B-party. This message exists because of historical reasons in development of GSM and ISDN standards. Actually it was specified to switch on CDR recording in the appropriate local exchange or serving MSC in earlier GSM/ISDN standard releases.

The following messages are used to release the voice call (see Figure 3.19):

UplinkDirectTransfer **DISC** – Disconnect message can be sent by A- or B-party. This indicates the beginning of call release procedure after a successful call setup. A cause value indicates the reason for disconnect request. This parameter is new compared to GSM/ISDN. The network is allowed in some cases to start call release procedure without having sent a Disconnect message before (see 3GPP 24.008 Ch. 5.4).

and/or

DownlinkDirectTransfer **RELEASE** – This request to release signaling resources for this call, also used to reject a call establishment. A cause value indicates the release cause, e.g. "normal call clearing".

UplinkDirectTransfer **RELCMP** – Release Complete message indicates complete release of signaling resources, e.g. stream identifier value can be used again.

The RRC connection is released with:

(RLC UMD) DL **rrcConnectionRelease**
(RLC UMD) UL **rrcConnectionReleaseComplete**

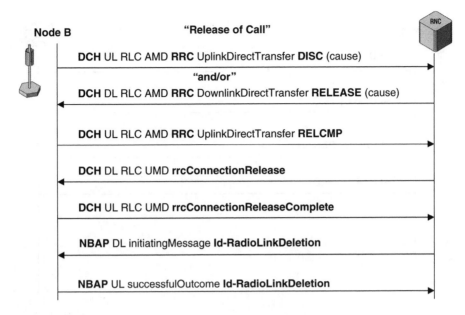

Figure 3.19   Iub MOC call flow 5/6.

Radio link deletion is continued by NBAP entity:

> **NBAP** DL initiatingMessage **Id-RadioLinkDeletion** (shortTransActionID=**j**,
> id-CRNC-CommunicationContextID=**d**, NodeBCommunicationsContext-ID=**p**)
> **NBAP** UL successfulOutcome **Id-RadioLinkDeletion** (shortTransActionID=**j**,
> id-CRNC-CommunicationContextID=**d**)

Finally, the dedicated transport channels for DCCH and DTCH are also released again (see Figure 3.20).

> **ALCAP** DL **REL** (Dest. Sign. Assoc. ID=**i**)
> **ALCAP** UL **RLC** (Dest. Sign. Assoc. ID=**f**)
> **ALCAP** DL **REL** (Dest. Sign. Assoc. ID=**q**)
> **ALCAP** UL **RLC** (Dest. Sign. Assoc. ID=**l**)

3.4   Iub CS – Mobile Terminated Call

This scenario describes the message flow for a user-terminated voice call (the mobile receives a call), which includes the allocation and release of an RAB.

3.4.1   Overview

The main difference between mobile-originated call (MOC) and mobile-terminated call (MTC) is that the MTC is initiated by a paging message sent from the CS core network domain (Figure 3.21).

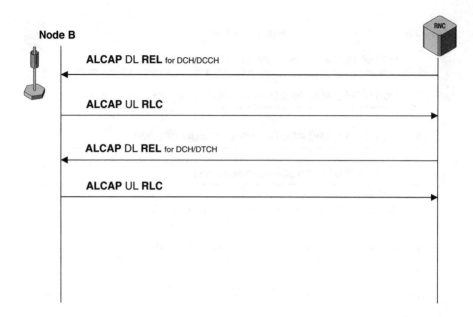

Figure 3.20   Iub MOC call flow 6/6.

There are two different types of paging messages defined.

Paging Type 1 message is used if the UE is in RRC_IDLE, RRC CELL_PCH, or URA_PCH state, that means, if no other connection using either DCH or common control channels is active before.

Paging Type 2 message is used if the UE is already in RRC CELL_FACH or CELL_DCH state, that means, if an RRC connection was already set up owing to request to establish a DCCH

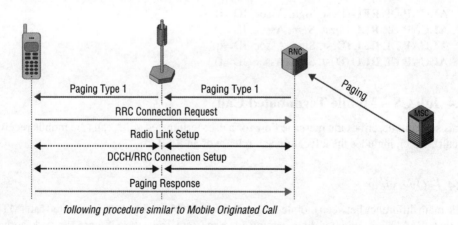

Figure 3.21   Iub mobile-terminated voice call overview.

for RRC/MMSMCC toward PS Domain. In addition, an active PDP context may be running using DCH/DTCH or in background on common transport channels RACH and FACH.

The paging messages are sent on the PCH in downlink direction. The logical channel related to this transport channel is the paging control channel (PCCH).

The paging message is answered with a paging response that indicates that the UE is able to accept the call request. After receiving paging response, the network(!) sends SETUP message. The following messages are similar to the messages seen in case of an MOC, but this time the calling party is on the network side.

3.4.2 Message Flow

Only the differences between MTC and MOC are highlighted because most parts of the general procedure, message parameters, and values are identical with MOC scenario.

The paging type 1 message is sent downlink in the PCH and contains a paging cause value that indicates the paging reason (Figure 3.22). In case of the MS in RRC_IDLE state, a signal on PICH is sent containing a paging indicator that "tells" all mobiles in RRC_IDLE state that belong to a specified paging group to listen to the PCH. The paging group number (=paging indicator) of the UE is derived from the UE's IMSI.

PCH: DL RLC TMD Paging Type 1 (**PagingCause=**
terminatingConversationalCall, IMSI or **TMSI**)

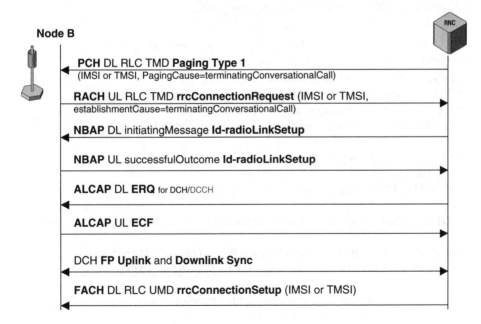

Figure 3.22 Iub MTC call flow 1/6.

```
|TS 29.331 PCCH (2002-06) (RRC_PCCH)  pagingType1 (= pagingType1)   |
|pCCH-Message                                                       |
|1 message                                                          |
|1.1 pagingType1                                                    |
|1.1.1 pagingRecordList                                             |
|1.1.1.1 pagingRecord                                               |
|1.1.1.1.1 cn-Identity                                              |
|000----- |1.1.1.1.1.1pagingCause  |terminatingConversationalCall   |
|---0---- |1.1.1.1.1.2 cn-DomainIdentity         |cs-domain          |
|1.1.1.1.1.3 cn-pagedUE-Identity                                    |
|**b32*** |1.1.1.1.1.3.1tmsi-GSM-MAP             |43 78 99 67        |
-------------------------------------------------------------------
```

Message Example 3.20   RRC Paging Type 1.

The UE answers after reception of the paging message with RRC connection request on the (uplink) RACH. The establishment cause indicates that RRC connection request was triggered by a paging message and so all the following messages belong to a terminating conversational call.

> **RACH**: UL RLC TMD rrcConnectionRequest (**IMSI or TMSI,
> establishmentCause=terminatingConversationalCall**)

No differences to MOC scenario are found in messages:

> **NBAP** DL initiatingMessage **Id-radioLinkSetup** (longTransActionID=**c**,
> id-CRNC-CommunicationContextID=**d**,
> ULscramblingCode/DLChannelizationCode=**b**/β)
> **NBAP** UL successfulOutcome **Id-radioLinkSetup** (longTransActionID=**c**,
> id-CRNC-CommunicationContextID=**d**, bindingID=**e**,
> NodeBCommunicationsContext-ID=**p**, CommunicationPortID=**o**)

> **ALCAP** DL **ERQ** (Originating Signal. Ass. ID=**f**, AAL2 Path=**g**, AAL2
> ChannelId=**h**, served user gen reference=**e**)
> **ALCAP** UL **ECF** (Originating Signal. Ass. ID=**i**, Destination Sign. Assoc. ID=**f**)

> **DCH** in AAL2 Path=**g'** and Channel=**h**: downlink and uplink FP synchronization

> **FACH**: DL RLC UMD **rrcConnectionSetup** (rrc-TransactionIdentifier=**a**,
> ULscramblingCode/DLChannelizationCode=**b**/β, **IMSI or TMSI**)

> **NBAP** UL initiating Message **Id-radioLinkRestoration** (shortTransActionID=**j**,
> id-CRNC-CommunicationContextID=**d**)

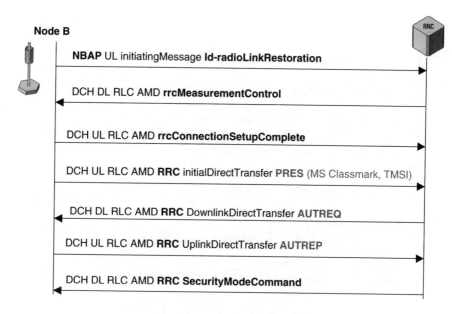

Figure 3.23   Iub MTC call flow 2/6.

RRC messages (RLC AMD) in this AAL2 channel (DCH)=**h** (see Figure 3.23):

DL RLC AMD **rrcMeasurementControl** (rrc-TransactionIdentifier=**a**)

UL RLC AMD **rrcConnectionSetupComplete** (rrc-TransactionIdentifier=**a**)

The following messages are: (see Figures 3.23 and 3.24):

UL initialDirect Transfer **PRES** – paging response message is sent by UE to the serving MSC to indicate that paging was successful and UE is able to continue with call setup. Enclosed in this message the mobile identity (as a rule TMSI) and information about the UEs technical capabilities (MS Classmark) are transmitted.

DownlinkDirectTransfer **AUTREQ**
UplinkDirectTransfer **AUTREP**
DL **RRC SecurityModeCommand**

UL **RRC SecurityModeComplete**

Opt.: DownlinkDirectTransfer **TRCMD** – TMSI reallocation command: new TMSI is assigned by CS core network domain
Opt.: UplinkDirectTransfer **TRCMP** – indicates successful TMSI reallocation
DownlinkDirectTransfer **SETUP** – called party number information element in this message is only optional, because the UE does not know its MSISDN. Hence, called party number is not necessary to route the call.

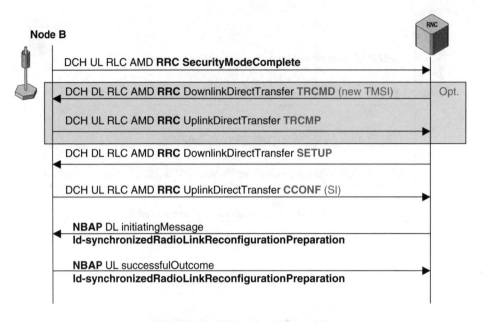

Figure 3.24   Iub MTC call flow 3/6.

UplinkDirectTransfer **CCONF** – UE confirms the incoming call setup request and assigns an SI

    NBAP DL: initiatingMessage
    Id-synchronisedRadioLinkReconfigurationPreparation (shortTransActionID=**j**,
    NodeBCommunicationsContext-ID=**p**,
    ULscramblingCode/DLChannelizationCode=**b/ϵ**)
    NBAP UL: successfulOutcome
    Id-synchronisedRadioLinkReconfigurationPreparation (shortTransActionID=**j**,
    id-CRNC-CommunicationContextID=**d**, bindID=**k**)

    ALCAP DL **ERQ** (Originating Signal. Ass. ID=**l**, AAL2 Path=**m**, AAL2
    ChannelId=**n**, served user gen reference=**k**)
    ALCAP UL **ECF** (Originating Signal. Ass. ID=**q**, Destination Sign. Assoc. ID=**l**)

Now the traffic channel on Iub (AAL2 Path=**m**′ and Channel=**n**) will be opened; downlink and uplink FP synchronization are sent (see Figure 3.25).

    NBAP DL initiatingMessage **Id-synchronisedRadioLinkReconfigurationCommit**
    (shortTransActionID=**j**, NodeBCommunicationsContext-ID=**p**)

The following are the RRC messages in the DCCH:

    DL **RRC RadioBearerSetup**
    UL **RRC RadioBearerSetupComplete**

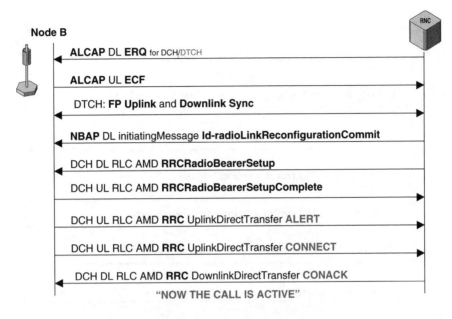

Node B

ALCAP DL **ERQ** for DCH/DTCH

ALCAP UL **ECF**

DTCH: **FP Uplink** and **Downlink Sync**

NBAP DL initiatingMessage **Id-radioLinkReconfigurationCommit**

DCH DL RLC AMD **RRCRadioBearerSetup**

DCH UL RLC AMD **RRCRadioBearerSetupComplete**

DCH UL RLC AMD **RRC** UplinkDirectTransfer ALERT

DCH UL RLC AMD **RRC** UplinkDirectTransfer CONNECT

DCH DL RLC AMD **RRC** DownlinkDirectTransfer CONACK

"NOW THE CALL IS ACTIVE"

Figure 3.25   Iub MTC call flow 4/6.

MM/SM/CC messages that follow:

UplinkDirectTransfer **ALERT**
UplinkDirectTransfer **CONNECT**
DownLinkDirectTransfer **CONACK**

Finally the call is active.
Release of the call (see Figures 3.26 and 3.27) is basically the same as in MOC:

UpLinkDirectTransfer **DISC**
                          and/or
DownlinkDirectTransfer **RELEASE**
UpLinkDirectTransfer **RELCMP**
(RLC UMD) DL **rrcConnectionRelease**
(RLC UMD) UL **rrcConnectionReleaseComplete**

NBAP DL initiatingMessage **Id-RadioLinkDeletion** (shortTransActionID=**j**,
id-CRNC-CommunicationContextID=**d**, NodeBCommunicationsContext-ID=**p**)
NBAP UL successfulOutcome **Id-RadioLinkDeletion** (shortTransActionID=**j**,
id-CRNC-CommunicationContextID=**d**)

ALCAP DL **REL** (Dest. Sign. Assoc. ID=**i**)
ALCAP UL **RLC** (Dest. Sign. Assoc. ID=**f**)
ALCAP DL **REL** (Dest. Sign. Assoc. ID=**q**)
ALCAP UL **RLC** (Dest. Sign. Assoc. ID=**l**)

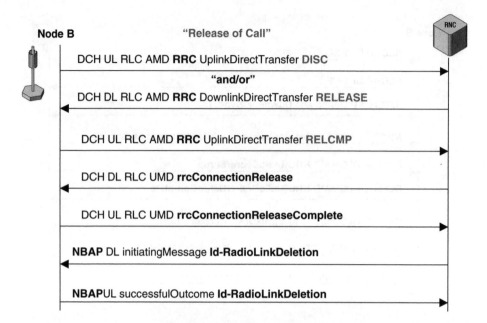

Figure 3.26   Iub MTC call flow 5/6.

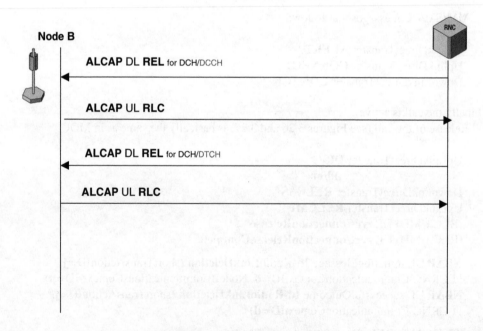

Figure 3.27   Iub MTC call flow 6/6.

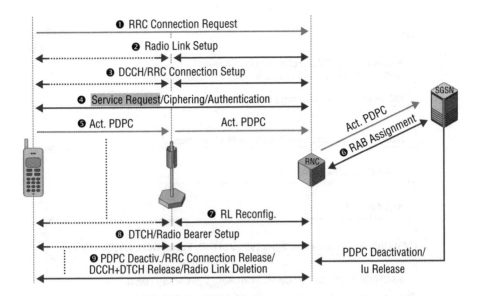

Figure 3.28 Iub PDP context activation/deactivation overview.

3.5 Iub PS – PDP Context Activation/Deactivation

If the user or the network activates any kind of data transmission, a PDP Context (PDPC) is activated and deactivated again after the transmission is finished or after a certain time period (see Figure 3.28).

3.5.1 Overview

In case of PDPC activation with the PDPC running in a DCH, Steps 1–3 are the same as in the case of mobile originated voice call (MOC) and IMSI/GPRS Attach procedure.

Step 4: A new mandatory session management message is the Service Request message. It shows if (and if yes, how many) PDP contexts are already running on the subscriber side. The optional ciphering/authentication procedure is the same as before.

Step 5: The PDPC activation starts with an Activate PDP Context Request message in MM/SM/CC layer. This message is forwarded by the RNC to the SGSN in the PS core network domain.

Step 6: The PS core network domain negotiates a QoS for the PDPC. RAB with this QoS is established. On IuPS interface the RAB is seen as GTP tunnel on user plane. On Iub interface the RAB is represented by an AAL2 SVC that is set up in Step 8.

Step 7: Again the NBAP Radio Link Reconfiguration procedure provides radio resources for the establishment of the Radio Bearer.

Step 8: On behalf of the parameter values negotiated in RAB Assignment procedure a new radio bearer is set up to carry the (logical) DTCH.

Step 9: The release procedures following a PDPC deactivation are the same as the release procedures in case of a voice call.

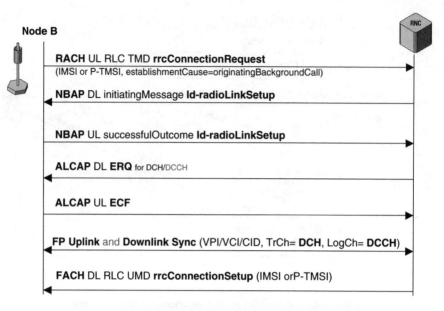

Figure 3.29   Iub PDP context activation/deactivation call flow 1/6.

3.5.2 Message Flow

The setup of radio link and RRC connection in case of PDPC activation are the same as in the case of an MOC. However, P-TMSI is used as temporary identifier (Figure 3.29). P-TMSI is the temporary identifier for SGSN-based services and is discriminated from circuit-switched TMSI by the value of the last two significant bits. Value 11 is used by SGSN, values 00, 01, and 10 are used by VLR.

> **RACH**: UL RLC TMD rrcConnectionRequest (**IMSI or P-TMSI**, establishmentCause=originatingBackgroundCall)

NBAP Radio link setup procedure is the same as in all cases described before, the same with AAL2 connection setup:

> **NBAP** DL initiatingMessage **Id-radioLinkSetup** (longTransActionID=**c**, id-CRNC-CommunicationContextID=**d**, ULscramblingCode/DLChannelizationCode=**b**/β)
> **NBAP** UL successfulOutcme **Id-radioLinkSetup** (longTransActionID=**c**, id-CRNC-CommunicationContextID=**d**, bindingID=**e**, NodeBCommunicationsContext-ID=**p**, CommunicationPortID=**o**)

> **ALCAP** DL **ERQ** (Originating Signal. Ass. ID=**f**, AAL2 Path=**g**, AAL2 Channel id=**h**, served user gen reference=**e**)

> **ALCAP** UL **ECF** (Originating Signal. Ass. ID=**i**, Destination Sign. Assoc. ID=**f**)
> **DCH Channel** in AAL2 Path=**g'** and Channel=**h**: downlink and uplink synchronization FP frames are exchanged.

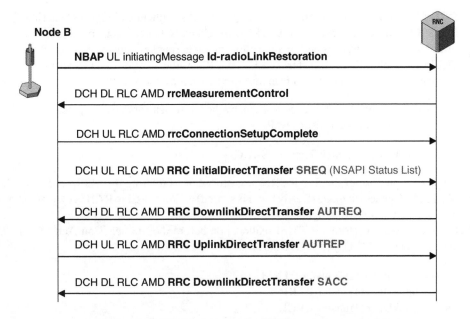

Figure 3.30 Iub PDP context activation/deactivation call flow 2/6.

RRC connection setup can once again contain the P-TMSI.

FACH: DL RLC UMD **rrcConnectionSetup** (rrc-TransactionIdentifier=**a**, ULscramblingCode/DLChannelizationCode=**b**/β, **IMSI or P-TMSI**)

NBAP UL initiatingMessage **Id-radioLinkRestoration** (shortTransActionID=**j**, id-CRNC-communicationContextID=**d**)

All following RRC messages (RLC AMD) are running in D(C)CH (AAL2 channel=**h**) (Figure 3.30):

DCH DL **rrcMeasurementControl** (rrc-TransactionIdentifier=**a**)
DCH UL **rrcConnectionSetupComplete** (rrc-TransactionIdentifier=**a**)

The Service Request message enclosed in an RRC Initial Direct Transfer message is sent to establish a logical association between mobile station (MS) and network. It contains a list of all available NSAPIs and whether they are in use by other PDP contexts or not. In addition, the P-TMSI is included as well.

DCH UL RRC initialDirectTransfer **SREQ (NSAPI Status List, P-TMSI)**

Ciphering/integrity protection is activated and the UE authenticated:

DCH DL **RRC SecurityModeCommand**
DCH UL **RRC SecurityModeComplete**

DCH RRC DownlinkDirectTransfer ACRQ — GPRS Authentication and Ciphering Request fulfills same function as AUTREQ for CS calls, but contains in addition information about used ciphering algorithm and ciphering key sequence number DCH RRC UplinkDirectTransfer ACRE — GPRS Authentication and Ciphering Response is used to send signed response (SRES) value back to RNC

The network accepts the service request sent by the UE with a Service Accept that optionally may contain an NSAPI status list as well.

DCH RRC DownlinkDirectTransfer **SACC**

The general purpose of the Service Request procedure is to bring the MS from PMM-Idle into the PMM-Connected mode (Ready timer is started) and/or to assign RAB in case of PDPCs are activated without RAB assigned.

Now the network performs P-TMSI reallocation before the MS sends an Activate PDPC Request message (APCR) (see Figure 3.31):

DCH DownlinkDirectTransfer **PTRM**
DCH UplinkDirectTransfer **PTRP**
DCH UplinkDirectTransfer **APCR**
(NSAPI, LLC SAPI, QoS req., PDP Add., opt. APN, PPP Dial-in info)

The Activate PDPC Request message contains the requested NSAPI, an LLC SAPI value (mandatory in case of handover to 2.5 G radio access network), the QoS requested by the user, and a PDP Address, e.g. an IPv4 or IPv6 Address, which is the target of the PDPC. In case that a dynamic PDP address will be assigned by the network, the PDP Address element is used as a placeholder and Access Point Name (APN) is included – this is the network

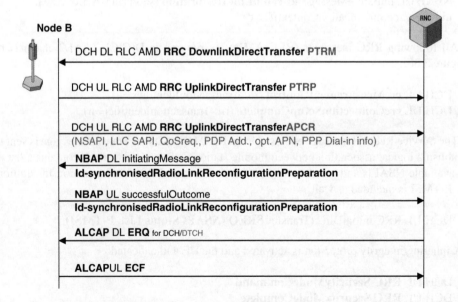

Figure 3.31 Iub PDP context activation/deactivation call flow 3/6.

server that will assign the dynamic PDP Address. In addition, this message often carries a PPP (Point-to-Point) protocol portion that encloses CHAP/PAP (Challenge Handshake Authentication Protocol/Password Authentication Protocol) information. PPP and CHAP/PAP are used to transmit username and password for dial-in service at Internet Service Provider (ISP) side. ISP can be the PLMN operator itself or any third-party outside the PLMN. In the latter case, this kind of PPP signaling will be transparently forwarded via Gi interface.

```
|TS 29.331 DCCH-UL (2002-03) (RRC_DCCH_UL)  uplinkDirectTransfer   |
|(= uplinkDirectTransfer)                                          |
|uL-DCCH-Message                                                   |
|1 integrityCheckInfo                                              |
|**b32*** |1.1 messageAuthenticationCode |'01110110011000111100   |
|11110010110'B                                                     |
|-0001--- |1.2 rrc-MessageSequenceNumber |1                        |
|2 message                                                         |
|2.1 uplinkDirectTransfer                                          |
|----1--- |2.1.1 cn-DomainIdentity      |ps-domain                 |
|**b936** |2.1.2 nas-Message   |0a 41 05 03 0b 00 00 00 00 00 00...|
|TS 24.008 GPRS Session Managemen t V3.11.0 (GSM-DMTAP)            |
|APCR (= Activate PDP context request)                             |
|Activate PDP context request                                      |
|----1010 |Protocol Discriminator       |GPRS session management   |
|         |                             | messages                 |
|-000---- |Transaction Id value (TIO)   |TI value 0                |
|0------- |Transaction Id flag          |message sent from orig TI |
|01000001 |Message Type                 |65                        |
|Network Service Access Point                                      |
|----0101 |NSAPI value                  |NSAPI 5                   |
|0000---- |Spare                        |0                         |
|LLC SAPI                                                          |
|----0011 |SAPI                         |SAPI 3                    |
|0000---- |Spare                        |0                         |
|Quality of Service                                                |
|00001011 |IE Length                    |11                        |
|-----000 |Reliability class          |Subscribed reliability class|
|--000--- |Delay class                |Subscribed delay class      |
|00------ |Spare                      |0                           |
|-----000 |Precedence class           |Subcribed prec.             |
|----0--- |Spare                      |0                           |
|0000---- |Peak throughput      |Subscr. peak throughput / reserved |
|---00000 |Mean throughput      |Subscr. mean throughput / Reserved |
|000----- |Spare                |0                                  |
|-----000 |Delivery erroneous SDU    |Subscr. delivery of err.      |
|         |                          | SDUs / r...                  |
|---00--- |Delivery order       |Subscr. delivery order / reserved  |
|000----- |Traffic Class        |Subscr. Traffic class / reserved   |
|00000000 |Maximum SDU Size     |Subscr. maximum SDU size / reserved|
```

```
|00000000 |Maximum BitRate Uplink           |Subscribed maximum bit   |
|                                            |rate for u...            |
|00000000 |Maximum BitRate DownLink         |Subscribed maximum bit   |
|                                            |rate for u...            |
|----0000 |SDU error ratio          |Subscr. SDU error ratio / reserved|
|0000---- |Residual BER             |Subscr. residual BER / reserved|
|------00 |Transfer Handling Prio.          |Subscr. Transfer Handling|
|                                            |Prio. /...               |
|000000-- |Transfer delay           |Reserved / Subscribed trnsfer delay|
|00000000 |Guaranteed BitRate UpLink        |0                        |
|00000000 |Guaranteed BitRate DownLink      |0                        |
|Packet Data Protocol Address_opt                                     |
|00000010 |IE Length                        |2                        |
|----0001 |Type of address                  |IETF specified address   |
|0000---- |Spare                            |0                        |
|00100001 |Packet data protocol type        |IPv4                     |
|Access Point Name                                                    |
|00101000 |IE Name                          |Access Point Name        |
|00001001 |IE Length                        |9                        |
|***B9*** |Access Point Name Value          |operator_name.           |
|                                            |operator_group.gprs      |
|Protocol Configuration Options                                       |
|00100111 |IE Name                          |Protocol Configuration   |
|                                            |Options                  |
|01010101 |IE Length                        |85                       |
|-----000 |Options format value             |PPP                      |
|-0000--- |Spare                            |0                        |
|1------- |Extension bit                    |Octet 3 is extended      |
|**B84*** |Address information              |username/password for ISP|
-----------------------------------------------------------------------
```

Message Example 3.21   RRC Uplink Direct Transfer with GPRS SM Activate PDP Context Request.

In the message example, RRC Uplink Direct Transfer is shown together with the embedded GPRS Session Management (unfortunately in some cases abbreviated GSM) Activate PDPC Request (APCR). It emerges that RRC message contains only the integrity protection message authentication code, CN domain identifier for proper message routing to SGSN, and message type. Then starts APCR that has TIO as already described for circuit-switched call control messages like SETUP. It will link all SM messages related to the same PDPC together. The NSAPI value 5 indicates that this is the first PDPC requested by this subscriber (that was identified by its P-TMSI in Service Request message before). NSAPI values 0–4 are reserved by international standards. Usable for mobile subscriber are values in the range from 5 to 15. So in total a single UE can have theoretically up to 11 PDPCs simultaneously. General idea of the standard people was to have 10 Internet connections plus one WAP connection running simultaneously.

LLC SAPI – as already mentioned – will only become important if the call is handed over to a 2G GSM cell. The quality of service parameters requested by the subscriber in the example indicate that the user will accept any QoS settings assigned by the network (as subscribed in HLR). We see a placeholder for a – still empty – IPv4 address, access point name, and PPP protocol configuration options. The access point name in the example follows a description given in 3GPP 23.003 (Numbering and Addressing). "Username/password for ISP" is just a comment inserted instead of any bit of real information.

The following procedures are necessary to activate the DTCH for this PDPC. For reasons we have explained in the MOC scenario, DL channelization code is changed:

NBAP DL initiatingMessage
Id-synchronisedRadioLinkReconfigurationPreparation(shortTransActionID=**j**,
NodeBCommunicationsContext-ID=**p**,
ULscramblingCode/DLChannelizationCode=**b**/ε)

NBAP UL successfulOutcome
Id-synchronisedRadioLinkReconfigurationPreparation(shortTransActionID=**j**,
id-CRNC-CommunicationContextID=**d**, bindingID=**k**)

ALCAP DL **ERQ** (Originating Signal. Ass. ID=**l**, AAL2 Path=**m**, AAL2
ChannelId=**n**, served user genreference=**k**)
ALCAP UL **ECF** (Originating Signal. Ass. ID=**q**, Destination Sign. Assoc. ID=**l**)

Dedicated Traffic Channel: Uplink and Downlink **FP Sync** indicate successful DTCH setup.

NBAP DL initiatingMessage **Id-synchronisedRadioLinkReconfigurationCommit**
(shortTransAction ID=**j**, NodeBCommunicationsContext-ID=**p**) indicates successful
synchronization on air interface (see Figure 3.32).

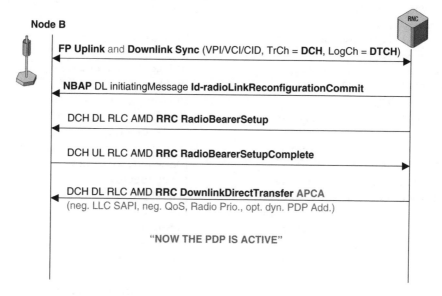

Figure 3.32   Iub PDP context activation/deactivation call flow 4/6.

The following RRC procedure with messages

DCH DL **RRC RadioBearerSetup**
DCH UL **RRC RadioBearerSetupComplete**

is used to inform UE about the reconfiguration of its dedicated radio link.

With DCH RRC DownlinkDirectTransfer **APCA (neg. LLC SAPI, neg. QoS, Radio Prio., opt. dyn. PDP Add.)** the network confirms the PDPC activation. The message contains the negotiated LLC SAPI, negotiated QoS (as registered for this user in the HLR subscription data), a radio priority to specify the priority level of data related to this PDPC on the lower layers, and optionally the dynamic PDP Address as assigned by APN server.

```
|TS 24.008 GPRS Session Managemen t V3.11.0 (GSM-DMTAP)   APCA       |
|(= Activate PDP context accept)                                      |
|Activate PDP context accept                                         |
|----1010 |Protocol Discriminator     |GPRS session management       |
|         |                           | messages                     |
|-000---- |Transaction Id value (TIO) |TI value 0                    |
|1------- |Transaction Id flag        |message sent to orig TI       |
|01000010 |Message Type               |66                            |
|LLC SAPI                                                             |
|----0011 |SAPI                       |SAPI 3                        |
|0000---- |Spare                      |0                             |
|Quality of Service                                                   |
|00001011 |IE Length                  |11                            |
|-----011 |Reliability class    |Unack. GTP&LLC,Ack.RLC, Prot. data|
|--100--- |Delay class                |Delay class 4 (best effort)   |
|00------ |Spare                      |0                             |
|-----011 |Precedence class           |Low priority                  |
|----0--- |Spare                      |0                             |
|0110---- |Peak throughput            |Up to 32000 octet/s           |
|---11111 |Mean throughput            |best effort                   |
|000----- |Spare                      |0                             |
|-----011 |Delivery erroneous SDU     |Erroneous SDUs are not        |
|         |                           | delivered...                 |
|---10--- |Delivery order         |Without delivery order ('no') |
|100----- |Traffic Class              |Background class              |
|10010110 |Maximum SDU Size           |1500 octets                   |
|01000000 |Maximum BitRate Uplink     |64 kbps                       |
|01101000 |Maximum BitRate DownLink   |384 kbps                      |
|----0100 |SDU error ratio            |1*10-4                        |
|0111---- |Residual BER               |1*10-5                        |
|------00 |Transfer Handling Prio.    |Subscr. Transfer Handling     |
|         |                           | Prio. /...                   |
|000000-- |Transfer delay     |Reserved / Subscribed trnsfer delay|
|00000000 |Guaranteed BitRate UpLink  |0                             |
```

```
|00000000 |Guaranteed BitRate DownLink |0                             |
|Radio Priority Level + Spare                                         |
|-----100 |Radio priority level value |priority level 4: lowest      |
|----0--- |Spare                       |0                             |
|0000---- |Spare                       |0                             |
|Packet Data Protocol Address_opt                                     |
|00101011 |IE Name                     |Packet Data Protocol Address  |
|00000110 |IE Length                   |6                             |
|----0001 |Type of address             |IETF specified address        |
|0000---- |Spare                       |0                             |
|00100001 |Packet data protocol type   |IPv4                          |
|***B4*** |IPv4-Address                |192.168.1.2                   |
|Protocol Configuration Options                                       |
|00100111 |IE Name                     |Protocol Configuration Options|
|00101000 |IE Length                   |40                            |
|-----000 |Options format value        |PPP                           |
|-0000--- |Spare                       |0                             |
|1------- |Extension bit               |Octet 3 is extended           |
|**B39*** |Address information         |CHAP/PAP answer from ISP      |
---------------------------------------------------------------------
```

Message Example 3.22   GPRS SM Activate PDP Context Accept.

When Activate PDPC Accept is received on the UE side, the PDPC becomes active and packet data can be transmitted in uplink and downlink direction (Figure 3.33).

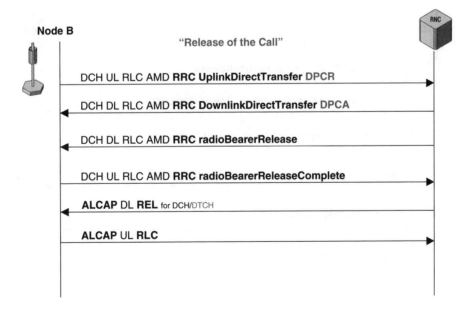

Figure 3.33   Iub PDP context activation/deactivation call flow 5/6.

The release of the PDPC starts with Deactivate PDPC procedure:

DCH RRC UpLinkDirectTransfer **DPCR – Deactivate PDPC** Request

The important information in this message is the transaction id value (TIO) which is the only information that links this message with the previous PDPC activation procedure and the SM cause value that informs about the reasons for deactivating the PDPC.

```
|TS 24.008 GPRS Session Management V3.8.0 (GSM-DMTAP)   DPCR       |
|(= Deactivate PDP context request)                                |
|Deactivate PDP context request                                    |
|----1010 |Protocol Discriminator      |GPRS session management    |
|         |                            | messages                  |
|-000---- |Transaction Id value (TIO)  |TI value 0                 |
|0------- |Transaction Id flag         |message sent from orig TI  |
|01000110 |Message Type                |70                         |
|SM Cause                                                          |
|00100100 |Reject cause value          |Regular deactivation       |
-------------------------------------------------------------------
```

Message Example 3.23   GPRS SM Deactivate PDP Context Request.

Deactivate PDPC Request is answered with
DCH RRC DownlinkDirectTransfer **DPCA – Deactivate PDPC Accept**
After this GPRS SM procedure, other release procedures follow (see Figure 3.34), as already seen in the case of MOC and MTC:

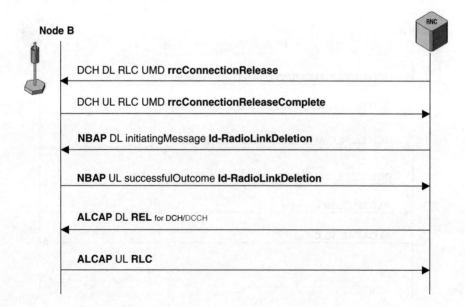

Figure 3.34   Iub PDP context activation/deactivation call flow 6/6.

DCH DL **RRC Radio Bearer Release**
DCH UL **RRC Radio Bearer Release Complete**

ALCAP DL **REL** (Dest. Sign. Assoc. ID=**q**) (DTCH)
ALCAP UL **RLC** (Dest. Sign. Assoc. ID=**l**)

DCH (RLC UMD) DL **rrcConnectionRelease**
DCH (RLC UMD) UL **rrcConnectionReleaseComplete**

NBAP DL initiatingMessage **Id-RadioLinkDeletion** (shortTransActionID=**j**,
id-CRNC-CommunicationContextID=**d**, NodeBCommunicationsContext-ID=**p**)

NBAP UL successfulOutcome **Id-RadioLinkDeletion** (shortTransActionID=**j**,
id-CRNC-CommunicationContextID=**d**)

ALCAP DL **REL** (Dest. Sign. Assoc. ID=**i**) (DCCH)
ALCAP UL **RLC** (Dest. Sign. Assoc. ID=**f**)

3.6 Iub – IMSI/GPRS Detach Procedure

3.6.1 Overview

Detach can be executed either toward CS Domain or toward PS Domain separately or as a
combined IMSI/GPRS Detach (Figure 3.35), e.g. in case the UE is switched off.

If the UE has no ongoing RRC connection, this means there are no ongoing CS or PS
connections to the network, and a new RRC connection needs to be set up in the same way as
shown for all other procedures before (Steps 1–3).

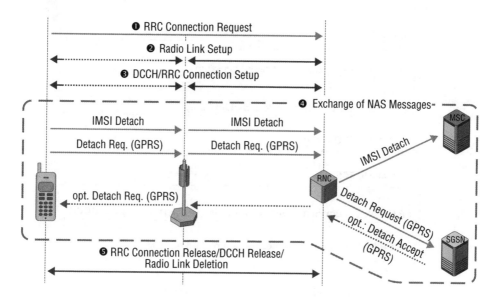

Figure 3.35 Iub IMSI/GPRS Detach procedure overview.

Step 4: IMSI Detach Indication message is sent by the UE to the network. No response is returned to the MS! GPRS detach is started by sending Detach Request message. The detach type information element in this message may indicate "GPRS detach with switching off," "GPRS detach without switching off," "IMSI detach," "GPRS/IMSI detach with switching off," or "GPRS/IMSI detach without switching off." If the MS is not switched off a Detach Accept message is sent by the network.

Step 5: Release of RRC connection, radio resources, and AAL2 SVC for DCCH.

3.6.2 Message Flow

Since there are no new radio network layer messages in this scenario compared to IMSI/GPRS, we just want to show the message flow as a reference for signaling analysis without any comments (see Figures 3.36 and 3.37).

RACH: UL RLC TMD **rrcConnectionRequest** (<u>**IMSI or P-TMSI**</u>, establishmentCause=Detach)

NBAP DL initiatingMessage **Id-radioLinkSetup** (longTransActionID=**c**, id-CRNC-CommunicationContextID=**d**, ULscramblingCode/ DLChannelizationCode=**b**/β)

NBAP UL successfulOutcome **Id-radioLinkSetup** (longTransActionID=**c**, id-CRNC-CommunicationContextID=**d**, bindingID=**e**, NodeBCommunicationsContext-ID=**p**, CommunicationPortID=**o**)

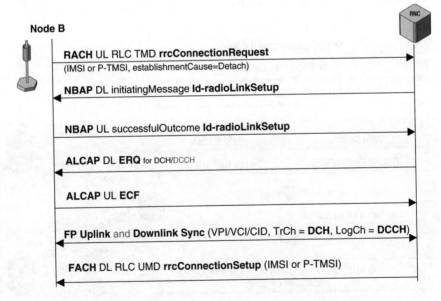

Figure 3.36 Iub IMSI/GPRS Detach call flow 1/3.

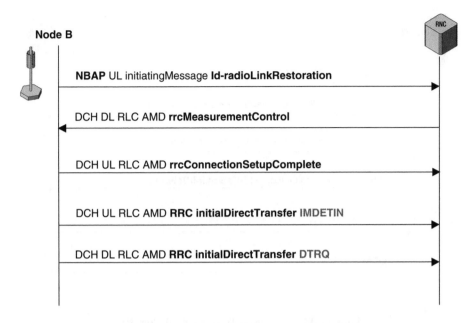

Figure 3.37   Iub IMSI/GPRS Detach call flow 2/3.

ALCAP DL **ERQ** (Originating Signal. Ass. ID=**f**, AAL2 Path=**g**, AAL2 ChannelId=**h**, served user gen reference=**e**)

ALCAP UL **ECF** (Originating Signal. Ass. ID=**i**, Destination Sign. Assoc. ID=**f**)

DCH Channel in AAL2 Path=**g** and Channel=**h**: downlink and uplink synchronization **FP frames**.

FACH: DL RLC UMD **rrcConnectionSetup** (rrc-TransactionIdentifier=**a**, ULscramblingCode/DLChannelizationCode=**b**/β, **IMSI or P-TMSI**)

NBAP UL initiatingMessage **id-radioLinkRestoration** (shortTransActionID=**j**, id-CRNC-communicationContextID=**d**)

"RRC messages (RLC AMD) in this DCH (AAL2 channel=h)"

DCH DL **rrcMeasurementControl** (rrc-TransactionIdentifier=**a**)

DCH UL **rrcConnectionSetupComplete** (rrc-TransactionIdentifier=**a**)

The following are the messages (Figure 3.38):

DCH UL RRC initialDirectTransfer (Domain Indicator=CS-Domain) **IMDETIN**
DCH UL RRC initialDirectTransfer (Domain Indicator=PS-Domain) **DTRQ**

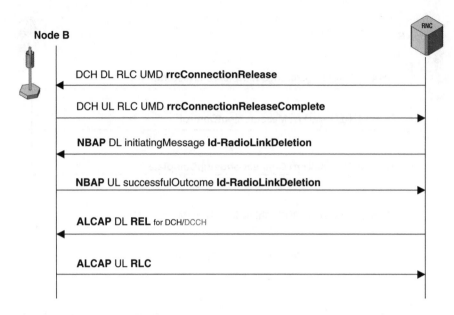

Figure 3.38 Iub IMSI/GPRS Detach call flow 3/3.

DCH (RLC UMD) DL **rrcConnectionRelease**
DCH (RLC UMD) UL **rrcConnectionReleaseComplete**

NBAP DL initiatingMessage **Id-RadioLinkDeletion** (shortTransActionID=**j**,
id-CRNC-CommunicationContextID=**d**, NodeBCommunicationsContext-ID=**p**)

NBAP UL successfulOutcome **Id-RadioLinkDeletion** (shortTransActionID=**j**,
id-CRNC-CommunicationContextID=**d**)

ALCAP DL **REL** (Dest. Sign. Assoc. ID=**i**)
ALCAP UL **RLC** (Dest. Sign. Assoc. ID=**f**)

3.7 RRC Measurement Procedures

To understand the following sections that deal with description of handover procedures, it is first
necessary to get a deeper knowledge about RRC measurement procedures. As already shown in
the previous call flows, the serving RNC sends an RRC Measurement Control message to the UE
after the establishment of RRC connection. This RRC Measurement Control message contains
information about which neighbor cells the UE shall monitor and what kind of measurement
reports the SRNC expects to receive from the mobile.

The different types of measurement are divided into different groups. Each measurement
type group can also be described by an appropriate group of event-IDs. These event-IDs are used
in case of event-triggered measurement reports. This means that a defined key measurement
parameter reaches a predefined threshold and so triggers the sending of a measurement report.
On the other hand, the SRNC may also order the UE to send RRC measurement reports

periodically, but typically this is not the default setting. However, it makes sense in selected scenarios, for instance if the SRNC is not able to add a strong cell to the active link set because of capacity shortage and a link addition shall be retried after a defined time period if the cell is still strong enough.

Note: In this book we will use the Event-ID names as defined in the ASN.1 specification of RRC protocol, e.g. "e1a," while in the RRC standard document general description, the same event is named "Event 1A." The meaning is the same.

3.7.1 RRC Measurement Types

Table 3.2 shows the RRC measurement types defined in 3GPP 25.331.

The evolution of UMTS standards may enhance the number of event-IDs and event-ID groups in the future step by step.

3.7.2 Cell Categories

The cells to be measured are categorized into three different sets:

Active Set Cells are all those FDD cells involved in softer and/or soft handover scenarios. In other words, all cells belonging to an active link set.

Note: In TDD mode there is always only ONE active cell, because there is no softer/soft handover in TDD!

Monitored Set Cells do not belong to the active set, but are monitored according to a neighbor list assigned by SRNC in the measurement control information. The UE can receive this list in two different ways: using an RRC Measurement Control message sent on a DCCH or reading SIB 11 or 12 (depending on radio infrastructure of cell) of Broadcast Control Channel (BCCH). And as shown in an earlier section, these SIBs have been sent during Node B Setup from CRNC to the Node B using the NBAP System Information Update procedure.

Table 3.2 RRC measurement types and Event-ID groups

Measurement type	Event-ID group	Typical tasks
Intrafrequency measurement	e1 . . .	Triggers softer or soft handover if necessary
Interfrequency measurement	e2 . . .	Triggers hard handover if necessary
Inter-RAT measurement	e3 . . .	Triggers handover from UTRAN (to, e.g., GSM) if necessary
Traffic volume measurement	e4 . . .	Triggers change of RRC State while PDPC stays active (channel type switching)
Quality reporting	e5 . . .	Informs SRNC that a predefined number of CRC errors is exceeded on UE side
UE internal measurement	e6 . . .	Delivers information about UE Tx Power (e.g. if maximum Tx Power is reached)
UE positioning reporting (3GPP Rel. 4 and higher)	e7 . . .	Informs network about problems with positioning accuracy (e.g. if position of UE is changing too often too fast to be reported)

Detected Set Cells have been detected by the UE despite the fact that they do neither belong to the active set nor they have been mentioned in the neighbor cell list. An intrafrequency measurement of these cells is done only if UE is in CELL_DCH state.

The following figures will introduce some typical procedures of how measurement settings are done and how the appropriate measurement reports are received.

3.7.3 Measurement Initiation for Intrafrequency Measurement

There are two different ways to initiate RRC measurement (Figure 3.39). The first way is that UE reads SIB 11 and/or 12 from Broadcast Channel of the current cell identified by a primary scrambling code that is also visible in RRC Connection Setup message. The second way is that SRNC sends an RRC Measurement Control message to the UE after successful establishment of RRC connection.

It is possible that several RRC Measurement Control messages are sent according to the different types of RRC measurement task. Also cell info lists (neighbor cell lists) and event criteria lists may be sent in separate messages. For instance, the first RRC Measurement Control message contains the Intrafrequency Cell Info List, and the second one the Event Criteria List with activated trigger conditions. In such a case, both messages may have different Measurement ID values and different RRC Transaction IDs.

The RRC Measurement Control messages are sent on a DCCH in downlink direction. A measurement identity is used as the setting identifier and will be used later in measurement reports related to these settings.

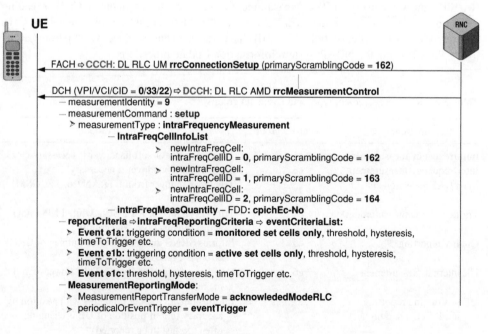

Figure 3.39 Initiation of RRC intrafrequency measurement.

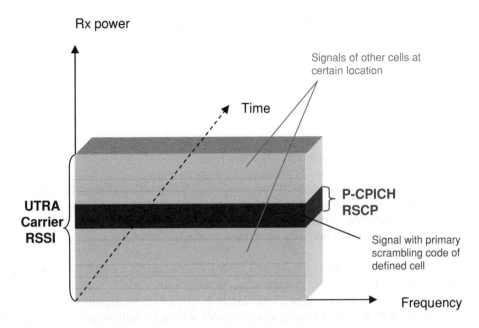

RSSI = Received Signal Strength Indicator   **RSCP** = Received Signal Code Power

Figure 3.40   RSSI and P-CPICH RSCP.

The Measurement Command value shows if settings are initial for the UE (setup), if previous settings are modified (modify), or if previous settings are deleted (deletion).

In case of intrafrequency measurement, an intrafrequency cell info list (neighbor cell list) is sent to the UE. It contains the primary scrambling code (PScrCd) of cells to be measured and assigns an appropriate Intrafrequency Cell ID to each cell of the list. Then the measurement quantity is defined – in the example, the Ec/No relation value of the primary CPICH signal of the listed FDD cells.

Ec/No (actually Ec/Io, but ASN.1 source code of RRC protocol uses Ec/No) stands for Energy per chip-to-total noise and interference power spectral density. In other words, it is the received signal code power (RSCP) of a P-CPICH in relation to the total signal strength (RSSI) of an UTRA carrier frequency, which must be imagined as the sum of all signals coming from cells received at a certain location (see Figure 3.40):

$$Ec/No = \frac{\text{P-CPICH RSCP}}{\text{UTRA Carrier RSSI}}$$

In the next section of the RRC Measurement Control message, the report criteria are defined in an event criteria list. In the example (Figure 3.39), the UE is requested to report triggered events e1a, e1b, and e1c to the SRNC if a defined CPICH Ec/No value threshold is reached. Additional parameters like hysteresis and time-to-trigger are used to optimize the number of measurement reports. It is further defined that RRC Measurement Report messages will be sent using acknowledged RLC mode and that measurement report sending is only event-triggered and not periodical.

3.7.4 Intrafrequency Measurement Events

3GPP 25.331 (Rel. 99) defines the following measurement events for intrafrequency measurement of FDD cells:

- e1a: A primary CPICH enters the reporting range
- e1b: A primary CPICH leaves the reporting range
- e1c: A nonactive primary CPICH becomes better than an active primary CPICH
- e1d: Change of best cell
- e1e: A primary CPICH becomes better than an absolute threshold
- e1f: A primary CPICH becomes worse than an absolute threshold

The Event-IDs e1a and e1b are used if Ec/No value of primary CPICH is to be measured, and e1e and e1f are used if absolute strength of primary CPICH is measured. Either the first or second possibility will be chosen. In both cases, e1a or e1e will trigger a radio link addition in a softer or soft handover situation, while e1b or e1f will trigger radio link deletion. e1c may trigger a radio link deletion with subsequent radio link addition in case the active link set already contains the maximum number of links. This action is also called radio link replacement. However, it should be noted that in any case the SRNC makes the decision if radio link additions/deletions are performed. e1d will not trigger any change in the radio link set necessarily.

Event-IDs e1g, e1h, and e1i describe events of TDD intrafrequency measurement.

The reporting range is a band bound to the level of the strongest cell of active set. Figure 3.41 shows the points when a neighbor cell primary CPICH level enters and leaves the reporting range and an appropriate measurement report is sent to SRNC.

As shown in Figure 3.42, the hysteresis is used to define a border area on top of the lowest reporting range level. Hysteresis ensures that only significant changes are reported.

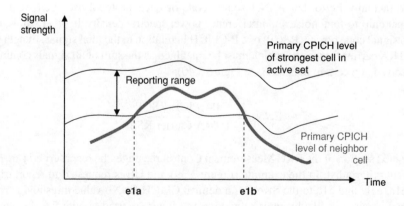

Figure 3.41   Reporting range and Event-ID.

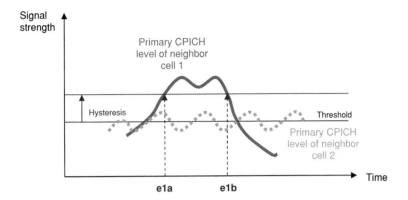

Figure 3.42   Hysteresis and influence on event report.

In the example, e1a and e1b for primary CPICH of cell 1 are reported, while change of cell 2 primary CPICH level does not become significant enough to trigger a reporting event.

Another parameter to limit the amount of RRC measurement reports is "time to trigger." It eliminates measurement reports that would be caused by short-time peaks of signal level. Only a cell that stays good over a longer time period (defined by "time to trigger") is added to the active link set as shown in Figure 3.43.

Another parameter worth to be discussed is the Offset. The Offset is a value added or subtracted from the originally measured signal level. It is unique for each cell to be measured. As a result, an event will be reported earlier – as shown in Figure 3.44 – or later than the threshold is reached in the given cell. This could be useful if the operator knows that a specific cell is interesting to monitor more carefully, even though it is not so good for the moment (positive offset). Our experience has shown that it is good to remove a defined cell earlier from the active set since it tends to lose strength very quickly (negative offset). As stated in 3GPP 25.331, this offset mechanism provides the network with an efficient tool to change the reporting of an individual primary CPICH.

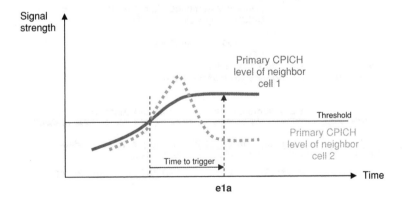

Figure 3.43   Time-to-trigger and influence on event report.

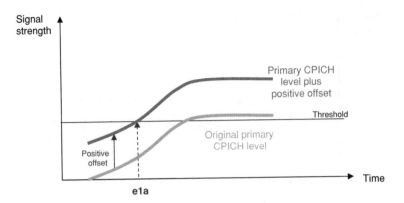

Figure 3.44   Offset and influence on event report.

Finally, it should be noted that it is also possible that in case of radio link replacement measurement (e.g. event-ID e1c), it is also possible to forbid defined primary CPICHs to affect the reporting range. This means that some cells can be excluded from RRC measurement.

3.7.5 Intrafrequency Measurement Report

Figure 3.45 shows an RRC Measurement Report that is related to the previous RRC Measurement Initiation by the same Measurement Identity = 9.

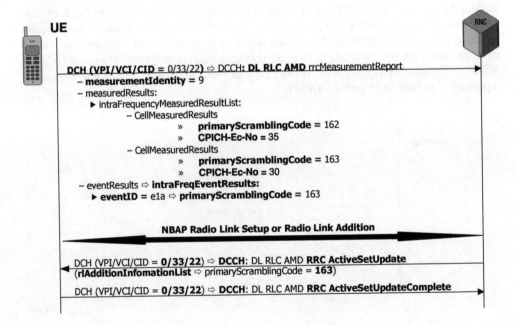

Figure 3.45   Intrafrequency measurement report.

The section Measured Results contains the intrafrequency measured result list. Here are CPICH Ec/No values reported for two cells that have been defined as cells to be monitored in the setup message before. An additional section contains the event result that was the reason why the measurement report was sent. In the example, the primary CPICH of the cell with PScrCd = 163 entered the reporting range, which led to an NBAP Radio Link Addition procedure in case of softer handover or NBAP Radio Link Setup procedure in case of soft handover procedure. In both cases the NBAP procedure was followed by an RRC Active Set Update procedure that contains once again the PScrCd of the cell that is added to the active set.

3.7.6 Intrafrequency Measurement Modification

Now after the SRNC receives the first measurement report, it may decide to change the RRC measurement configuration. In this case, a new RRC Measurement Control message is sent from SRNC to UE containing the same measurement identity (= 9) as in measurement setup message, but Measurement Command this time is set to "modify" (Figure 3.46).

The cell that was ordered to be measured, but not found by UE, can be removed from intrafrequency cell info list. It is indicated in this procedure using the intrafrequency cell ID defined in the measurement setup message. In addition, the event criteria list is included in the message as well, either having the same or different parameter settings for the defined trigger events.

A new measurement report (see Figure 3.47) will be received after this modification, depending on changing position of UE or changing transmission conditions on radio interface. In the example it is reported that the cell with PScrCd = 163 becomes weak, which is also

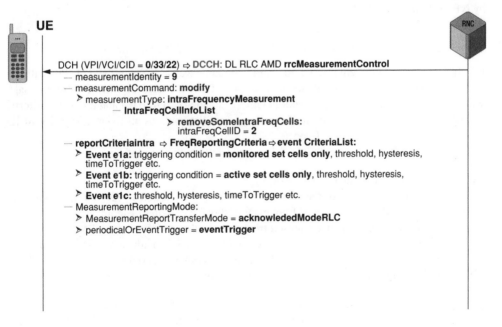

Figure 3.46 Intrafrequency measurement modification.

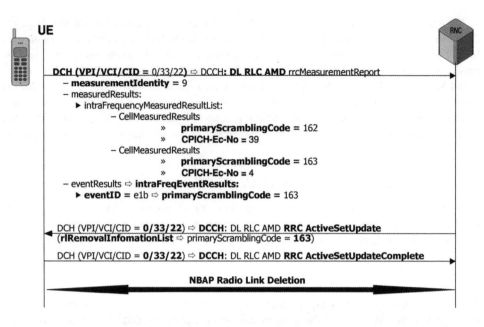

Figure 3.47   New intrafrequency measurement report.

documented with the low CPICH Ec/No value from the intrafrequency measured result list. As a result, the radio link from this cell will be removed from active link set with RRC Active Set Update (Radio link Removal List: PScrCd = 163). This procedure will be followed now by NBAP Radio Link Deletion.

3.7.7 Measurement Initiation for Interfrequency Measurement

Interfrequency measurement is initiated in quite the same way as measurement of cells with the same frequency as the present cell (see Figure 3.48). There is a Measurement Control message sent by SRNC to UE containing an optional interfrequency cell info list and reporting criteria, e.g. primary CPICH Ec/No. The interfrequency event list contains all Event-IDs of Event-ID group e2 . . . to be activated on UE side.

3GPP 25.331 (Rel. 99) specifies the following events in this group:

- e2a: Change of best frequency
- e2b: The estimated quality of the currently used frequency is below a certain threshold *and* the estimated quality of a nonused frequency is above a certain threshold
- e2c: The estimated quality of a nonused frequency is above a certain threshold
- e2d: The estimated quality of the currently used frequency is below a certain threshold
- e2e: The estimated quality of a nonused frequency is below a certain threshold
- e2f: The estimated quality of the currently used frequency is above a certain threshold

Often it is observed in network environments that all interfrequency measurement is set up and reported separately from other measurements, which means that this kind of measurement is related to a dedicated measurement identity. In the example, measurementID = 12.

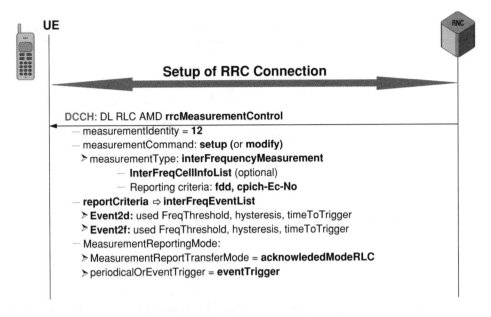

Figure 3.48   RRC Interfrequency measurement initiation.

The appropriate Measurement Report message is received later in the same way and based on the same conditions as in case of intrafrequency measurement.

3.7.8 Further RRC Measurement Groups

Event-ID group **e3**...deals with inter-RAT (Radio Access Technology) measurement. For instance, cells of GSM, CDMA2000, or TDMA networks can be monitored and reported to SRNC to change over to a different RAT if UTRAN operation conditions become too bad for the UE.

The events defined for inter-RAT measurement are:

- e3a: The estimated quality of the currently used UTRAN frequency is below a certain threshold *and* the estimated quality of the other system is above a certain threshold
- e3b: The estimated quality of the other system is below a certain threshold
- e3c: The estimated quality of the other system is above a certain threshold
- e3d: Change of best cell in other system

Event-ID group e4...allows SRNC to decide on behalf of UE's RLC payload buffer in which RRC state the UE shall operate during an active PDPC. The type of hard handover that is executed when measurement reports with e4...events are received is also known as channel type switching. A full scenario of such a procedure is shown in a following section.

The events defined for RLC payload buffer measurement (Figure 3.49) are:

- e4a: RLC buffer payload exceeds an absolute threshold
- e4b: RLC buffer payload becomes smaller than an absolute threshold

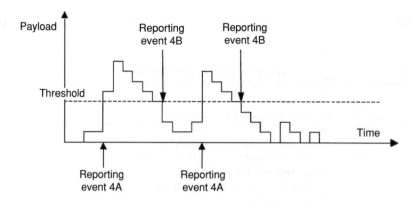

Figure 3.49 Traffic volume measurement and events (from 3GPP 25.331).

A parameter to limit and optimize the number of measurement reports in this group is Pending-Time-After-Trigger (Figure 3.50). It is a timer that ensures that positive or negative short-time buffer peaks will not result in consecutive measurement reports.

Event-ID **e5a** is used to report that the number of CRC errors on a certain transport channel (downlink transport channel block error rate [BLER]) in downlink direction exceeds a threshold. Once again the Pending-Time-After-Trigger parameter is used in case of this event to limit the number of measurement reports.

For Event-ID group **e6** ... the following trigger events are defined:

- e6a: The UE Tx power becomes larger than an absolute threshold
- e6b: The UE Tx power becomes less than an absolute threshold
- e6c: The UE Tx power reaches its minimum value
- e6d: The UE Tx power reaches its maximum value
- e6e: The UE RSSI reaches the UE's dynamic receiver range

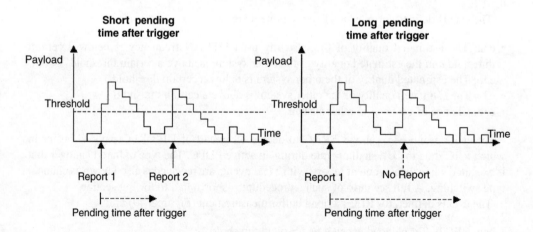

Figure 3.50 Pending time after trigger (from 3GPP 25.331).

- e6f: The UE Rx–Tx time difference for an RL included in the active set becomes larger than an absolute threshold
- e6g: The UE Rx–Tx time difference for an RL included in the active set becomes less than an absolute threshold

3.7.9 Changing Reporting Conditions After Transition to CELL_FACH

After transition to RRC CELL_FACH state, UE *stops sending intrafrequency, interfrequency, and intersystem measurement reports* according to previous definition in Measurement Control messages. Then it *starts monitoring neighboring cells listed in BCH System Information Block 12 or 11* and stores reporting criteria received from the same SIB.

Traffic volume measurement reporting (Event-ID Group e4...) is continued if *UE state of reporting* = "all states" or "all states except CELL_DCH" in previous Measurement Control. If no traffic volume measurement was defined in previous Measurement Control, UE starts traffic volume measurement according to info in BCH SIB 12 or 11.

Previous measurement definitions (received in former CELL_DCH state) can be resumed when UE changes back to CELL_DCH. In this case it should be noted that reporting criteria defined in Measurement Control message have a higher priority in comparison to the same measurement definitions sent on BCH SIB 12 or 11.

3.8 Iub – Physical Channel Reconfiguration (PDPC)

This scenario describes the case of PDPC activation with physical channel reconfiguration (Figure 3.51). After regular setup of PDPC with DCHs for DCCH and DTCH (Steps 1–4), the UE sends a measurement report including an event-ID (Step 5).

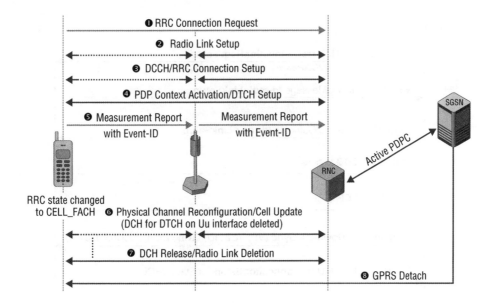

Figure 3.51   Iub physical channel reconfiguration during active PDPC overview.

The event-ID indicates that there is only a small amount of data to be transported from UE to the network. Hence, it made the decision to change into CELL_FACH state and perform user data transmission on RACH/FACH.

Step 6: The Physical Channel Reconfiguration and Cell Update procedures are performed to release the radio resources provided for the DCH that carried the DTCH. However, the AAL2 SVC for the DTCH stays active on Iub interface.

Step 7: Since the DCHs are not used anymore, they as well as their appropriate AAL2 SVCs and radio resources are deleted.

Step 8: The PDPC will finally be deactivated by a GPRS Detach procedure.

3.8.1 Message Flow

The following description will only highlight the messages and parameters that are important to understand the physical channel reconfiguration (Figure 3.52). All other messages/parameters are as described in PDPC activation/deactivation.

3.8.1.1 Iub PS PDP Activation and Physical Channel Reconfiguration

To analyze this scenario it is recommended to set an emphasis on watching some parameters already shown in the RRC Connection Setup message example earlier, especially RNTI and RRC state indicator. The mapping info for SRBs will show from the beginning two mapping options: either DCH can be used to exchange signaling or DCCHs can alternatively be mapped onto RACH (uplink) and FACH (downlink).

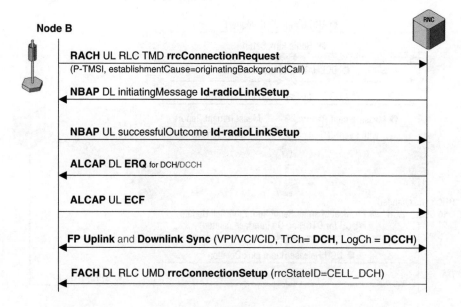

Figure 3.52   Iub physical channel reconfiguration (PDPC) call flow 1/6.

FACH DL RLC UMD **rrcConnectionSetup** (rrc-TransactionIdentifier=**a**, ULscramblingCode/DLChannelizationCode=**b**/β, **P-TMSI**, U-RNTI: SRNC-ID + S-RNTI, SRB Mapping Options for DCH and RACH/FACH, RRC State Indicator = CELL_DCH).

```
|TS 29.331 CCCH-DL (2002-03) (RRC_CCCH_DL)  rrcConnectionSetup     |
|(= rrcConnectionSetup)                                            |
|dL-CCCH-Message                                                   |
|1 message                                                         |
|1.1 rrcConnectionSetup                                            |
|1.1.1 r3                                                          |
|1.1.1.1 rrcConnectionSetup-r3                                     |
|1.1.1.1.1 initialUE-Identity                                      |
|1.1.1.1.1.1 tmsi-and-LAI                                          |
|**b32***  |1.1.1.1.1.1.1 tmsi                          |c7 38 56 98|
```
〰〰
```
|1.1.1.1.3 new-U-RNTI                                              |
|**b12***  |1.1.1.1.3.1 srnc-IDentity                  |44         |
|**b20***  |1.1.1.1.3.2 S-RNTI                         |13567      |
|--00----  |1.1.1.1.4 rrc-StateIndicator               |cell-DCH   |
```
〰〰
```
|1.1.1.1.7 srb-InformationSetupList                                |
|1.1.1.1.7.1 sRB-InformationSetup                                  |
|00000---  |1.1.1.1.7.1.1 rb-Identity                  |1          |
|1.1.1.1.7.1.2 rlc-InfoChoice                                      |
|1.1.1.1.7.1.2.1 rlc-Info                                          |
|1.1.1.1.7.1.2.1.1 ul-RLC-Mode                                     |
|1.1.1.1.7.1.2.1.1.1 ul-UM-RLC-Mode                                |
|1.1.1.1.7.1.2.1.2 dl-RLC-Mode                                     |
|          |1.1.1.1.7.1.2.1.2.1 dl-UM-RLC-Mode         |0          |
|1.1.1.1.7.1.3 rb-MappingInfo                                      |
|1.1.1.1.7.1.3.1 rB-MappingOption                                  |
|1.1.1.1.7.1.3.1.1 ul-LogicalChannelMappings                       |
|1.1.1.1.7.1.3.1.1.1 oneLogicalChannel                             |
|1.1.1.1.7.1.3.1.1.1.1 ul-TransportChannelType                     |
|***b5***  |1.1.1.1.7.1.3.1.1.1.1.1 dch                |32         |
|---0001-  |1.1.1.1.7.1.3.1.1.1.2 logicalChannelIdentity|1         |
```
〰〰
```
|1.1.1.1.7.1.3.1.2.1 dL-LogicalChannelMapping                      |
|1.1.1.1.7.1.3.1.2.1.1 dl-TransportChannelType                     |
|11111---  |1.1.1.1.7.1.3.1.2.1.1.1 dch                |32         |
|***b4***  |1.1.1.1.7.1.3.1.2.1.2 logicalChannelIdentity|1         |
|1.1.1.1.7.1.3.2 rB-MappingOption                                  |
|1.1.1.1.7.1.3.2.1 ul-LogicalChannelMappings                       |
|1.1.1.1.7.1.3.2.1.1 oneLogicalChannel                             |
|1.1.1.1.7.1.3.2.1.1.1 ul-TransportChannelType                     |
|          |1.1.1.1.7.1.3.2.1.1.1.1 rach               |0          |
```

```
|***b4*** |1.1.1.1.7.1.3.2.1.1.2 logicalChannelIdentity |1         |
|1.1.1.1.7.1.3.2.1.1.3 rlc-SizeList                                |
|1.1.1.1.7.1.3.2.1.1.3.1 explicitList                              |
|1.1.1.1.7.1.3.2.1.1.3.1.1 rLC-SizeInfo                            |
|--00000- |1.1.1.1.7.1.3.2.1.1.3.1.1.1 rlc-SizeIndex       |1        |
|***b3*** |1.1.1.1.7.1.3.2.1.1.4 mac-LogicalChannelPri..|1         |
|1.1.1.1.7.1.3.2.2 dl-LogicalChannelMappingList                    |
|1.1.1.1.7.1.3.2.2.1 dL-LogicalChannelMapping                      |
|1.1.1.1.7.1.3.2.2.1.1 dl-TransportChannelType                     |
|         |1.1.1.1.7.1.3.2.2.1.1.1 fach                    |0        |
|***b4*** |1.1.1.1.7.1.3.2.2.1.2 logicalChannelIdentity |1         |
```

Message Example 3.24   RRC Connection Setup with RB Mapping Options.

The message example shows only the RB mapping options for SRB 1. All other SRBs have similar mapping options.

On behalf of the Measurement Control message (Figure 3.53), RRC-related measurement in the UE is activated (see previous section).

DCH DL rrc measurementControl (rrc-TransactionIdentifier=**a**, measurementIdentity=**r**, Eventtrigger enabled for event-ID: **e4a, e4b**)

A DCH for a DTCH is installed by the RNC after receiving Activate PDPC Request message from the mobile (Figures 3.54 and 3.55).

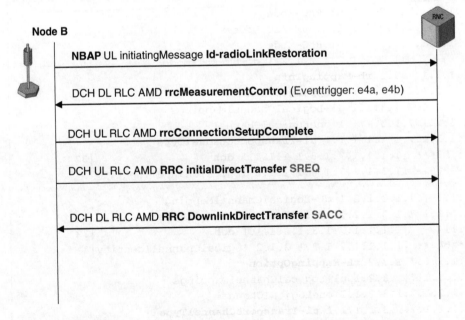

Figure 3.53   Iub physical channel reconfiguration (PDPC) call flow 2/6.

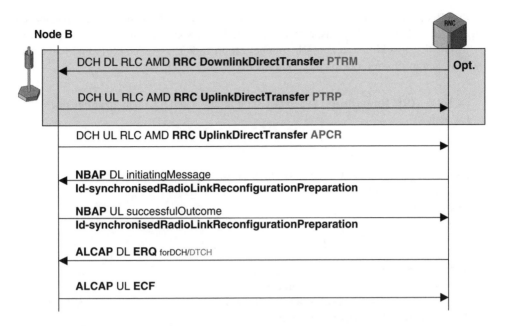

Figure 3.54   Iub physical channel reconfiguration (PDPC) call flow 3/6.

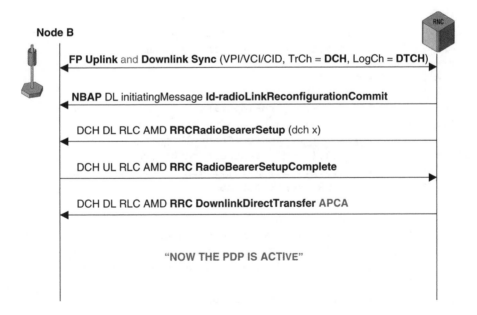

Figure 3.55   Iub physical channel reconfiguration (PDPC) call flow 4/6.

In the Radio Bearer Setup message the transport channel used for the payload transmission is identified as **dch x** (RRC Transport Channel ID). Later, RRC measurement reports will refer to this identity. In addition, a new DL channelization code is assigned according to expected higher data transmission rate on radio interface.

DCH DL **RRC RadioBearerSetup** (ULscramblingCode/DLChannelizationCode= b/ε, dch x)

```
TS 29.331 DCCH-DL (2002-03) (RRC_DCCH_DL)  radioBearerSetup      |
|(= radioBearerSetup)                                            |
|dL-DCCH-Message                                                 |
|1 message                                                       |
|1.1 radioBearerSetup                                            |
|1.1.1 r3                                                        |
|1.1.1.1 radioBearerSetup-r3                                     |
|00------ |1.1.1.1.1 rrc-TransactionIdentifier         |0         |
|--00---- |1.1.1.1.3 rrc-StateIndicator                |cell-DCH  |
|1.1.1.1.4 rab-InformationSetupList                              |
|1.1.1.1.4.1 rAB-InformationSetup                                |
|1.1.1.1.4.1.1 rab-Info                                          |
|1.1.1.1.4.1.1.1 rab-Identity                                    |
|***b8*** |1.1.1.1.4.1.1.1.1 gsm-MAP-RAB-Identity      |5         |
|--1----- |1.1.1.1.4.1.1.2 cn-DomainIdentity           |ps-domain |
|---0---- |1.1.1.1.4.1.1.3 re-EstablishmentTimer       |useT314   |
```
〰〰
```
|1.1.1.1.4.1.2 rb-InformationSetupList                           |
|1.1.1.1.4.1.2.1 rB-InformationSetup                             |
|00101--- |1.1.1.1.4.1.2.1.1 rb-Identity               |6         |
|1.1.1.1.4.1.2.1.3.1.2 dl-LogicalChannelMappingList              |
|1.1.1.1.4.1.2.1.3.1.2.1 dL-LogicalChannelMapping                |
|1.1.1.1.4.1.2.1.3.1.2.1.1 dl-TransportChannelType               |
|***b5*** |1.1.1.1.4.1.2.1.3.1.2.1.1.1 dch             |11        |
|1.1.1.1.4.1.2.1.3.2 rB-MappingOption                            |
|1.1.1.1.4.1.2.1.3.2.1 ul-LogicalChannelMappings                 |
|1.1.1.1.4.1.2.1.3.2.1.1 oneLogicalChannel                       |
|1.1.1.1.4.1.2.1.3.2.1.1.1 ul-TransportChannelType               |
|         |1.1.1.1.4.1.2.1.3.2.1.1.1.1 rach           |0         |
|0110---- |1.1.1.1.4.1.2.1.3.2.1.1.2 logicalChannelIde..|6        |
|1.1.1.1.4.1.2.1.3.2.1.1.3 rlc-SizeList                          |
|1.1.1.1.4.1.2.1.3.2.1.1.3.1 explicitList                        |
|1.1.1.1.4.1.2.1.3.2.1.1.3.1.1 rLC-SizeInfo                      |
|---00001 |1.1.1.1.4.1.2.1.3.2.1.1.3.1.1.1 rlc-SizeIndex|2       |
|001----- |1.1.1.1.4.1.2.1.3.2.1.1.4 mac-LogicalChanne..|2        |
```

```
|1.1.1.1.4.1.2.1.3.2.2 dl-LogicalChannelMappingList            |
|1.1.1.1.4.1.2.1.3.2.2.1 dL-LogicalChannelMapping              |
|1.1.1.1.4.1.2.1.3.2.2.1.1 dl-TransportChannelType             |
|          |1.1.1.1.4.1.2.1.3.2.2.1.1.1 fach           |0       |
|***b4*** |1.1.1.1.4.1.2.1.3.2.2.1.2 logicalChannelIde..|6       |
```
\/

Message Example 3.25   RRC Radio Bearer Setup with RB mapping options.

The message example shows that there are RB mapping options for the DTCH (for IP services like Web-browsing only one RB is necessary) similar to those for the DCCHs seen in the RRC Connection Setup message example. Scrambling codes and channelization codes are not shown in the message example.

When the UE receives the Activate PDPC Accept message, the PDPC becomes active.

An RRC Measurement Report including event-ID = e4b triggers the physical channel reconfiguration (Figure 3.56):

DCH UL RLC AMD **rrcMeasurementReport** (measurementIdentity=**r**, **dch x**,
Event-ID=e4b)

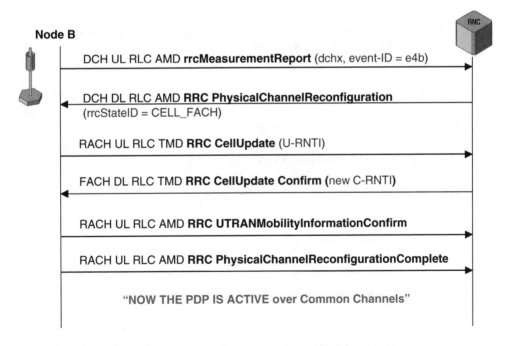

Figure 3.56   Iub physical channel reconfiguration (PDPC) call flow 5/6.

```
|TS 29.331 DCCH-UL (2002-03) (RRC_DCCH_UL)   measurementReport   |
|(= measurementReport)                                           |
|uL-DCCH-Message                                                 |
|1 message                                                       |
|1.1 measurementReport                                           |
|---1011- |1.1.1 measurementIdentity                  |12        |
|1.1.2 eventResults                                              |
|1.1.2.1 trafficVolumeEventResults                               |
|1.1.2.1.1 ul-transportChannelCausingEvent                       |
|***b5*** |1.1.2.1.1.1 dch                            |11        |
|-1------ |1.1.2.1.2 trafficVolumeEventIdentity       |e4b       |
-----------------------------------------------------------------
```

Message Example 3.26 RRC Measurement Report (Event-ID = e4b).

It is indicated by Event-ID = e4b that RLC buffer payload became smaller than an absolute threshold. Hence, the decision is made by the RNC to switch the UE into the CELL_FACH state and release the radio resources for the DCH with identifier **dch x** that carried the DTCH.

DCH DL RLC AMD **RRC PhysicalCannelReconfiguration** (rrc-TID=**a**, rrcStateIndicator=CELL_FACH)

```
|TS 29.331 DCCH-DL (2002-09) (RRC_DCCH_DL) physicalChannelRecon- |
|figuration (= physicalChannelReconfiguration)                   |
|dL-DCCH-Message                                                 |
|1 integrityCheckInfo                                            |
|**b32*** |1.1 messageAuthenticationCode     |'101101101110010   |
|                                             00001100100001001'B|
|-0111--- |1.2 rrc-MessageSequenceNumber     |7                 |
|2 message                                                       |
|2.1 physicalChannelReconfiguration                              |
|2.1.1 r3                                                        |
|2.1.1.1 physicalChannelReconfiguration-r3                       |
|--00---- |2.1.1.1.1 rrc-TransactionIdentifier |0               |
|----01-- |2.1.1.1.2 rrc-StateIndicator        |cell-FACH       |
|2.1.1.1.3 modeSpecificInfo                                      |
|2.1.1.1.3.1 fdd                                                 |
-----------------------------------------------------------------
```

Message Example 3.27 RRC physical channel reconfiguration.

After transition into CELL_FACH state, UE sends a Cell Update message to confirm the successful change of RRC state and to request a C-RNTI that was not assigned so far. The Cell Update message contains U-RNTI as UE identifier, start values to continue the active ciphering of IP payload on RACH/FACH, and a cell update cause = "cell reselection."

RACH UL RLC TMD **RRC CellUpdate** (U-RNTI: srnc-IDentity=**c** + S-RNTI=λ, start values for ciphering, cell update cause=**"cell reselection"**)

```
|TS 29.331 CCCH-UL (2002-09) (RRC_CCCH_UL)  cellUpdate            |
|(= cellUpdate)                                                   |
|uL-CCCH-Message                                                  |
~~~~~~~~~~~~~~~~~~~~~~~~~~~~~~~~~~~~~~~~~~~~~~~~~~~~~~~~~~~~~~~~~~~~~~~~~~~
|2.1 cellUpdate                                                   |
|2.1.1 U-RNTI                                                     |
|**b12*** |2.1.1.1 srnc-IDentity            |44                  |
|**b20*** |2.1.1.2 S-RNTI                   |13567               |
|2.1.2 startList                                                  |
|2.1.2.1 sTARTSingle                                              |
|----0--- |2.1.2.1.1 cn-DomainIdentity      |cs-domain           |
|**b20*** |2.1.2.1.2 start-Value            |'00000000000000000110'B|
|2.1.2.2 sTARTSingle                                              |
|-1------ |2.1.2.2.1 cn-DomainIdentity      |ps-domain           |
|**b20*** |2.1.2.2.2 start-Value            |'00000000000000010110'B|
|000----- |2.1.5 cellUpdateCause            |cellReselection     |
-------------------------------------------------------------------
```

Message Example 3.28   RRC Cell Update.

With Cell Update confirm message the UE receives the requested new C-RNTI that will be a valid UE identifier for all RLC/MAC frames containing IP payload as long as UE stays in the present cell.

RACH UL RLC TMD **RRC CellUpdateConfirm** (new C-RNTI=ω, RRC State Indicator = CELL_FACH)

```
|TS 29.331 DCCH-DL (2002-09) (RRC_DCCH_DL)  cellUpdateConfirm     |
|(= cellUpdateConfirm)                                            |
|dL-DCCH-Message                                                  |
~~~~~~~~~~~~~~~~~~~~~~~~~~~~~~~~~~~~~~~~~~~~~~~~~~~~~~~~~~~~~~~~~~~~~~~~~~~
|2.1 cellUpdateConfirm                                            |
|2.1.1 r3                                                         |
|2.1.1.1 cellUpdateConfirm-r3                                     |
|---00--- |2.1.1.1.1 rrc-TransactionIdentifier   |0             |
|**b16*** |2.1.1.1.2 new-C-RNTI                  |1689          |
|-----01- |2.1.1.1.3 rrc-StateIndicator          |cell-FACH     |
-------------------------------------------------------------------
```

Message Example 3.29   RRC Cell Update confirm.

RACH UL RLC AMD **RRC UTRANMobilityInformationConfirm** (rrc-TID=**a**) confirms that the new C-RNTI was received and is valid. RRC Transaction Identifier (rrc-TID) value links this message to the Physical Channel Reconfiguration procedure.

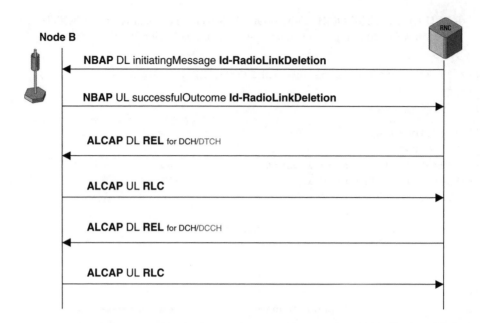

Figure 3.57 Iub physical channel reconfiguration (PDPC) call flow 6/6.

After the next message

> **RACH** UL RLC AMD **RRC PhysicalChannelReconfigurationComplete**
> (rrc-TID=**a**)

the PDPC is active over common channels and IP packets will be transmitted using RACH and FACH channels. All user plane packets on common transport channels will be identified by C-RNTI.

All radio and AAL2 resources for DCHs, which are not used anymore, are deleted (as shown in Figure 3.57), but the PDPC is still active over common channels. *The release of this connection will happen with a GPRS Detach procedure if there is no change back to CELL_DCH RRC state!*

3.9 Channel Type Switching

3.9.1 Overview

The procedure that uses physical channel reconfiguration to change the state of RRC connection is also called Channel Type Switching (Figure 3.58). Channel Type Switching can be triggered by events detected by the UE or by the network (SRNC). If during an active PDPC the RLC buffer of the connection stays for a defined time period below a certain threshold, DCCHs as well as DTCHs will be mapped onto common transport channels (RACH and FACH), which are used to transport signaling and IP payload in uplink or downlink direction. If then the RLC buffer is filled up again, dedicated resources will be assigned once again to serve the requested QoS.

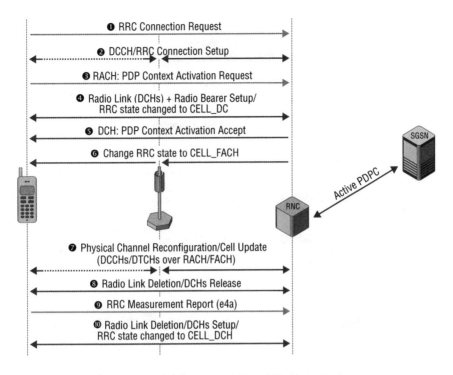

Figure 3.58   PDPC Channel Type Switching overview.

From network optimization point of view, Channel Type Switching is one of today's biggest problems especially for Web-browsing. The source of the problem is located in the user plane. Here HTTP (HyperText Transfer Protocol) uses connection-oriented transport services of TCP (Transmission Control Protocol). TCP is designed in a way that every TCP frame sent needs to be acknowledged by the peer entity. The window size parameter that defines after how many sent frames the connection is set on hold by sending entity to wait for the acknowledgement of the peer entity cannot be changed by users. (Well, experts say TCP window size could be changed if Windows™ registry files are edited, but this is not what normal Windows™ users will do.)

It is a pretty long distance between the computer connected to UE and the Web server, which is its TCP peer entity. In other words, the delay time for a TCP acknowledgment frame under these conditions is very long – so long that RLC buffer for either uplink or downlink data transfer becomes empty while waiting for TCP acknowledgment and this triggers change to RRC state CELL_FACH, because UTRAN does not expect a high data transfer rate for the connection after the long delay time.

But suddenly after successful acknowledgment there is another "wave" of TCP data entering the RLC buffer and now packet scheduler expects a very high data transmission rate on user plane. Hence, dedicated resources are assigned to guarantee high data transmission rate, but the next delay caused by waiting for TCP acknowledgment comes soon ...

There is only one true solution to overcome this problem in the future: design a new TCP standard with a new flexible window size parameter. However, to define a new protocol standard will take some years. Meanwhile network operators try to reduce the TCP delay times by installing proxy servers on Gi interface close to GGSNs to speed up at least the connections

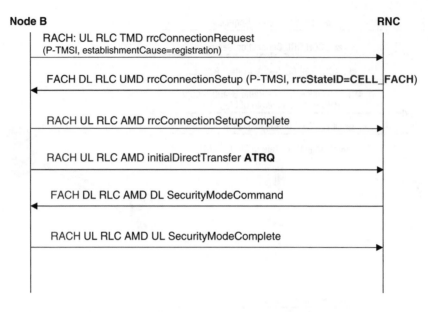

Figure 3.59   PDPC channel type switching call flow 1/9.

for frequently visited Web sites. Although there are still far too many, channel type switching procedures are monitored in the UTRAN. An interesting example that is more sophisticated than the one described in the previous scenario is given as follows (Figure 3.59):

Step 1: Setup of an RRC Connection is requested by UE. This message is sent on RACH.

Step 2: An RRC Connection that does not use dedicated resources is set up. All signaling messages will be transported on RACH (uplink) and FACH (downlink). The DCHs are mapped onto these CTrCHs.

Step 3: A PDPC Activation Request message is sent from UE side.

Step 4: Before the PDPC is activated, a dedicated radio link that will carry and also set up radio bearer is established. SRNC orders UE to switch into the CELL_DCH state.

Step 5: Using the transport capabilities of the established DCHs, PDPC activation is confirmed by the network.

Step 6: Caused by first TCP acknowledgment delay, the RLC buffer on network side is below a certain threshold. Hence, SRNC orders UE to change to CELL_FACH state.

Step 7: Physical channel reconfiguration is performed to map DCCHs and DTCHs onto common transport channels RACH and FACH. Cell Update procedure is embedded to assign new UE indentity (new C-RNTI).

Step 8: Dedicated resources for radio links and AAL2 SVCs on Iub are deleted.

Step 9: UE sends RRC measurement report to announce that RLC buffer is full again. This triggers SRNC to order change back to CELL_DCH state and to reassign dedicated resources.

Step 10: New radio link is set up. DCCHs and DTCHs are mapped onto DCHs, RRC state is changed to CELL_DCH, but it is predictable that RLC buffer either on UE or on network side will soon be empty again and channel type switching is triggered once again.

3.9.2 Message Flow

Stage 1 – Attach and PDPC Activation

As usual, the procedure starts with an RRC Connection Request sent by UE. The establishment cause is "registration" because this UE is not attached to the PS domain yet.

> RACH: UL RLC TMD **rrcConnectionRequest** (<u>**P-TMSI**</u>,
> establishmentCause=registration)

The RNC decided that for the exchange of signaling messages the common transport channels will be used. This decision can be driven by present traffic situation in the cell. Maybe there are already many dedicated resources in use. In RRC Connection Setup message sent on FACH, the RRC state indicator orders the UE to continue the RRC connection in CELL_FACH state. This is the reason why not only a U-RNTI, but also a C-RNTI is assigned. Uplink channel type of the SRBs is RACH, downlink channel type of SRBs is FACH. The transport channel identities are those assigned during CTrCH Setup (see Node B Setup scenario).

> FACH: DL RLC UMD **rrcConnectionSetup** (rrc-TransactionIdentifier=**a**, **P-TMSI**,
> U-RNTI: SRNC-ID=**c** + S-RNTI=**d**, C-RNTI=**e**, rrc-StateIndicator=**cell-FACH**,
> ul-transportChannelType=**RACH**, dl-transportChannelType=**FACH**,
> ul/dl-transportChannelIdentity)

```
|TS 29.331 CCCH-DL (2002-09) (RRC_CCCH_DL)   rrcConnectionSetup      |
|(= rrcConnectionSetup)                                              |
|dL-CCCH-Message                                                     |
|1 message                                                          |
|1.1 rrcConnectionSetup                                             |
|1.1.1 r3                                                           |
|1.1.1.1 rrcConnectionSetup-r3                                     |
|1.1.1.1.1 initialUE-Identity                                      |
|1.1.1.1.1.1 p-TMSI-and-RAI                                        |
|**b32*** |1.1.1.1.1.1.1 p-TMSI             |e0 87 99 53 |
```
~~~~~~
```
|1.1.1.1.3 new-U-RNTI                                               |
|**b12*** |1.1.1.1.3.1 srnc-IDentity        |44          |
|**b20*** |1.1.1.1.3.2 S-RNTI               |19774       |
|**b16*** |1.1.1.1.4 new-C-RNTI             |8           |
|--01---- |1.1.1.1.5 rrc-StateIndicator     |cell-FACH   |
```
~~~~~~
```
|1.1.1.1.8 srb-InformationSetupList                                 |
|1.1.1.1.8.1 sRB-InformationSetup                                  |
|00000--- |1.1.1.1.8.1.1 rb-Identity        |1           |
|1.1.1.1.8.1.2 rlc-InfoChoice                                      |
|1.1.1.1.8.1.2.1 rlc-Info                                          |
|1.1.1.1.8.1.2.1.1 ul-RLC-Mode                                     |
|1.1.1.1.8.1.2.1.1.1 ul-UM-RLC-Mode                                |
```

```
|1.1.1.1.8.1.2.1.2 dl-RLC-Mode                                          |
|           |1.1.1.1.8.1.2.1.2.1 dl-UM-RLC-Mode              |0          |
|1.1.1.1.8.1.3 rb-MappingInfo                                           |
|1.1.1.1.8.1.3.1 rB-MappingOption                                       |
|1.1.1.1.8.1.3.1.1 ul-LogicalChannelMappings                            |
|1.1.1.1.8.1.3.1.1.1 oneLogicalChannel                                  |
|1.1.1.1.8.1.3.1.1.1.1 ul-TransportChannelType                          |
|***b5*** |1.1.1.1.8.1.3.1.1.1.1.1 dch                      |32          |
|---0001- |1.1.1.1.8.1.3.1.1.1.2 logicalChannelIdentity |1          |
|1.1.1.1.8.1.3.1.1.1.3 rlc-SizeList                                     |
|           |1.1.1.1.8.1.3.1.1.1.3.1 configured           |0          |
|-000---- |1.1.1.1.8.1.3.1.1.1.4 mac-LogicalChannelPri..|1          |
|1.1.1.1.8.1.3.1.2 dl-LogicalChannelMappingList                         |
|1.1.1.1.8.1.3.1.2.1 dL-LogicalChannelMapping                           |
|1.1.1.1.8.1.3.1.2.1.1 dl-TransportChannelType                          |
|11111--- |1.1.1.1.8.1.3.1.2.1.1.1 dch                      |32          |
|***b4*** |1.1.1.1.8.1.3.1.2.1.2 logicalChannelIdentity |1          |
|1.1.1.1.8.1.3.2 rB-MappingOption                                       |
|1.1.1.1.8.1.3.2.1 ul-LogicalChannelMappings                            |
|1.1.1.1.8.1.3.2.1.1 oneLogicalChannel                                  |
|1.1.1.1.8.1.3.2.1.1.1 ul-TransportChannelType                          |
|           |1.1.1.1.8.1.3.2.1.1.1.1 rach                     |0          |
|***b4*** |1.1.1.1.8.1.3.2.1.1.2 logicalChannelIdentity |1          |
|1.1.1.1.8.1.3.2.1.1.3 rlc-SizeList                                     |
|1.1.1.1.8.1.3.2.1.1.3.1 explicitList                                   |
|1.1.1.1.8.1.3.2.1.1.3.1.1 rLC-SizeInfo                                 |
|--00000- |1.1.1.1.8.1.3.2.1.1.3.1.1.1 rlc-SizeIndex       |1          |
|***b3*** |1.1.1.1.8.1.3.2.1.1.4 mac-LogicalChannelPri..|1          |
|1.1.1.1.8.1.3.2.2 dl-LogicalChannelMappingList                         |
|1.1.1.1.8.1.3.2.2.1 dL-LogicalChannelMapping                           |
|1.1.1.1.8.1.3.2.2.1.1 dl-TransportChannelType                          |
|           |1.1.1.1.8.1.3.2.2.1.1.1 fach                     |0          |
|***b4*** |1.1.1.1.8.1.3.2.2.1.2 logicalChannelIdentity |1          |
---------------------------------------------------------------------
```

Message Example 3.30   RRC Connection Setup for PDP Context in CELL_FACH.

The message example shows that the same RB mapping options are given as in the case of PDPC activation in CELL_DCH state, but besides the different RRC state indicators there is also no uplink scrambling code and downlink channelization code sent to UE, because there is no dedicated physical channel assigned.

On RACH, UE sends RRC Connection Setup Complete. The RRC message itself is linked to the previous RRC Connection Setup by RRC Transaction Identifier, but it does not contain a UE identifier. While in CELL_FACH state the UE is known by its C-RNTI that is part of the MAC frame. It is possible that the RRC message is transported in several segments by RLC/MAC and finally reassembled by RNC or measurement equipment that monitors Iub

From	2. Prot	2. MSG	3. Prot	3. MSG	MAC: UE-ID Type	MAC: UE-ID
E2 RACH Cell0	RLC/MAC	FP DATA RACH				
E2 RACH Cell0	RLC reasm.	TM DATA RACH	RRC_CCCH_UL	rrcConnectionRequest		
E2 FACH1 Cell0	RLC/MAC	FP DATA FACH				
E2 FACH1 Cell0	RLC/MAC	FP DATA FACH				
E2 FACH1 Cell0	RLC/MAC	FP DATA FACH				
E2 FACH1 Cell0	RLC reasm.	UM DATA FACH	RRC_CCCH_DL	rrcConnectionSetup		
E2 RACH Cell0	RLC/MAC	FP DATA RACH			C-RNTI (Cell Radio Network Temporary Identity) 0	
E2 RACH Cell0	RLC/MAC	FP DATA RACH			C-RNTI (Cell Radio Network Temporary Identity) 0	
E2 RACH Cell0	RLC/MAC	FP DATA RACH			C-RNTI (Cell Radio Network Temporary Identity) 0	
E2 RACH Cell0	RLC reasm.	AM DATA RACH	RRC_DCCH_UL	rrcConnectionSetupComplete		
E2 FACH1 Cell0	RLC/MAC	FP DATA FACH			C-RNTI (Cell Radio Network Temporary Identity) 0	
E2 RACH Cell0	RLC/MAC	FP DATA RACH			C-RNTI (Cell Radio Network Temporary Identity) 0	
E2 RACH Cell0	RLC/MAC	FP DATA RACH			C-RNTI (Cell Radio Network Temporary Identity) 0	
E2 RACH Cell0	RLC reasm.	AM DATA RACH	RRC_DCCH_UL	initialDirectTransfer		

Figure 3.60 RRC Connection Setup: Segmented and reassembled RLC frames.

interface (Figure 3.60). Depending on used measurement software it is possible that C-RNTI is only shown for the segmented frames, but not for the reassembled RRC message.

RACH: UL MAC: UE-ID Type=C-RNTI, MAC: UE-ID=**e**, RLC AMD
rrcConnectionSetupComplete (rrc-TransactionIdentifier=**a**,
UE-RadioAccessCapability).

With an RRC Initial Direct Transfer, the Attach Request (ATRQ) message is sent on RACH to RNC and forwarded to SGSN and after successfully switching on of the security functions the Attach is accepted and a new P-TMSI is assigned with ATAC message. Reception of new P-TMSI is confirmed with Attach Complete (ACOM) (see Figure 3.61):

RACH: UL MAC: UE-ID Type=C-RNTI, MAC: UE-ID=**e**, RLC AMD RRC
initialDirectTransfer **ATRQ**

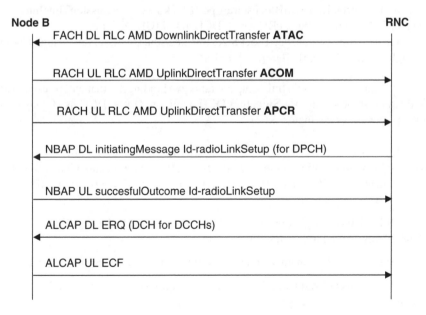

Figure 3.61 PDPC channel type switching 2/9.

FACH: DL MAC: UE-ID Type=C-RNTI, MAC: UE-ID=**e**, RRC
SecurityModeCommand

RACH: UL MAC: UE-ID Type=C-RNTI, MAC: UE-ID=**e**, RRC
SecurityModeComplete

FACH: DL MAC: UE-ID Type=C-RNTI, MAC: UE-ID=**e**, DownlinkDirectTransfer
ATAC

RACH: UL MAC: UE-ID Type=C-RNTI, MAC: UE-ID=**e**, UplinkDirectTransfer
ACOM

Now the subscriber wants to activate a PDPC. Still on RACH, Activate PDPC Request
(APCR) is sent to the network containing Access Point Name (APN) to get a dynamic IP
address assigned by this server. In addition, the requested QoS attributes are included.

RACH: UL MAC: UE-ID Type=C-RNTI, MAC: UE-ID=**e**, RRC
UplinkDirectTransfer **APCR** (APN, requested Quality of Service)

Reception of APCR in SGSN triggers RANAP RAB assignment procedure on IuPS. Based
on the QoS requested for the RAB, SRNC packet scheduler function decides to set up DCHs
for signaling and user traffic. In the next step, SRNC (that is also CRNC for this Node B)
performs NBAP Radio Link Setup procedure. In NBAP Initiating Message of this procedure,
for each DCH the uplink and downlink TFS is defined as shown in former scenarios.

NBAP DL initiatingMessage **Id-radioLinkSetup** (longTransActionID=**f**,
id-CRNC-CommunicationContextID=**g**, ULscramblingCode=**b**, DCH-ID=**h**
[UL/DL TFS: nrOfTransportBlocks, transportBlockSize, transmissionTimeInterval,
channelCoding, codingRate, cRC-Size], DCH-ID=**i** [UL/DL TFS:
nrOfTransportBlocks, transportBlockSize, transmissionTimeInterval, channelCoding,
codingRate, cRC-Size], rL-ID=**j**, C-ID=**k**)

Successful Outcome for radio link setup contains two bindingIDs that will be used to identify
ALCAP Establish procedure for Signaling DCH and User Traffic DCH as already described
in previous scenarios (see Figure 3.62).

NBAP UL successfulOutcome **Id-radioLinkSetup** (longTransActionID=**f**,
id-CRNC-CommunicationContextID=**g**, id-NodeB-CommunicationContextID=**l**,
CommunicationPortID=**m**, rL-ID=**j**, rL-Set-ID=**n**, dCH-ID=**h** ⇨ bindingID=**o**,
dCH-ID=**i** ⇨ bind-ID=**p**)

ALCAP DL **ERQ** (Originating Signal. Ass. ID=**q**, AAL2 Path=**r**, AAL2
ChannelId=**s**, served user gen reference=**o**)

ALCAP UL **ECF** (Originating Signal. Ass. ID =**t**, Destination Sign. Assoc. ID=**q**)

ALCAP DL **ERQ** (Originating Signal. Ass. ID=**u**, AAL2 Path=**r**, AAL2
ChannelId=**v**, served user generated reference=**p**)

ALCAP UL **ECF** (Originating Signal. Ass. ID=**w**, Destination Sign. Assoc. ID=**u**)

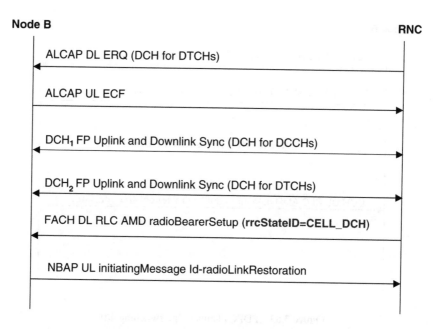

Figure 3.62 PDPC channel type switching 3/9.

Then follows the synchronization procedures of FP on both AAL2 SVCs.

DCH₁ for DCCHs in AAL2 Path=**r′** and Channel=**s**: downlink and uplink synchronization FP frames

DCH₂ for DTCHs in AAL2 Path=**r′** and Channel=**v**: downlink and uplink synchronization FP frames

Now RRC Radio Bearer Setup is sent to UE using FACH, because the UE still does not know anything about the provided dedicated resources. The message also contains RRC State Indicator that orders UE to change into CELL_DCH state.

FACH: DL MAC: UE-ID Type=C-RNTI, MAC: UE-ID=**e**, RLC AMD **RRC radioBearerSetup** (rrc-Transaction Identifier=**a**, rrc-StateIndicator=**cell-DCH**, gsm-MAP-RAB-Identity=**5**, ul/dl-transportChannelType=DCH, ul/dl-transportChannelIdentity=**h,i**, UL_scramblingCode/DL_channelizationCode=**b**/β, primaryScramblingCode=**x**)

NBAP Radio Link Restoration indicates that the dedicated physical channels have been found by UE on radio interface.

NBAP UL initiatingMessage **id-radioLinkRestoration** (shortTransActionID=**y**, id-CRNC-CommunicationContextID=**g**, rL-Set-ID=**n**)

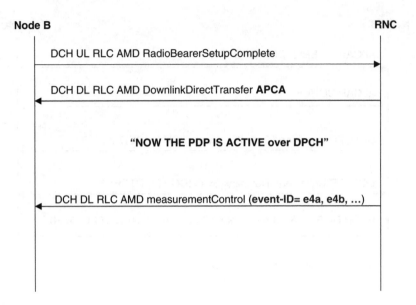

Figure 3.63   PDPC channel type switching 4/9.

After Radio Bearer Setup completion (Figure 3.63), activation of PDPCs is confirmed including the negotiated QoS that is based on the subscribed QoS values stored in HLR.

DCH UL RLC AMD **RRC RadioBearerSetupComplete**
(rrc-TransactionIdentifier=**a**)

DCH DL RLC AMD RRC DownlinkDirectTransfer **APCA** (subscribed Quality of
Service)

Now the PDPC is active and RRC traffic volume measurement on UE side is initialized by
SRNC.

DCH DL RLC AMD **RRC measurementControl** (trafficVolumeMeasurement,
dCH-ID=**i**, event-ID= **e4a**, event-ID=**e4b**, ...)

Stage 2 – Change Back to CELL_FACH (Figures 3.64 and 3.65)
The delay in IP payload transmission caused by TCP acknowledgment procedure leads to
an empty RLC buffer on RNC side. Because of the obviously low data tranmission rate,
RNC decides to continue the connection in CELL_FACH and release the previously assigned
dedicated radio resources. An RRC Cell Update procedure is performed to assign a new
C-RNTI:

DCH DL RLC AMD RRC PhysicalChannelReconfiguration
(rrc-StateIndicator=cell-FACH)

RACH UL RLC TMD RRC cellUpdate (U-RNTI: srnc-IDentity=**c** + S-RNTI=**d**,
cellUpdateCause = cellReselection)

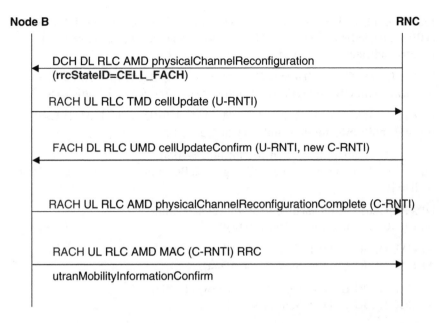

Figure 3.64   PDPC channel type switching overview 5/9.

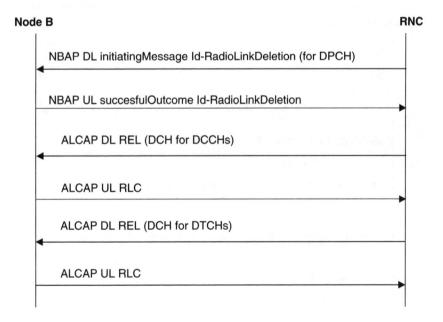

Figure 3.65   PDPC channel type switching 6/9.

FACH DL MAC: UE-ID Type=U-RNTI, MAC: UE-ID=**c+d**, RLC UMD **RRC cellUpdateConfirm** (new C-RNTI=**z** {in following message example z=**9**}, rrc-StateIndicator=cell-FACH)

RACH UL MAC: UE-ID Type=C-RNTI, MAC: UE-ID=**z**, RLC AMD **RRC physicalChannelReconfigurationComplete** (rrc-Transaction Identifier=**a**)

RACH UL MAC: UE-ID Type=C-RNTI, MAC: UE-ID=**z**, RLC AMD **RRC utranMobilityInformationConfirm** (rrc-TransactionIdentifier=**a**)

NBAP DL initiatingMessage **Id-RadioLinkDeletion** (id-CRNC-CommunicationContextID=**g**, NodeBCommunicationsContexr-ID=**l**, rL-ID=**j**)

NBAP UL successfulOutcome **Id-RadioLinkDeletion** (id-CRNC-CommunicationContextID=**g**)

ALCAP DL **REL** (Dest. Sign. Assoc. ID=**t**) (DCH)
ALCAP UL **RLC** (Dest. Sign. Assoc. ID=**q**)

ALCAP DL **REL** (Dest. Sign. Assoc. ID=**w**) (Traffic)
ALCAP UL **RLC** (Dest. Sign. Assoc. ID=**u**)

After release of all dedicated resources IP payload as well as signaling messages are transported using RACH and FACH. This goes on until an RRC Measurement Report informs the network that RLC buffer on UE side is quite full (Event-ID **e4a**) and higher data transmission rate can be expected (Figure 3.66).

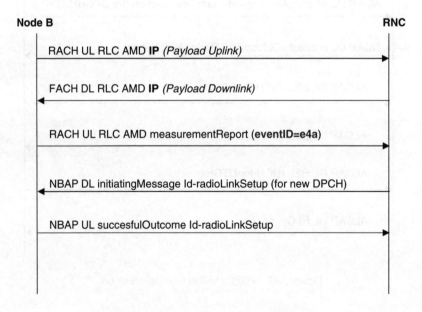

Figure 3.66   PDPC channel type switching 7/9.

RACH: UL MAC: UE-ID Type=C-RNTI, MAC: UE-ID=**z**, RLC AMD **IP payload**

```
|"10/26/9" E2 RACH Cell0 RLC reasm. AM DATA RACH IP IPv4 UDP DTGR  |
|TS 25.322 V3.10.0 (2002-03) reassembled (RLC reasm.)  AM DATA RACH|
|(= Acknowledged Mode Data RACH)                                   |
|Acknowledged Mode Data RACH                                       |
|          |FP:  VPI/VCI/CID              |"10/26/9"                |
|          |FP:  Direction               |Uplink                   |
|          |FP:  Transport Channel Type  |RACH (Random Access      |
|          |                             |Channel)                 |
|          |MAC: UE-ID                   |9                        |
|          |MAC: Target Channel Type     |DTCH (Dedicated          |
|          |                             |Traffic Channel)         |
|          |MAC: C/T Field               |Logical Channel 5        |
|          |MAC: RLC Mode                |Acknowledge Mode         |
|**B161**  |RLC: Whole Data              |45 00 00 a1 0c a         |
|          |                             | 7 00 00 01 11 dd...     |
|IP - Internet Protocol (v4/v6), RFC791/2460 (IP) IPv4             |
|(= Internet Protocol version 4)                                   |
|Internet Protocol version 4                                       |
|0100----  |Version                      |4                        |
```
~~~~~~
```
|00010001  |Protocol                     |UDP   User Datagram      |
|          |                             |[RFC768,JBP]             |
|***B2***  |Header Checksum              |dd9c                     |
|***B4***  |Source Address (IPv4)        |172.xxx.xxx.xxx          |
|***B4***  |Destination Address (IPv4)   |239.xxx.xxx.xxx          |
|**B141**  |Data                         |40 c3 07 c8 00 8d 7e|
|          |                             |93 4d 2d 53...           |
```

Message Example 3.31 Uplink IPv4 payload frame sent over RACH.

In the message example we see an IPv4 frame sent in uplink direction over RACH. C-RNTI of the UE is 9 and higher layer protocol on top of IPv4 is UDP. IP source address identifies the IP terminal connected to UE; destination IP address identifies, e.g., a server in the Internet (expressions "xxx" are used to hide real address values).

FACH: DL MAC: UE-ID Type=C-RNTI, MAC: UE-ID=**z**, RLC AMD **IP payload**

Stage 3 - Change Back to CELL_DCH

RACH: UL MAC: UE-ID Type=C-RNTI, MAC: UE-ID=**z**, RLC AMD **RRC measurementReport** (trafficVolumeEventIdentity=**e4a**)

Based on the event in the received RRC measurement report, SRNC decides once again to assign dedicated resources and continue connection in RRC CELL_DCH state (Figures 3.66 and 3.67).

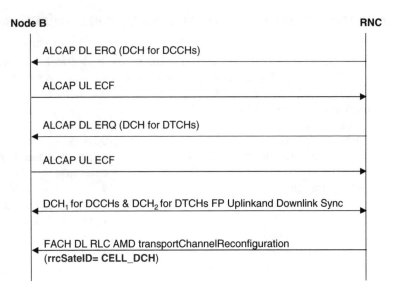

Figure 3.67   PDPC channel type switching 8/9.

NBAP DL initiatingMessage **Id-radioLinkSetup** (longTransActionID=**aa**,
id-CRNC-CommunicationContextID=**bb**, scramblingCode=**b**,
DCH-ID=**h** + TFS (nrOfTransportBlocks, transportBlockSize,
transmissionTimeInterval, channelCoding, codingRate, cRC-Size), DCH-ID=**I** +
TFS (nrOfTransportBlocks, transportBlockSize, transmissionTimeInterval,
channelCoding, codingRate, cRC-Size), rL-ID=**j**, C-ID=**k**)

NBAP UL successfulOutcome **Id-radioLinkSetup** (longTransActionID=**aa**,
id-CRNC-CommunicationContextID=**bb**, id-NodeB-CommunicationContextID=**cc**,
CommunicationPortID=**m**, rL-ID=**j**, rL-Set-ID=**n**, dCH-ID=**h**, bindingID=**dd**,
dCH-ID=**i**, bindingID=**ee**)

ALCAP DL **ERQ** (Originating Signal. Ass. ID=**ff**, AAL2 Path=**r**, AAL2
ChannelId=**hh**, served user gen reference=**dd**)

ALCAP UL **ECF** (Originating Signal. Ass. ID=**gg**, Destination Sign. Assoc. ID=**ff**)

ALCAP DL **ERQ** (Originating Signal. Ass. ID=**ii**, AAL2 Path=**r**, AAL2
ChannelId=**jj**, served user gen reference=**ee**)

ALCAP UL **ECF** (Originating Signal. Ass. ID=**kk**, Destination Sign. Assoc. ID=**gg**)

DCH$_1$ for DCCHs in AAL2 Path=**r'** and Channel=**hh**: downlink and uplink
synchronization FP frames

DCH$_2$ for DTCHs in AAL2 Path=**r'** and Channel=**jj**: downlink and uplink
synchronization FP frames

Because the previously used scrambling codes and channelization codes are still reserved
for this connection, this time a Transport Channel Reconfiguration is performed (Figures 3.67
and 3.68):

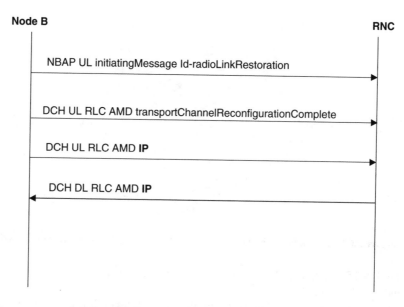

Figure 3.68   PDPC channel type switching 9/9.

FACH: DL MAC: UE-ID Type=C-RNTI, MAC: UE-ID=**z**, RLC AMD **RRC
transportChannelReconfiguration** (rrc-TransactionIdentifier= **a**,
rrc-StateIndicator=**cell-DCH**, ul/dl-transportChanelType=DCH,
ul/dl-transportChanelIdentity=**h,i**, primaryScramblingCode=**x**,
UL_scramblingCode/DL_channelizationcode=**b/**β, **P-TMSI**)

NBAP UL initiatingMessage **Id-radioLinkRestoration** (shortTransActionID=**j**,
id-CRNC-CommunicationContextID=**bb**, rL-Set-ID=**n**)

DCH$_1$ UL RLC AMD **RRC transportChannelReconfigurationComplete**
(rrc-TransactionIdentifier=**a**)

DCH$_2$ UL RLC AMD **IP payload**

DCH$_2$ DL RLC AMD **IP payload**

Further channel type switching procedures can be expected during this connection because
of further TCP acknowledgment delays.

3.10   Iub – Mobile-Originated Call with Soft Handover
(Inter-Node B, Intra-RNC)

3.10.1   Overview

The Soft Handover of a mobile-originated call (MOC) consists of three phases (see Figure 3.69).
Each phase is related to the position of the MS, also named User Equipment (UE). Both names

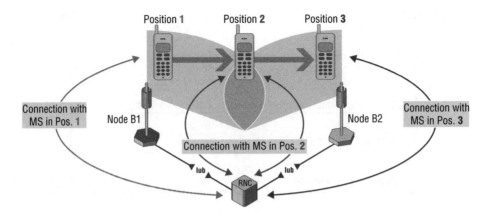

Figure 3.69   Iub mobile-originated voice call (MOC) with Soft Handover overview.

are used in UMTS standards. The example described in this chapter is an Inter-Node B/Intra-RNC handover procedure.

Note: A soft handover may not only appear during a voice call. It is also seen if plain RRC signaling is exchanged and during active PDPC. This is because the main purpose of soft handover is not to update the network with the current location of the subscriber, but to keep transmission power levels low also when UE is moving along cell borders. Less necessary transmission power is less interference in cell and a lower interference level grants the cell a higher capacity: more users and services can be served.

Position 1: The call is after setup in an active state. DCCHs for exchange of signaling messages and DTCHs for transport of user data (voice packets) are active between the MS and the RNC. For each of them, DCCH and DTCH, there is an Iub transport bearer (AAL2 SVC) running between SRNC and Node B1. The MS is only able to have radio contact with a cell of Node B1.

Position 2: The MS changes its position and is now able to have radio contact with both, the cell of Node B1 and another cell belonging to Node B2. The MS detects the availability of the new cell and sends a measurement report including an Event-ID to the RNC. Triggered by this measurement report SRNC decides to set up a second connection (radio link plus appropriate Iub bearers) to the MS via Node B2. Finally, two different connections between RNC and MS are active; both belong to the same active set. An active set can handle up to six connections simultaneously.

Position 3: The MS loses radio contact with Node B1 and sends an appropriate measurement report again. Iub transport bearers for DCCH and DTCH(s) on Iub to Node B1 are released.

3.10.2 Message Flow (Figure 3.70)

Iub CS MOC and Handover from Node B1 to Node B2

To make the whole process better understandable and show the complete signaling, first the messages of the MOC setup are described as already done in scenario MOC:

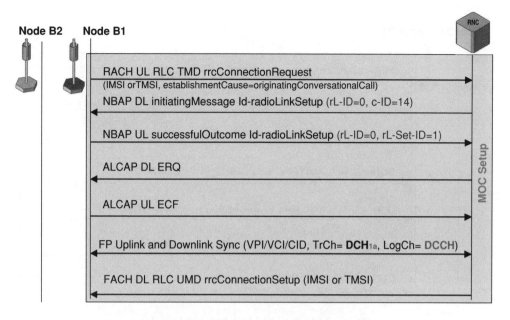

Figure 3.70   Iub MOC with Soft Handover 1/9.

RACH: UL RLC TMD rrcConnectionRequest (**IMSI or TMSI**, establishmentCause=**originatingConversationalCall**)

NBAP DL initiatingMessage **Id-radioLinkSetup** (longTransActionID=**c**, id-CRNC-CommunicationContextID=**d**, ULscramblingCode/DLChannelizationCode=**b**/β, rL-ID=**0**, c-ID=**14**)

Besides other parameters (already discussed in location update and MOC scenarios) this message (as shown in Message Example 3.11) contains radio link ID (rL-ID) and cell-ID (c-ID) of the cell where the radio link is originally set up.

In NBAP Successful Outcome message of Radio Link Setup procedure there is another mandatory information element which is very important for the monitoring of handover procedures: radio link set ID (rL-Set-ID) – related to rL-ID previously assigned by SRNC function of RNC. Values of rL-ID and rL-Set-ID in the example are decimal integer numbers to prevent a growing number of variables (Figure 3.71):

NBAP UL successfulOutcome **Id-radioLinkSetup** (longTransActionID=**c**, id-CRNC-CommunicationContextID=**d**, bindingID=**e**, NodeBCommunicationsContext-ID=**p**, CommunicationPortID=**o**, rL-ID=**0**, rL-Set-ID=**1**)

ALCAP DL **ERQ** (Originating Signal. Ass. ID=**f**, AAL2 Path=**g**, AAL2 ChannelId=**h**, served user gen reference=**e**)

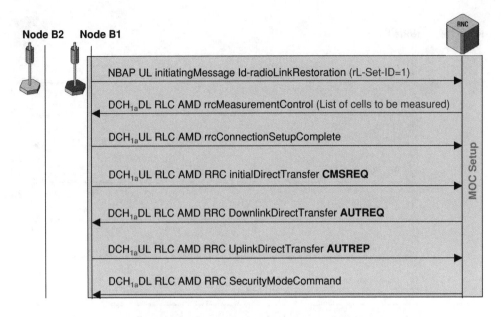

Figure 3.71 Iub MOC with Soft Handover 2/9.

ALCAP UL **ECF** (Originating Signal. Ass. ID=**i**, Destination Sign. Assoc. ID=**f**)

DCH in AAL2 Path=**g′** and Channel=**h**: downlink and uplink FP synchronization

FACH: DL RLC UMD **rrcConnectionSetup** (rrc-TransactionIdentifier=**a**, ULscramblingCode/DLChannelizationCode=**b**/β, **IMSI or TMSI**)

Also NBAP Radio Link Restoration message contains the rL-Set-ID, but neither rL-ID nor c-ID:

NBAP UL initiating Message **id-radioLinkRestoration** (shortTransActionID=**j**, id-CRNC-CommunicationContextID=**d**, rL-Set-ID=**1**)

DCH₁ₐ RRC messages (RLC AMD) in this AAL2 channel=h

The RRC Measurement Control message in this call flow contains definitions of intrafrequency cell measurement: a list of neighbor cells that use the same frequency as the cell used for the setup of the present radio link and a list of event-IDs out of event-ID group **e1** . . . that need to be reported to SRNC (Figure 3.72).

DCH DL RLC AMD **rrcMeasurementControl** (rrc-**TransactionIdentifier**=**a**, measurementIdentity=**r**, List of cells to be measured – **identified by their Primary Scrambling Codes [PScrCd]**, List of **event-IDs** to be reported)

DCH UL RLC AMD **rrcConnectionSetupComplete** (rrc-TransactionIdentifier=**a**)

DCH UL RLC AMD RRC initialDirect Transfer **CMSREQ**
DCH DL RLC AMD RRC DownlinkDirectTransfer **AUTREQ**

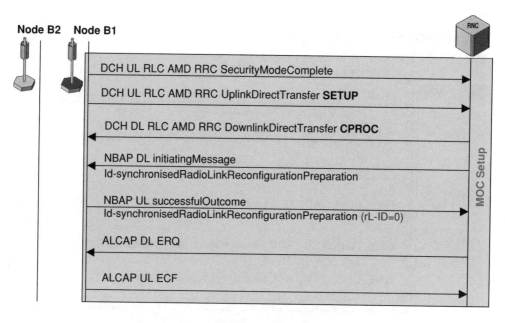

Node B2 Node B1

DCH UL RLC AMD RRC SecurityModeComplete

DCH UL RLC AMD RRC UplinkDirectTransfer **SETUP**

DCH DL RLC AMD RRC DownlinkDirectTransfer **CPROC**

NBAP DL initiatingMessage
Id-synchronisedRadioLinkReconfigurationPreparation

NBAP UL successfulOutcome
Id-synchronisedRadioLinkReconfigurationPreparation (rL-ID=0)

ALCAP DL ERQ

ALCAP UL ECF

MOC Setup

RNC

Figure 3.72 Iub MOC with Soft Handover 3/9.

DCH UL RLC AMD RRC UplinkDirectTransfer **AUTREP**
DCH DL RLC AMD **RRC SecurityModeCommand**

DCH UL RLC AMD **RRC SecurityModeComplete**
DCH UL RLC AMD RRC UplinkDirectTransfer **SETUP**
DCH DL RLC AMD RRC DownlinkDirectTransfer **CPROC**

NBAP DL: initiatingMessage**Id-synchronisedRadioLink
ReconfigurationPreparation**(shortTransActionID=**j**,
NodeBCommunicationsContext-ID=**p**,
ULscramblingCode/DLChannelizationCode=**b**/ε)

NBAP UL: successfulOutcome **Id-synchronisedRadioLink
ReconfigurationPreparation**(shortTransActionID=**j**,
id-CRNC-CommunicationContextID=**d**, bindingID=**k**, rL-ID=**0**)

ALCAP DL **ERQ** (Originating Signal. Ass. ID=**l**, AAL2 Path=**m**, AAL2 Channel
id=**n**, served user gen reference=**k**)

ALCAP UL **ECF** (Originating Signal. Ass. ID=**q**, Destination Sign. Assoc. ID=**l**)

*Now the transport bearer for traffic channels (Path= m' and Channel = n) on Iub are opened,
and downlink and uplink FP synchronization frames are transmitted (Figure 3.73).*

NBAP DL initiatingMessage **Id-synchronisedRadioLinkReconfigurationCommit**
(shortTransAction ID=**j**, NodeBCommunicationsContext-ID=**p**)

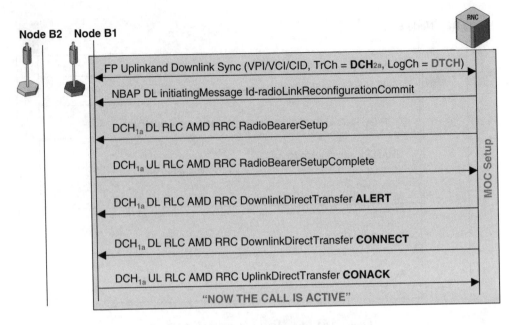

Node B2   Node B1

FP Uplinkand Downlink Sync (VPI/VCI/CID, TrCh = **DCH₂ₐ**, LogCh = **DTCH**)

NBAP DL initiatingMessage Id-radioLinkReconfigurationCommit

DCH₁ₐ DL RLC AMD RRC RadioBearerSetup

DCH₁ₐ UL RLC AMD RRC RadioBearerSetupComplete

DCH₁ₐ DL RLC AMD RRC DownlinkDirectTransfer **ALERT**

DCH₁ₐ DL RLC AMD RRC DownlinkDirectTransfer **CONNECT**

DCH₁ₐ UL RLC AMD RRC UplinkDirectTransfer **CONACK**

"NOW THE CALL IS ACTIVE"

MOC Setup

Figure 3.73   Iub MOC with Soft Handover 4/9.

The following are the messages in DCH AAL2 Path=**g′** and Channel=**h**:

DCH DL **RRC RadioBearerSetup** (ULscramblingCode/DLChannelizationCode=**b/ε**)
DCH UL **RRC RadioBearerSetupComplete**
DCH RRC DownlinkDirectTransfer **ALERT**
DCH RRC DownlinkDirectTransfer **CONNECT**
DCH RRC UplinkDirectTransfer **CONACK**

The handover procedure is triggered by a measurement report sent by the MS via Node B1. This measurement report message is related to the Measurement Control sent during call setup on behalf of the same measurement identity. The report indicates the occurrence of an event that is described by its Event-ID. Following RRC protocol specification TS 25.331 report, event 1a (e1a) means the mobile detected a new primary Common Pilot Channel (CPICH). This means a new cell (from the list of cells to be measured in the Measurement Control message) has come close enough and can now be used for communication between MS and network. The new cell is identified on radio interface by its primary scrambling code (Figure 3.74):

DCH₁ₐ UL RLC AMD **rrcMeasurementReport** (measurementIdentity=**r**, Event Results: Event-ID=*e1a*= "a primary CPICH enters the reporting range", PrimaryScramblingCode=**hh**)

Based on the measurement report the RNC decides to hand over the call to the new cell. First radio resources must be provided using NBAP Radio Link Setup Request message sent to

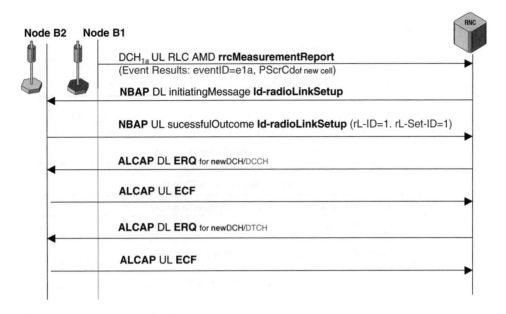

Figure 3.74   Iub MOC with Soft Handover 5/9.

Node B2, then AAL2 SVCs for DCCH and DTCH are set up on the Iub interface between RNC and Node B2. Since in the Intra-RNC Soft Handover case all radio resources are administrated by the same RNC, the second NBAP Radio Link Setup Request message will most likely contain the same CRNC Communication Context value as in the first message of this type during call setup. However, in some cases it was also seen that a new CRNC communication context value was assigned.

Node B Communication Context will definitely have a new value, because Node B and hence Iub interface is a different one compared to first radio link setup. rL-ID and rl-Set-ID in Successful Outcome message of Radio link setup procedure show that a new link is set up, but belongs to the same link set as the already existing one.

The downlink channelization code of the new radio link does not need to be the same as for the first link in link set. Often a new downlink channelization code is assigned for the second radio link in addition to the one used for the first link of active set and UE is informed about the additional code number during RRC Active Set Update procedure.

NBAP DL initiatingMessage **Id-RadioLinkSetup** (longTransActionID=**s**,
id-CRNC-communicationsContext-ID=**d** [in some cases: **t**]),
ULscramblingCode/DLChannelizationCode=**b**/δ[new!!!], DCH-Ids [Transport
Format Sets]

NBAP UL successfulOutcome **Id-radioLinkSetup** (longTransActionID=**s**,
id-CRNC-CommunicationContextID=**t**, NodeBCommunicationContext-ID=**u**,
CommunicationPortID=**v**, DCH-Id, bindingID=**w**, DCH-Id, bindingID=**x**,
rL-ID=**1**, rL-Set-ID=**1**)

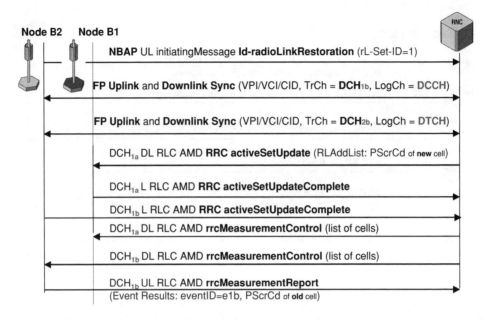

Figure 3.75   Iub MOC with Soft Handover 6/9.

ALCAP DL **ERQ** (Originating Signal. Ass. ID=**y**, AAL2 Path=**z**, AAL2 ChannelId=**aa**, served user gen reference=**w**)
ALCAP UL **ECF** (Originating Signal. Ass. ID=**bb**, Destination Sign. Assoc. ID=**y**)

ALCAP DL **ERQ** (Originating Signal. Ass. ID=**cc**, AAL2 Path=**dd**, AAL2 ChannelId=**ee**, served user gen reference=**x**)

ALCAP UL **ECF** (Originating Signal. Ass. ID=**ff**, Destination Sign. Assoc. ID=**cc**)

The NBAP message Synchronized Radio Link Restoration (see Figure 3.75) indicates that Node B2 became uplink synchronic with UE/MS. This is possible because the mobile still uses the same uplink scrambling code that was given with the second Radio Link Setup procedure to Node B2 as well. So Node B2 is able to listen to UE while the mobile has not received any information yet on how to detect the provided new physical resources.

Note: Specific synchronization and reporting criterias – as far as defined in international standards – can be found in 3GPP 25.214 (Physical Layer Procedures [FDD]).

NBAP UL initiatingMessage **Id-RadioLinkRestoration** (shortTransActionID=**gg**, id-CRNC-communicationsContext-ID=**d** [or **t**])

FP Uplink and Downlink Synchronization messages are used for initial alignment of physical transport bearers for DCCH and DTCH toward Node B2:

Dedicated Control Channel DCH$_{1b}$ **FP Uplink** and **Downlink Sync** Messages in the AAL2 Path=**z′** and AAL2 Channel=**aa**

Dedicated Traffic Channel DCH$_{2b}$ **FP Uplink** and **Downlink Sync** Messages in the AAL2 Path=**dd′** and AAL2 Channel=**ee**

Then an RRC Active Set Update message is sent from RNC to the MS still via the old Node B. This is because the MS still does not know that a second radio link in a new cell was successfully established. After receiving RRC Active Set Update with included radio link addition information list, the UE/MS knows in which cell (identified by a primary scrambling code) the additional link was established.

DCH$_{1a}$ DL RLC AMD **RRC activSetUpdate** (rrc-TransactionIdentifier=**a**, RLAdditionInformationList: PrimaryScramblingCode= **hh**)

```
|TS 29.331 DCCH-DL (2002-06) (RRC_DCCH_DL)  activeSetUpdate     |
|(= activeSetUpdate)                                            |
|dL-DCCH-Message                                               |
|1 integrityCheckInfo                                          |
/\/\/\/\/\/\/\/\/\/\/\/\/\/\/\/\/\/\/\/\/\/\/\/\/\/\/\/\/\/\/\/\/\/\/\/\/\/\/\
|2 message                                                    |
|2.1 activeSetUpdate                                          |
|2.1.1 r3                                                     |
|2.1.1.1 activeSetUpdate-r3                                   |
|***b2*** |2.1.1.1.1 rrc-TransactionIdentifier      |0        |
|2.1.1.1.3 rl-AdditionInformationList                        |
|2.1.1.1.3.1 rL-AdditionInformation                          |
|2.1.1.1.3.1.1 primaryCPICH-Info                             |
|***b9*** |2.1.1.1.3.1.1.1 primaryScramblingCode   |291      |
/\/\/\/\/\/\/\/\/\/\/\/\/\/\/\/\/\/\/\/\/\/\/\/\/\/\/\/\/\/\/\/\/\/\/\/\/\/\/\
|2.1.1.1.3.1.2.1.3.1 dL-ChannelizationCode                   |
|2.1.1.1.3.1.2.1.3.1.1 sf-AndCodeNumber                      |
|***b7*** |2.1.1.1.3.1.2.1.3.1.1.1 sf128           |4        |
|-1------ |2.1.1.1.3.1.2.1.3.1.2 scramblingCodeChange |noCode- |
|                                                   Change    |
```

Message Example 3.32   RRC Active Set Update (radio link addition).

In the message example we see further that a new downlink channelization code was assigned during NBAP Radio Link Setup procedure on Iub 2. The parameter scramblingCodeChange "indicates whether the alternative scrambling code is used for compressed mode method 'SF/2'" (3GPP 25.331 [RRC]). This means the downlink, not the uplink scrambling code!

After UE adds the new radio link to its radio link set, it sends RRC Active Set Update Complete message using DCCH on both old and new Iub interface. That the message on both interfaces is indeed the same and has the same origin can be proved by checking the message authentication code that is equal:

DCH$_{1a}$ UL RLC AMD **RRC activSetUpdateComplete** (messageAuthenticationCode=**xx**, rrc-TransactionIdentifier=**a**)

DCH$_{1b}$ UL RLC AMD **RRC activSetUpdateComplete** (messageAuthenticationCode=**xx**, rrc-TransactionIdentifier=**a**)

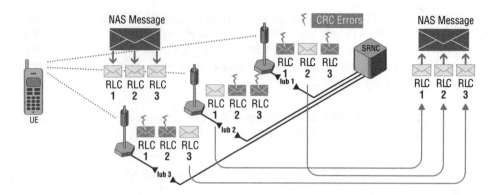

Figure 3.76   RLC reassembly and macrodiversity.

In general, all downlink messages can be monitored now on both Iub interfaces simultaneously as long as there is no site selection diversity (SSDT) activated. Uplink RLC frames can also be found on both Iub interfaces, but they are not necessarily error-free.

Soft Handover and Macrodiversity

When monitoring a UE in soft handover the physical links FP connections on Iub/Iur interface(s) are terminated in SRNC and following the rules of macrodiversity. SRNC compares FP frames (each FP frame contains one single RLC frame) incoming from different Iub interfaces. Good frames are taken to reassemble the original sent uplink messages, bad frames are deleted.

Figure 3.76 explains this for an active link set with three radio links. CRC errors occur during transmission on radio interface, but can also be monitored while looking at FP frames on different Iub interfaces.

Whether a frame is good or bad is indicated by quality estimate (QE) parameter that is part of the FP trailer. QE value is bound to radio interface bit error ratio (BER). How BER is mapped onto QE depends on manufacturer implementation; there is no general rule given by 3GPP standards. Often the best quality estimate value is 0, the worst 255 (= CRC error). SRNC will always use the frame with lowest QE value for reassembly of higher layer messages.

QE values are found in all uplink frames that do not contain RLC status information (control messages of RLC acknowledged mode), as shown in Figure 3.77 for RLC frames that carry RRC signaling messages of a single call. However, QE is also used in case RLC encloses user plane traffic. These statements made for RRC signaling messages in soft handover situations are also valid (and probably much more important) for user plane voice/data.

To give an overview of how QEs develop during a call, a diagram can be used (see Figure 3.78). It will show that especially in Soft Handover situations, QE and hence the BER on radio interface are not very good – but this is what is expected in Soft Handover. A bad BER is accepted as long as transmission power of UE and cell antenna remains low so that any unnecessary interference and excessive cell breathing effects are prevented. However, in a well-optimized radio network it is for sure also a target to keep BER/QE as good as possible.

Qual_Estimate_1call.html

VPI/VCI/CID	2. Prot	2. MSG	3. Prot	3. MSG	4. Prot	4. MSG	FP: Quality Estimate	eventID
"10/26/38"	RLC reasm.	AM DATA DCH	RRC_DCCH_UL	rrcConnectionSetupCompl..				
"10/26/38"	RLC/MAC	FP DATA DCH						
"10/26/38"	RLC/MAC	FP DATA DCH						
"10/26/38"	RLC/MAC	FP DATA DCH						
"10/26/38"	RLC/MAC	FP DATA DCH					0	
"10/26/38"	RLC/MAC	FP DATA DCH						
"10/26/38"	RLC/MAC	FP DATA DCH						
"10/26/38"	RLC/MAC	FP DATA DCH						
"10/26/38"	RLC/MAC	FP DATA DCH						
"10/26/38"	RLC/MAC	FP DATA DCH						
"10/26/38"	RLC reasm.	AM DATA DCH	RRC_DCCH_DL	measurementControl				
"10/26/38"	RLC reasm.	AM DATA DCH	RRC_DCCH_DL	measurementControl				
"10/26/38"	RLC/MAC	FP DATA DCH					75	
"10/26/38"	RLC/MAC	FP DATA DCH					33	
"10/26/38"	RLC reasm.	AM DATA DCH	RRC_DCCH_UL	initialDirectTransfer	MM-DMTAP	CMSREQ		
"10/26/38"	RLC/MAC	FP DATA DCH					75	
"10/26/38"	RLC/MAC	FP DATA DCH						
"10/26/38"	RLC/MAC	FP DATA DCH						
"10/26/38"	RLC/MAC	FP DATA DCH						
"10/26/38"	RLC reasm.	AM DATA DCH	RRC_DCCH_DL	downlinkDirectTransfer	MM-DMTAP	AUTREQ		
"10/26/38"	RLC/MAC	FP DATA DCH					111	
"10/26/38"	RLC/MAC	FP DATA DCH					0	
"10/26/38"	RLC reasm.	AM DATA DCH	RRC_DCCH_UL	uplinkDirectTransfer	MM-DMTAP	AUTREP		
"10/26/38"	RLC/MAC	FP DATA DCH						

Figure 3.77   Quality estimate values for uplink RLC frames.

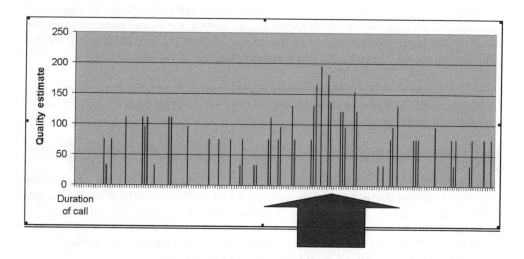

Figure 3.78   Quality estimate values indicating Soft Handover situation.

After link set information is updated, the same RRC Measurement Control message is sent twice, via the old and via the new radio link for reasons mentioned above:

DCH$_{1a}$ DL RLC AMD **rrcMeasurementControl** (measurementIdentity=**r**)
DCH$_{1b}$ DL RLC AMD **rrcMeasurementControl** (measurementIdentity=**r**)

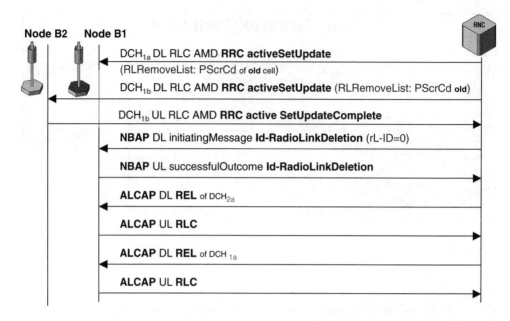

Figure 3.79   Iub MOC with Soft Handover 7/9.

When the mobile moves further and loses contact with the old cell of Node B1 a new measurement report is sent. Once again an Event-ID indicates that the old cell became too bad (on behalf of CPICH measurement result) and the primary scrambling code identifies the old cell (different one than in Active Set Update message before!):

> DCH$_{1b}$ UL RLC AMD **rrcMeasurementReport** (measurementIdentity=**r**, Event
> Results: eventID=e1b= "a primary CPICH leaves the reporting range,"
> PrimaryScramblingCode=**jj**)

The reception of the new measurement report triggers the release procedures between RNC and Node B1.

Still Active Set Update message is sent via both the old and the new DCCH, but in comparison to the first Active Set Update this time the message contains a Radio Link Removal Information List with the primary scrambling code as identifier of the old cell on Node B1 (Figure 3.79).

> DCH$_{1a}$ DL RLC AMD **RRC activSetUpdate** (rrc-TransactionIdentifier=**a**, Radio
> Link Removal Information List: PrimaryScramblingCode= **jj**)

```
|TS 29.331 DCCH-DL (2002-06) (RRC_DCCH_DL)  activeSetUpdate          |
|(= activeSetUpdate)                                                 |
|dL-DCCH-Message                                                     |
wwwwwwwwwwwwwwwwwwwwwwwwwwwwwwwwwwwwwwwwwwwwwwwwwwwwwwwwwwwwwwwwwwwwwwwww
|2 message                                                          |
|2.1 activeSetUpdate                                                |
```

```
|2.1.1 r3                                                                |
|2.1.1.1 activeSetUpdate-r3                                              |
|***b2*** |2.1.1.1.1 rrc-TransactionIdentifier          |0              |
|-1010011 |2.1.1.1.2 maxAllowedUL-TX-Power              |33             |
|2.1.1.1.3 rl-RemovalInformationList                                     |
|2.1.1.1.3.1 primaryCPICH-Info                                           |
|***b9*** |2.1.1.1.3.1.1 primaryScramblingCode          |285            |
-------------------------------------------------------------------------
```

Message Example 3.33   RRC Active Set Update (Radio Link Removal).

Same message on second Iub:

DCH$_{1b}$ DL RLC AMD **RRC activeSetUpdate** (rrc-TransactionIdentifier=**a**, Radio Link Remove Information List: PrimaryScramblingCode= **jj**)

The UE that removed the link toward Node B1 answers via Node B2 with an Active Set Update Complete message:

DCH$_{1b}$ UL RLC AMD **RRC activeSetUpdateComplete** (rrc-TransactionIdentifier=**a**)

Now the radio links and the AAL2 SVCs between RNC and Node B1 are deleted. Radio link Deletion Request message contains the Radio Link ID=0, so this is the old link.

NBAP DL initiatingMessage **Id-RadioLinkDeletion** (shortTransActionID=**ii**, id-CRNC-CommunicationContextID=**d**, NodeBCommunicationsContext-ID=**p**, rL-ID=**0**)
NBAP UL successfulOutcome **Id-RadioLinkDeletion** (shortTransActionID=**ii**, id-CRNC-CommunicationContextID=**d**)

ALCAP DL **REL** (Dest. Sign. Assoc. ID=**q**) (Traffic Channel to Node B1)
ALCAP UL **RLC** (Dest. Sign. Assoc. ID=**l**)

ALCAP DL **REL** (Dest. Sign. Assoc. ID=**i**) (DCCH to Node B1)
ALCAP UL **RLC** (Dest. Sign. Assoc. ID=**f**)

The release procedure (Figure 3.80), starting with MM/CC/SM Disconnect message, is also already known from the MOC scenario. However, since the call was handed over to Node B2, now all signaling messages are exchanged via the channels on the new Iub interface:

DCH$_{1b}$ UL RLC AMD RRC UpLinkDirectTransfer **DISC**
and/or
DCH$_{1b}$ DL RLC AMD RRC DownlinkDirectTransfer **RELEASE**
DCH$_{1b}$ UL RLC AMD RRC UpLinkDirectTransfer **RELCMP**
DCH$_{1b}$ DL RLC UMD DL **rrcConnectionRelease** (rrc-TransactionIdentifier=**a**)
DCH$_{1b}$ UL RLC UMD UL **rrcConnectionReleaseComplete** (rrc-TransactionIdentifier=**a**)

NBAP DL initiatingMessage **Id-RadioLinkDeletion** (shortTransActionID=**jj**, id-CRNC-CommunicationContextID=**t**, NodeBCommunicationsContext-ID=**u**)

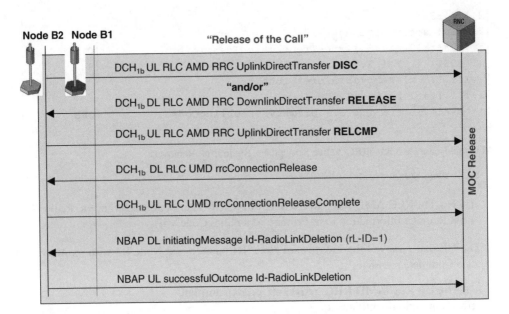

Figure 3.80   Iub MOC with Soft Handover 8/9.

NBAP UL successfulOutcome **Id-RadioLinkDeletion** (shortTransActionID=**jj**,
id-CRNC-CommunicationContextID=**t**)

ALCAP DL **REL** (Dest. Sign. Assoc. ID=**bb**) (Figure 3.81)
ALCAP UL **RLC** (Dest. Sign. Assoc. ID=**y**)
ALCAP DL **REL** (Dest. Sign. Assoc. ID=**ff**)
ALCAP UL **RLC** (Dest. Sign. Assoc. ID=**cc**)

3.11 Iub – Softer Handover

3.11.1 Overview

The Softer Handover (Figure 3.82) is based on the same moving steps as the Soft Handover
procedure, but in case of Softer Handover the neighbor cells belong to the same Node B. So
there is no need to set up additional physical transport bearers (AAL2 SVCs). All signaling
messages can be monitored on the same Iub interface.

To illustrate better the interactions between radio links and cells, c-IDs are shown as integer
numbers in the call flow diagram (Figure 3.83). The focus in the following call trace example
is on messages and parameters directly related to the Softer Handover procedure. That is the
reason why ALCAP messages will not be shown and MM/CC/SM messages can just be found
in a kind of shortcut.

3.11.2 Message Flow

MOC with Softer Handover

During call establishment phase of the first radio link, important link identifiers like rL-ID and
rL-Set-ID are assigned to the NBAP connection represented by CRNC/Node B Communication

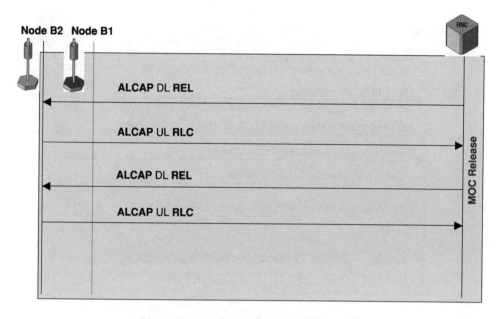

Figure 3.81   Iub MOC with Soft Handover 9/9.

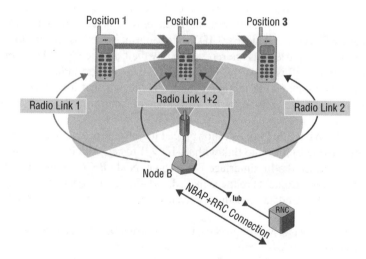

Figure 3.82   Iub Softer Handover overview.

Context and to the UE represented by its uplink scrambling code ScrCd. An optional indicator says that it is the first radio link set for this UE, and C-ID names the target cell that is addressed by RNC.

In RRC Connection Setup message the rL-Set-ID binds the RRC Connection to the radio link set. After RRC Connection Setup Complete, all RRC messages will be transmitted in DCCH as long as the UE is in CELL_DCH state.

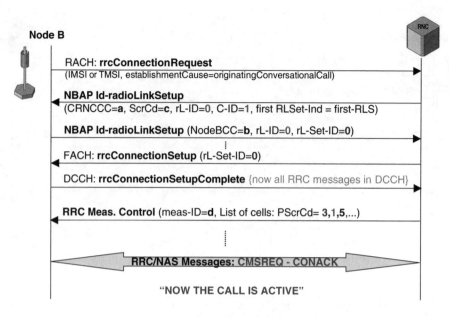

Figure 3.83 Iub Softer Handover call flow 1/4.

RRC Measurement Control Message is sent by RNC and used to request measurements by the UE. The message contains a measurement ID. Later reports referring to this Measurement. Control message will have the same meas-ID. In the same message a list of cells to be monitored by the UE is found. All these cells are represented by their unique PScrCd (see Figure 3.84).

Some time after the call is active, an RRC Measurement Report indicates that the primary CPICH of a new cell entered the reporting range, this means that it became strong enough. This new cell is one of the cells from the Measurement Control message. It is identified by PScrCd = 1.

RNC starts an NBAP Radio Link Addition procedure (instead of Radio Link Setup procedure in case of Soft Handover) for radio link with rL-ID = 1 to cell with c-ID=0, which is the cell with PScrCd = 1. Since the Iub interface is the same, Node B Communication Context and CRNC Communication Context are the same. Also rL-Set-ID is constant.

The RRC Active Set Update procedure activates the new radio link for the UE and adds it to the link set.

Measurement Control is sent to request the UE to inform RNC about latest updates, especially if events are triggered.

The next measurement report contains such an event report again. This time the primary CPICH of cell with primary scrambling code=3 is outside the reporting range; the beam is too weak. The RNC decides to clear all links running over this cell to this UE (Figure 3.85).

The cell with PScrCd=3 is taken out of the UE's active set.

The radio link (rL-ID=0) is deleted by NBAP procedure.

Obviously the mobile is moving back now, because with the third measurement report in this trace the old cell with PScrCd = 3 becomes available again, the radio link is added by NBAP and the active set of the UE updated by RRC.

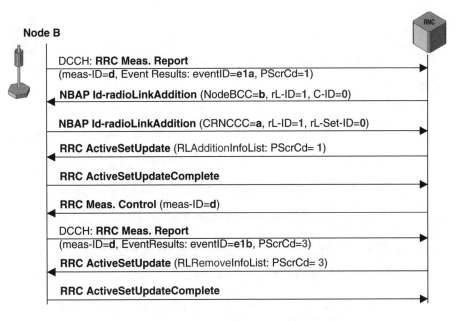

Figure 3.84 Iub Softer Handover call flow 2/4.

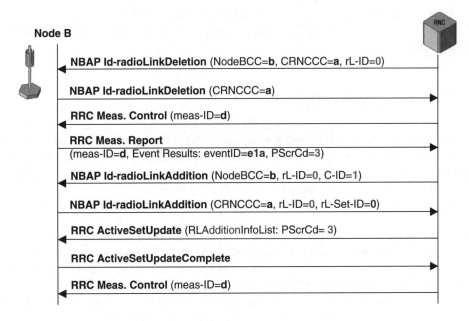

Figure 3.85 Iub Softer Handover call flow 3/4.

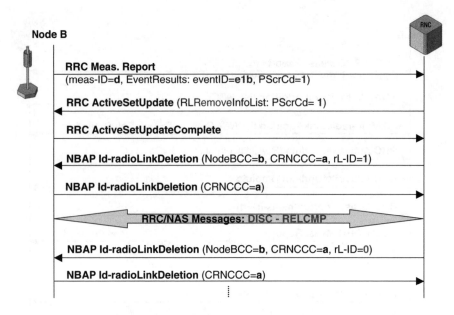

Figure 3.86   Iub Softer Handover call flow 4/4.

Finally the UE goes back into its very first position (before voice call was active) – so now the cell with PScrCd=1 goes out of range and the radio link is deleted again.

After release of the voice call the last active radio link (with rL-ID=0) is deleted as well (Figure 3.86).

3.12 Iub-Iu – Location Update

Now we will have a more detailed look at the signaling procedures on Iu interfaces. First we will have look at basic procedures on IuCS and IuPS before Inter-RNC handover and relocation via Iur interface will be explained. To understand Iu procedures it is also necessary to look back on what is running on Iub – as described in the call flow examples before. However, the focus will be on those Iub messages that trigger Iu activities.

The start is the already well-known Location Update (LUP) procedure (Figure 3.87).

Step 1: Set up the DCCH for the RRC connection on Iub interface.

Step 2: MM/CC/SM (Mobility Management/Call Control/Session Management) messages are transparently forwarded to the RNC on behalf of RRC direct transfer messages, in this case, Location Update Request (LUREQ) message.

Step 3: The reception of the LUREQ message on RNC triggers the setup of an SCCP/RANAP connection on IuCS interface toward MSC/VLR. The LUREQ is embedded in a RANAP Initial Message, which is also embedded in an SCCP Connection Request. The answer can be Location Update Accept (LUACC) or Location Update Reject (LUREJ).

Step 4: After sending the answer message the SCCP/RANAP connection on IuCS is released.

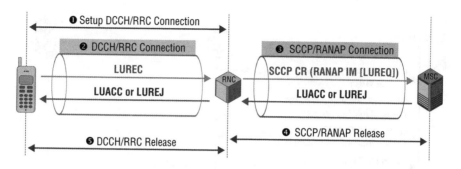

Figure 3.87 Iub-Iu Location Update procedure overview.

Step 5: Triggered by the release messages from the IuCS the RRC connection and its DCCH are also released.

3.12.1 Message Flow

Iu-LUP

First the DCCH on Iub interface is set up (Figure 3.88).

After RRC connection is established, MM/CC/SM messages can be exchanged embedded in RRC Direct Transfer messages. The mobile sends a LUREQ.

When RNC receives the NAS message, it starts setting up SCCP connection on IuCS interface on behalf of SCCP Connection Request message. This CR message includes a RANAP Initial_UE_Message that carries the embedded NAS message LUREQ. The Source Local

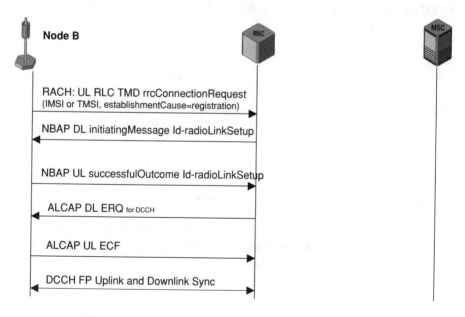

Figure 3.88 Iub-Iu Location Update procedure call flow 1/4.

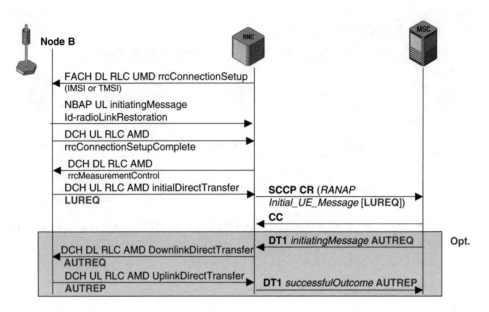

Figure 3.89 Iub-Iu Location Update procedure call flow 2/4.

Reference Number in the CR message identifies the calling party of this SCCP connection. It will be used as destination local reference number in all messages sent by the other side (called party) of the SCCP connection; in this case the other party is the MSC/VLR:

SCCP CR (source local reference=a, *RANAP Initial_UE_Message*, NAS message=**LUREQ**)

```
+---------+-----------------------------+-----------------------+
|BITMASK  |ID Name                      |Comment or Value       |
+---------+-----------------------------+-----------------------+
|  frm RNC  SSCOP  SD  RL  RL  id-InitialUE-Message              |
|  SCCP  CR  RANAP  initiating..  MM-DMTAP  LUREQ                |
|                                                               |
/\/\/\/\/\/\/\/\/\/\/\/\/\/\/\/\/\/\/\/\/\/\/\/\/\/\/\/\/\/\/\/\/\/\/\/\
|ITU-T Routing Label (RL)  RL (= Routing Label)                 |
|Routing Label                                                  |
|----0011 |Service Indicator            |SCCP                   |
|--00---- |Sub-Service: Priority        |Spare/priority 0 (U.S.A. only)|
|10------ |Sub-Service: Network Ind     |National message       |
|**b14*** |Destination Point Code       |1000                   |
|**b14*** |Originating Point Code       |2000                   |
|ITU-T WB SCCP (SCCP)  CR (= Connection Request)                |
|Connection Request                                             |
|0001---- |Signalling Link Selection|1                          |
```

```
|00000001 |SCCP Message Type         |1                                |
|***B3*** |Source Local Reference     |131073                          |
|----0010 |Protocol Class            |Class 2                          |
|0000---- |Spare                     |0                                |
|00000010 |Pointer to parameter      |2                                |
|00000110 |Pointer to parameter      |6                                |
|Called address parameter                                              |
|00000100 |Parameter Length          |4                                |
|-------1 |Point Code Indicator      |PC present                       |
|------1- |Subsystem No. Indicator   |SSN present                      |
|--0000-- |Global Title Indicator    |No global title included         |
|-1------ |Routing Indicator         |Route on DPC + Subsystem No.     |
|0------- |For national use          |0                                |
|**b14*** |Called Party SPC          |1000                             |
|00------ |Spare                     |0                                |
|10001110 |Subsystem number          |RANAP                            |
```
〰〰〰
```
|TS 25.413 V3.6.0 (2001-06) (RANAP)  initiatingMessage                 |
|(= initiatingMessage)                                                 |
|ranapPDU                                                              |
|1 initiatingMessage                                                   |
|00010011 |1.1 procedureCode         |id-InitialUE-Message             |
|01------ |1.2 criticality           |ignore                           |
|1.3 value                                                             |
|1.3.1 protocolIEs                                                     |
|1.3.1.1 sequence                                                      |
|***B2*** |1.3.1.1.1 id              |id-CN-DomainIndicator            |
|01------ |1.3.1.1.2 criticality     |ignore                           |
|0------- |1.3.1.1.3 value           |cs-domain                        |
|1.3.1.2 sequence                                                      |
|***B2*** |1.3.1.2.1 id              |id-LAI                           |
|01------ |1.3.1.2.2 criticality     |ignore                           |
|1.3.1.2.3 value                                                       |
|***B3*** |1.3.1.2.3.1 pLMNidentity  |92 f9 00                         |
|***B2*** |1.3.1.2.3.2 lAC           |00 01                            |
|1.3.1.3 sequence                                                      |
|***B2*** |1.3.1.3.1 id              |id-SAI                           |
|01------ |1.3.1.3.2 criticality     |ignore                           |
|1.3.1.3.3 value                                                       |
|***B3*** |1.3.1.3.3.1 pLMNidentity  |92 f9 00                         |
|***B2*** |1.3.1.3.3.2 lAC           |00 01                            |
|***B2*** |1.3.1.3.3.3 sAC           |00 02                            |
```
〰〰〰
```
|***B2*** |1.3.1.6.1 id              |id-GlobalRNC-ID                  |
|01------ |1.3.1.6.2 criticality     |ignore                           |
|1.3.1.6.3 value                                                       |
```

```
|***B3*** |1.3.1.6.3.1 pLMNidentity     |92 f9 00                  |
|***B2*** |1.3.1.6.3.2 rNC-ID           |99                        |
|TS 24.008 Mobility Management V3.8.0 (MM-DMTAP)                    |
|LUREQ (= Location updating req.)                                  |
|Location updating req.                                            |
|----0101 |Protocol Discriminator      |mobility management       |
|         |                            | messages                 |
|0000---- |Sub-protocol discriminator  |Skip Indicator            |
|--001000 |Message Type                |8                         |
|00------ |Send Sequence Number        |Message sent              |
|         |                            | from the Network         |
|------10 |LUT                         |IMSI attach               |
|-----0-- |Spare                       |0                         |
|----0--- |Follow-On Request           |No follow-on              |
|         |                            | request pending          |
|0000---- |Key sequence                |0                         |
|Location Area identification                                      |
|----1111 |MCC digit 1                 |15                        |
|1111---- |MCC digit 2                 |15                        |
|----1111 |MCC digit 3                 |15                        |
|1111---- |MNC digit 3                 |15                        |
|----1111 |MNC digit 1                 |15                        |
|1111---- |MNC digit 2                 |15                        |
|***B2*** |LAC                         |0                         |
|Mobile Station Classmark 1                                        |
~~~~~~~~~~~~~~~~~~~~~~~~~~~~~~~~~~~~~~~~~~~~~~~~~~~~~~~~~~~~~~~~~~~~~~~~~
|Mobile IDentity                                                   |
|00001000 |IE Length                   |8                         |
|-----001 |Type of identity            |IMSI                      |
|----1--- |Odd/Even Indicator          |Odd no of digits          |
|**b60*** |Identity digits             |299001800094051           |
|Mobile Station Classmark for UMTS                                 |
------------------------------------------------------------------
```

Message Example 3.34   *Location Update Request including RANAP and SCCP transport.*

LUREQ message example shows SCCP Connection Request and RANAP Initial Direct Tranfer messages as well. From the routing label it can be detected that SS#7 Signaling Point Code (SPC) of MSC is 1000 while SPC of sending RNC is 2000. SCCP Connection Request message shows the source local reference number and a called party number based on SPC addressing. Called party on SCCP level is once again the MSC (SPC = 1000). A short description of global title addressing can be found in core network signaling procedures part of this book (Chapter 4).

The RANAP Initial Direct Transfer message further includes domain identifier to indicate the core network domain to which the message is routed to. Then current location area information (LAI) and service area information (SAI) of the subscriber that sent LUREQ is part of the

RANAP message as well as global RNC identity that leaves no doubt as to which RNC in the world sent this message.

LUREQ message in the example shows location update type (LUT) = IMSI attach and because it is an IMSI attach, IMSI is included and (old) location area code as stored on USIM shows default values of digits. This is the default LAI found on every new USIM when it is bought for instance in a shop without having been used for attachment to a network before.

When the RNC receives the SCCP Connection Confirm message from MSC the SCCP connection is established successfully:

SCCP CC (source local reference=**b**, destination local reference=**a**)

```
+---------+-----------------------------------+-----------------------+
|BITMASK  |ID Name                            |Comment or Value       |
+---------+-----------------------------------+-----------------------+
|16:53:36,034,105  frm RNC  SSCOP  SD  RL  RL  SCCP  CC               |
|NNI SSCOP (SSCOP)  SD (= Seq. Conn.mode Data)                        |
/\/\/\/\/\/\/\/\/\/\/\/\/\/\/\/\/\/\/\/\/\/\/\/\/\/\/\/\/\/\/\/\/\/\/\/\
|ITU-T Routing Label (RL)  RL (= Routing Label)                       |
|Routing Label                                                        |
|----0011 |Service Indicator                  |SCCP                   |
|--00---- |Sub-Service: Priority              |Spare/priority 0       |
|         |                                   |(U.S.A. only)          |
|10------ |Sub-Service: Network Ind           |National message       |
|**b14*** |Destination Point Code             |2000                   |
|**b14*** |Originating Point Code             |1000                   |
|ITU-T WB SCCP (SCCP)  CC (= Connection Confirm)                      |
|Connection Confirm                                                   |
|0010---- |Signalling Link Selection          |2                      |
|00000010 |SCCP Message Type                  |2                      |
|***B3*** |Destination Local Ref.             |131073                 |
|***B3*** |Source Local Reference             |12390716               |
|----0010 |Protocol Class                     |Class 2                |
-----------------------------------------------------------------------
```

Message Example 3.35 *SCCP Connection Confirm of previous shown SCCP Connection Request.*

This SCCP CC example message shows routing label when message is sent from MSC to RNC as well as both SLR and DLR values

For exchange of user data, SCCP provides Data Format 1 (DT1) messages in case of an SCCP Class 2 connection like this. In these DT1 messages once again RANAP messages and NAS messages (MM/CC/SM) are embedded, but only destination local reference (DLR) is used on SCCP level to identify the receiver of the message:

SCCP DT1 (destination local reference=**a**, *RANAP initiatingMessage*, NASmessage=**AUTREQ**)

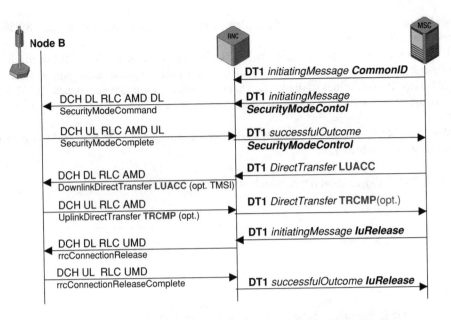

Figure 3.90 Iub-Iu Location Update procedure call flow 3/4.

SCCP DT1 (destination local reference=**b**, *RANAP successfulOutcome*,
NASmessage=**AUTREP**)

The Authentication procedure shown in this call flow example is optional.

With the RANAP Initiating Message that contains the Common ID procedure code the true
identity (IMSI) is sent to RNC, which needs this information and a frequent update of it to
ensure proper paging services for the user identified by this IMSI (see Figure 3.90).

SCCP DT1 (destination local reference=**a**, *RANAP initiatingMessage*, *Common ID*)

With the RANAP Security Mode Control procedure, ciphering and/or integrity protection
between RNC and UE are activated:

SCCP DT1 (destination local reference=**a**, *RANAP initiatingMessage*,
SecurityModeControl)
SCCP DT1 (destination local reference=**b**, *RANAP successfulOutcome*,
SecurityModeControl)

The LUACC message confirms the registration of the new UE location in VLR/HLR
databases. Optionally a new TMSI may be assigned by the VLR:

SCCP DT1 (destination local reference=**a**, *RANAP DirectTransfer*,
NASmessage=**LUACC** (opt. TMSI))

```
|TS 24.008 Mobility Management V3.8.0 (MM-DMTAP)                |
|LUACC (= Location updating accept)                            |
|Location updating accept                                      |
|----0101 |Protocol Discriminator      |mobility management   |
|         |                            | messages             |
|0000---- |Sub-protocol discriminator  |Skip Indicator        |
|--000010 |Message Type                |2                     |
|00------ |Send Sequence Number        |Message sent from the |
|         |                            | Network              |
|Location Area identification                                  |
|----0010 |MCC digit 1                 |2                     |
|1001---- |MCC digit 2                 |9                     |
|----9001 |MCC digit 3                 |9                     |
|1111---- |MNC digit 3                 |15                    |
|----0000 |MNC digit 1                 |0                     |
|0000---- |MNC digit 2                 |0                     |
|***B2*** |LAC                         |1                     |
|Mobile IDentity                                               |
|00010111 |IE Name                     |Mobile IDentity       |
|00000101 |IE Length                   |5                     |
|-----100 |Type of identity            |TMSI/P-TMSI           |
|----0--- |Odd/Even Indicator          |Even no of digits     |
|1111---- |Filler                      |15                    |
|***B4*** |MID TMSI                    |a1 14 3c b4           |
----------------------------------------------------------------
```

Message Example 3.36   Location Update Accept.

Location Update Accept message example shows the answer to the previous LUREQ example. LAI digits have now the same values as in RANAP Initial Direct Transfer and a TMSI is assigned.

In case of new TMSI assignment a TMSI Reallocation Complete (TRCMP) message is sent back by the UE:

> Opt. **SCCP DT1** (destination local reference=**b**, *RANAP DirectTransfer*, NASmessage=**TRCMP**)

Now the Location Update procedure is complete and the SCCP/RANAP connection can be released. The first IuRelease contains a release cause for the RANAP layer.

> **SCCP DT1** (destination local reference=**a**, *RANAP initiatingMessage*, **IuRelease** [Id Cause])
> **SCCP DT1** (destination local reference=**a**, *RANAP successfulOutcome*, **IuRelease**)

RANAP IuRelease message triggers the RRC Connection Release on Iub interface.

While on Iub the radio links are deleted the MSC sends an SCCP Released (RLSD) message that triggers the sending of ALCAP Release Request (for the DCCH) on Iub interface. The

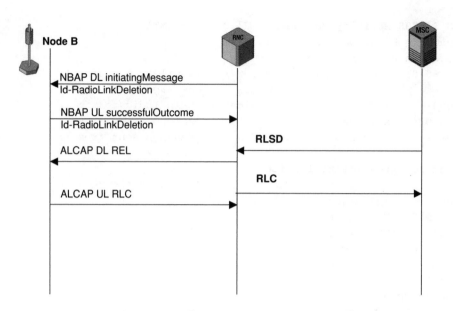

Figure 3.91   Iub-Iu Location Update procedure call flow 4/4.

SCCP Released message contains a release cause for the SCCP layer (Figure 3.91):

> **SCCP RLSD** (source local reference=**b**, destination local reference=**a**,
> Release Cause)
> **SCCP RLC** (source local reference=**a**, destination local reference=**b**)

3.13 Iub-Iu – Mobile-Originated Call

3.13.1 Overview (Figure 3.92)

Step 1: Set up the DCCH for the RRC connection on Iub interface.

Step 2: NAS messages of MM/CC/SM protocol are transparently forwarded to the RNC on behalf of RRC direct transfer messages.

Step 3: The reception of the first MM/CC/SM message at RNC triggers the setup of a SCCP/RANAP connection on IuCS interface toward MSC. The NAS message is embedded in an RANAP Initial Message, which is also embedded in an SCCP Connection Request. The further NAS messages for call setup and release are exchanged in the same SCCP/RANAP connection.

Step 4: After the call setup process reaches a defined state, when it is necessary to have traffic channel an RANAP RAB Assignment message is sent by the MSC. RAB includes the Iu Bearer and the Radio Bearer.

Step 5: Triggered by reception of RAB Assignment message, the RNC starts the setup of a DTCH, which follows the UMTS Bearer concept seen as the Radio Bearer.

Step 6: Iu User Plane protocol packets that contain voice information are exchanged between the MS and the MSC using the RAB as the traffic channel.

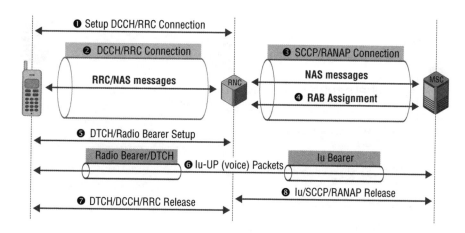

Figure 3.92   Iub-Iu mobile-originated voice call (MOC) procedure overview.

Step 7: After release of the call in MM/CC/SM layer (messages are exchanged via SCCP/RANAP), the Iu Bearer and the SCCP/RANAP connection on IuCS are released as well.

Step 8: Triggered by the release messages from the IuCS the RRC connection and its DCCH are also released.

3.13.2 Message Flow

Until the first NAS message is sent by the mobile, the already described procedures on the Iub interface can be monitored (Figure 3.93).

The first NAS message is Connection Management Service Request, which is sent by the MS to the network to request ciphering for the following connection management procedure (Figure 3.94). The message is forwarded by the RNC using an SCCP/RANAP connection toward the MSC. The SCCP Connection Request (CR) message includes both the first RANAP and the first NAS message. The RANAP Initial_UE_Message contains a domain indicator to enable the RNC to make a decision as to which core network domain (in this case, CS domain) the CMSREQ message should be forwarded. CMSREQ contains an MS identifier, either IMSI or TMSI (Figure 3.94).

SCCP CR (source local reference=**a**, *RANAP Initial_U E_Message*
(Id-CN-DomainIndicator), NASmessage=**CMSREQ** (IMSI or TMSI))

SCCP CC indicates that the SCCP connection was set up successfully. Source and destination local reference stand for called and calling party of the SCCP connection. Value "a" identifies the calling party (RNC), value "b" the called party (MSC):

SCCP CC (source local reference=**b**, destination local reference=**a**)

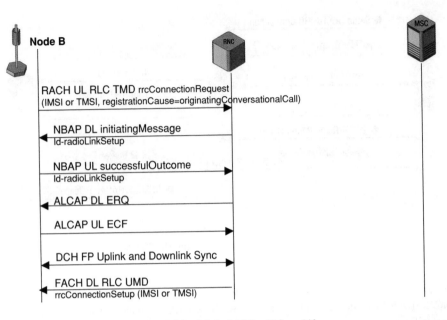

Figure 3.93   Iub-Iu MOC call flow 1/6.

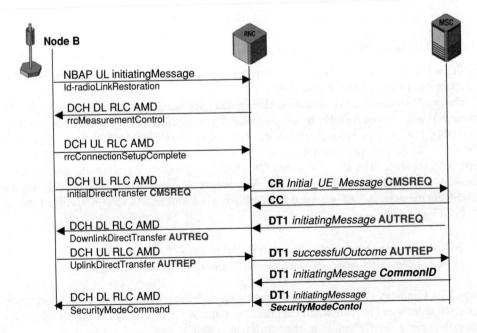

Figure 3.94   Iub-Iu MOC call flow 2/6.

Now the already described Authentication procedure follows (optional):

SCCP DT1 (destination local reference=**a**, *RANAP initiatingMessage*, NASmessage=**AUTREQ**)
SCCP DT1 (destination local reference=**b**, *RANAP successfulOutcome*, NASmessage=**AUTREP**)

As in the case of Location Update procedure the MSC sends the true MS identity to the RNC using an RANAP Initiating Message including a Common ID procedure code:

SCCP DT1 (destination local reference=**a**, *RANAP initiatingMessage*, **Common ID (IMSI)**)

Security Mode Control procedure of RANAP triggers activation of ciphering and/or integrity protection between RNC and MS (see Figure 3.95):

SCCP DT1 (destination local reference=**a**, *RANAP initiatingMessage*, *SecurityModeControl*)

SCCP DT1 (destination local reference=**b**, *RANAP successfulOutcome*, *SecurityModeControl*)

The SETUP message contains the Called Party Address, the number dialed by the user of the MS.

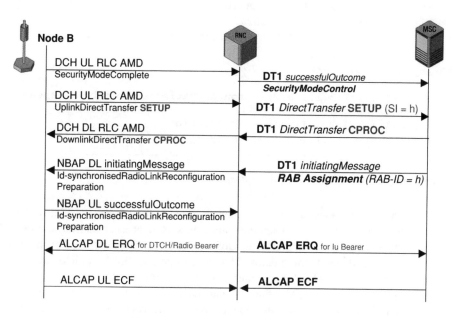

Figure 3.95 Iub-Iu MOC call flow 3/6.

SCCP DT1 (destination local reference=b, *RANAP DirectTransfer*,
NASmessage=**SETUP** (Called Party Address, Stream Identifier [SI]=**h**))

Call Proceeding message indicates that the call request was accepted and no more call establishment information for this call will be accepted by MSC. The call request is now forwarded on the E interface.

SCCP DT1 (destination local reference=a, *RANAP DirectTransfer*,
NASmessage=**CPROC**)

Now it is necessary to set up a traffic channel or – following the wording of UMTS bearer concept – a Radio Access Bearer. Sending an RANAP RAB Assignment message from the MSC to the RNC starts this procedure. The RAB ID is the identifier of the traffic channel for this connection. The RAB ID has the same value as the SI discussed in Iub MOC scenario.

SCCP DT1 (destination local reference=**a**, *RANAP initiatingMessage*, **RAB**
Assignment (bindingID=**c**, RAB ID=**h**))

```
|TS 25.413 V3.9.0 (2002-03) (RANAP)  initiatingMessage              |
|(= initiatingMessage)                                              |
|ranapPDU                                                           |
|1 initiatingMessage                                                |
|00000000 |1.1 procedureCode               |id-RAB-Assignment      |
/\/\/\/\/\/\/\/\/\/\/\/\/\/\/\/\/\/\/\/\/\/\/\/\/\/\/\/\/\/\/\/\/\/\/\/\/\/\/\/\/\/\/\/\/\/\/\/\/\/\/\
|***b8*** |1.3.1.1.3.1.1.3.1 rAB-ID        |1                      |
|1.3.1.1.3.1.1.3.2 rAB-Parameters                                  |
|00------ |1.3.1.1.3.1.1.3.2.1 trafficClass |conversational        |
|---00--- |1.3.1.1.3.1.1.3.2.2             |symmetric-             |
|rAB-AsymmetryIndicator                    bidirectional          |
/\/\/\/\/\/\/\/\/\/\/\/\/\/\/\/\/\/\/\/\/\/\/\/\/\/\/\/\/\/\/\/\/\/\/\/\/\/\/\/\/\/\/\/\/\/\/\/\/\/\/\
|--0----- |1.3.1.1.3.1.1.3.2.10 sourceStatisticsDescri.. |speech   |
/\/\/\/\/\/\/\/\/\/\/\/\/\/\/\/\/\/\/\/\/\/\/\/\/\/\/\/\/\/\/\/\/\/\/\/\/\/\/\/\/\/\/\/\/\/\/\/\/\/\/\
|1.3.1.1.3.1.1.3.4.2 iuTransportAssociation                        |
|***B4*** |1.3.1.1.3.1.1.3.4.2.1 bindingID  |68 c0 b0 00           |
-------------------------------------------------------------------
```

Message Example 3.37 *RANAP RAB Assignment.*

This RAB Assignment message example shows parameters mentioned in call flow diagram. Not shown, but important, are RAB subflow parameters. Here is especially the subflow SDU size defined that is necessary to guarantee the QoS of the call. RNC receives subflow parameters from core network and uses them to define TFSs transmitted in NBAP and RRC messages.

As already known from ALCAP procedures on Iub, the bindingID will be used as SUGR in the ALCAP ERQ message. AAL2 Path ID and AAL2 CID (channel ID) are used to select appropriate AAL2 parameters. It should be noted that it is always the RNC that sends ALCAP ERQ messages.

ALCAP ERQ (orig.Sign.Asso.ID=**d**, served user generated reference=**c**, AAL2
Path ID=**f**, AAL2_CID=**g**)
ALCAP ECF (orig.Sign.Asso.ID=**e**, dest.Sign.Asso.ID=**d**)

Now the Iu Bearer (IuCS traffic channel) on AAL2 Path=**f**' and AAL2_CID=**g** is available.
This Iu Bearer is used to exchange Iu User Protocol (IuUP) frames between RNC and MSC.

Before user data can be exchanged the IuUP connection must be initialized. This is done by
sending a Control Procedure Frame (PROCOD) with initialization procedure indicator and an
Initialization Acknowledge (ACK). The initialization frame (PROCOD) includes RAB sub-
Flow Combination Indicators (RFCIs) containing information about different attributes of this
peer-to-peer connection between IuUP instances, e.g. the size of exchanged IuUP frames in
the connection (Figure 3.96).

IuUP PROCOD (Type 14, Control procedure frame, Initialization: RFCI Formats)
IuUP ACK (Type 14, Positive Acknowledgement, Initialization)

The AMR voice frames will be transmitted later embedded in IuUP Type 0 messages, which
are sent constantly every 20 ms:

IuUP Type 0 (RFCI Number, Payload (AMR Speech every 20 ms))

After the IuUP initiation, the RAB Assignment can be completed:

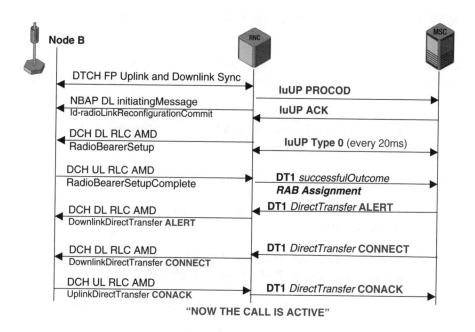

Figure 3.96   Iub-Iu MOC call flow 4/6.

SCCP DT1 (destination local reference=**b**, *RANAP successfulOutcome*, ***RAB Assignment*** (RAB ID=**h**))

ALERT message indicates that the call was successfully forwarded to the B-party; it can be seen as a ringing indicator.

SCCP DT1 (destination local reference=**a**, RANAP DirectTransfer, NASmessage=**ALERT**)

CONNECT is sent when the B-party accepts the call, e.g. picks up the handset.

SCCP DT1 (destination local reference=**a**, RANAP DirectTransfer, NASmessage=**CONNECT**)

With sending CONNECT ACKNOWLEDGE, A-party (MS) confirms the successful call setup.

SCCP DT1 (destination local reference=**b**, RANAP DirectTransfer, NASmessage=**CONACK**)

Now the call is active.

Call release starts when A- or B-party sends a DISCONNECT message including a cause, e.g. "normal call clearing".

SCCP DT1 (destination local reference=**b**, *RANAP DirectTransfer*, NASmessage=**DISC** (cause))

In some cases also a RELEASE message can be sent to stop the call (Figure 3.97). The RELEASE is used to set free all used transaction identifiers for this connection, e.g. the used SI.

SCCP DT1 (destination local reference=**a**, *RANAP DirectTransfer*, NASmessage=**RELEASE** (cause))

The peer entity sends RELEASE COMPLETE to confirm:

SCCP DT1 (destination local reference=**b**, *RANAP DirectTransfer*, NASmessage=**RELCMP**)

Now the release of the Iu bearer starts with RANAP procedure IuRelease. The Initiating Message contains a release cause for RANAP layer (Figure 3.98).

SCCP DT1 (destination local reference=**a**, *RANAP initiatingMessage*, ***IuRelease*** (Id cause))

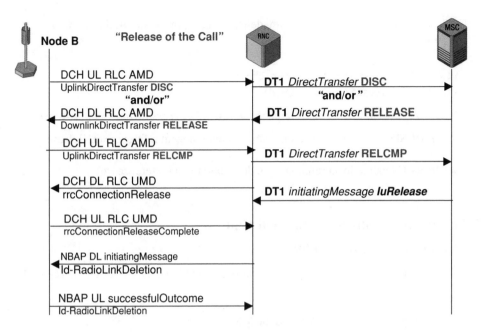

Figure 3.97   Iub-Iu MOC call flow 5/6.

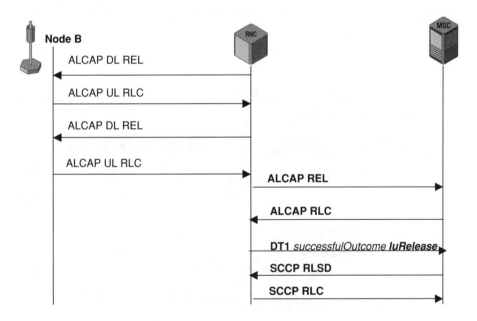

Figure 3.98   Iub-Iu MOC call flow 6/6.

Iu Bearer release is completed with deleting the appropriate AAL2 SVC and SCCP Connection:

ALCAP REL (dest.Sign.Asso.ID=**e**, cause)
ALCAP RLC (dest.Sign.Asso.ID=**d**)

SCCP DT1 (destination local reference=**a**, *RANAP successfulOutcome*, *Iu Release*)

SCCP RLSD (source local reference=**b**, destination local reference=**a**, Release Cause)
SCCP RLC (source local reference=**a**, destination local reference=**b**)

3.14 Iub-Iu – Mobile-Terminated Call

3.14.1 Overview (Figure 3.99)

The main difference between mobile-terminated call (MTC) and mobile-originated call (MOC) is the paging procedure.

Step 0: The paging is sent from MSC to the MS.

All other steps are identical to the steps discussed in the MOC scenario.

It should be noted that Paging Response is one of the NAS messages exchanged in DCCH/RRC Connection on Iub interface and later embedded in SCCP/RANAP messages on the IuCS interface.

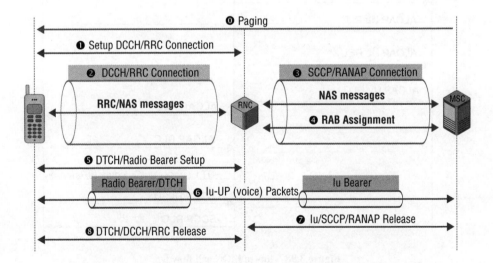

Figure 3.99 Iub-Iu mobile-terminated voice call (MTC) procedure overview.

3.14.2 Message Flow

Since most messages in the MTC scenario are identical or quite similar to the ones described in MOC scenario, only those messages will be explained in detail that make the main differences.

The call scenario starts with a Paging message that is sent in RANAP layer from MSC to RNC (Figure 3.100). The message contains a core network indicator and the subscriber identity (TMSI *and* IMSI). It should be noted that the IMSI is mandatory in this message because RNC needs to derive the paging indicator or paging group number used on PICH from IMSI.

Since there is no defined connection between MS and network in this phase of the call, the Paging is sent enclosed in an SCCP Unitdata (UDT) message. In case there is no paging response received within a defined time interval controlled by a counter, the Paging will be repeated.

> **SCCP UDT** (*RANAP Id-Paging* (Id-CN-Indicator, IMSI, TMSI)

The Paging Response (PRES) message triggers the setup of the SCCP/RANAP connection on IuCS interface. It contains the MS identity (Figure 3.101).

> **SCCP CR** (source local reference=**a**, *RANAP Initial_UE_Message*
> (Id-CN-DomainIndicator), NAS message=**PRES** (IMSI or TMSI))

Next messages are known from the MOC scenario:

> **SCCP CC** (source local reference=**b**, destination local reference=**a**)

> **SCCP DT1** (destination local reference=**a**, *RANAP initiatingMessage*,
> NASmessage=**AUTREQ**)

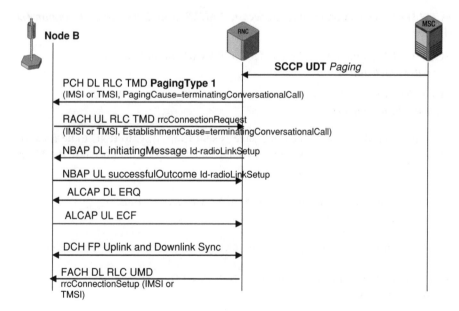

Figure 3.100   Iub-Iu MTC call flow 1/6.

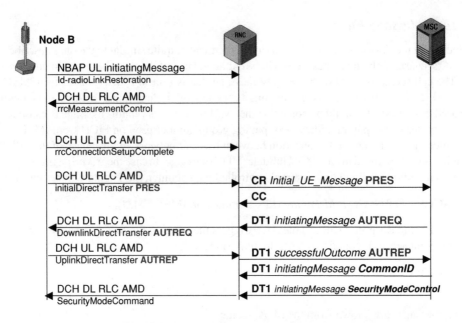

Figure 3.101   Iub-Iu MTC call flow 2/6.

SCCP DT1 (destination local reference=**b**, *RANAP successfulOutcome*,
NASmessage=**AUTREP**)

SCCP DT1 (destination local reference=**a**, *RANAP initiatingMessage, Common ID*
(IMSI))

SCCP DT1 (destination local reference=**a**, *RANAP initiatingMessage,
SecurityModeControl*)

SCCP DT1 (destination local reference=**b**, *RANAP successfulOutcome,
SecurityModeControl*)

Since this is an MTC, the SETUP message is sent from MSC to RNC to be forwarded to
the MS (Figure 3.102). It also does not contain a called party number because the MS has no
idea about the MSISDN related to its IMSI.

SCCP DT1 (destination local reference=**a**, *RANAP DirectTransfer*,
NASmessage=**SETUP**)

The MS confirms the incoming call request by sending CALL CONFIRMING message:

SCCP DT1 (destination local reference=**b**, *RANAP DirectTransfer*,
NASmessage=**CCONF**)

Setup of traffic channels (Radio Access Bearer), IuUP initialization, and NAS messages
until call is active are the same procedures as explained in MOC scenario (Figure 3.103):

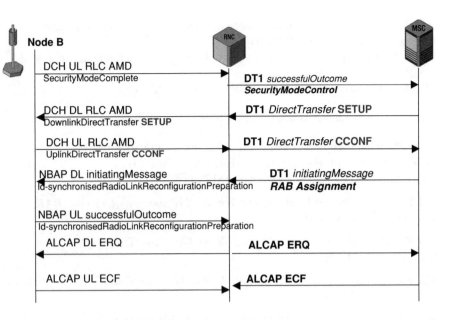

Figure 3.102 Iub-Iu MTC call flow 3/6.

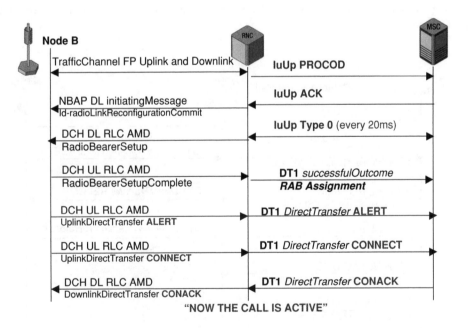

Figure 3.103 Iub-Iu MTC call flow 4/6.

SCCP DT1 (destination local reference=**a**, *RANAP initiatingMessage*, **RAB Assignment** (bindingID=**c**, RAB ID=**h**))
ALCAP ERQ (orig.Sign.Asso.ID=**d**, served user generated reference=**c**, AAL2 Path ID=**f**, AAL2_CID=**g**)
ALCAP ECF (orig.Sign.Asso.ID=**e**, dest.Sign.Asso.ID=**d**)

In the AAL2 Path=**f′** and AAL2_CID=**g** the traffic channel (IuUP) will start:

IuUP PROCOD (Type 14, Control procedure frame, Initiation of RFCIs)
IuUP ACK (Type 14, Positive Acknowledgment, Initiation)
IuUP Type 0 (RFCI Number, Payload (AMR Speech every 20 ms))

SCCP DT1 (destination local reference=**b**, *RANAP successfulOutcome*, **RAB Assignment** (RAB ID=**h**)

SCCP DT1 (destination local reference=**b**, *RANAP DirectTransfer*, NASmessage=**ALERT**)

SCCP DT1 (destination local reference=**b**, *RANAP DirectTransfer*, NASmessage=**CONNECT**)

SCCP DT1 (destination local reference=**a**, *RANAP DirectTransfer*, NASmessage=**CONACK**)

The procedures following DISCONNECT message are also the same as in the case of an MOC (Figure 3.104):

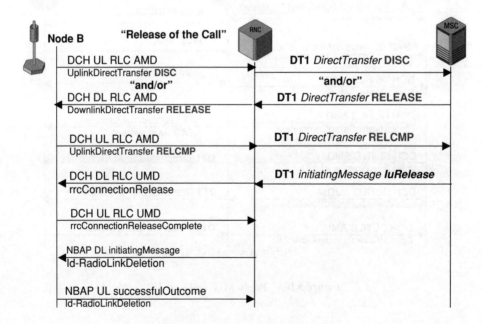

Figure 3.104   Iub-Iu MTC call flow 5/6.

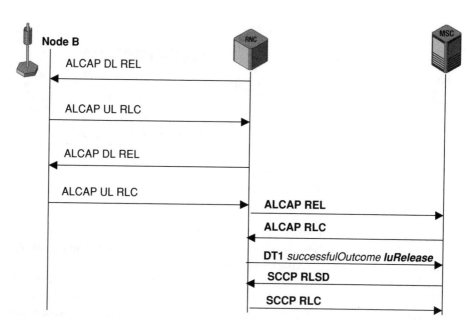

Figure 3.105 Iub-Iu MTC call flow 6/6.

SCCP DT1 (destination local reference=**b**, *RANAP DirectTransfer*, NASmessage=**DISC** (cause))

And/or

SCCP DT1 (destination local reference=**a**, *RANAP DirectTransfer*, NASmessage=**RELEASE** (cause))
SCCP DT1 (destination local reference=**b**, *RANAP DirectTransfer*, NASmessage=**RELCMP**)
SCCP DT1 (destination local reference=**a**, *RANAP initiatingMessage*, IuRelease (Id cause))

ALCAP REL (dest.Sign.Asso.ID=**e**, cause) (Figure 3.105)
ALCAP RLC (dest.Sign.Asso.ID=**d**)
SCCP DT1 (destination local reference=**a**, *RANAP successfulOutcome, **Iu Release***)
SCCP RLSD (source local reference=**b**, destination local reference=**a**, Release Cause)
SCCP RLC (source local reference=**a**, destination local reference=**b**)

3.15 Iub-Iu – Attach

3.15.1 Overview

The Attach procedure on IuPS interface (Figure 3.106) is quite similar to the Location Update procedure on IuCS (see Figure 3.87). Only NAS messages and the SGSN as core network element are different.

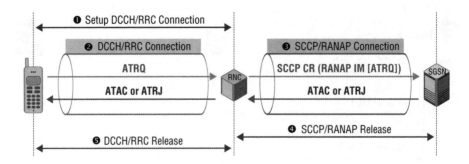

Figure 3.106 Iub-Iu GPRS Attach procedure overview.

Step 1: Set up the DCCH for the RRC connection on Iub interface.

Step 2: MM/CC/SM messages are transparently forwarded to the RNC on behalf of RRC direct transfer messages, in this case Attach (ATRQ) message.

Step 3: The reception of the ATRQ message triggers the setup of an SCCP/RANAP connection on IuPS interface toward the SGSN. The ATRQ is embedded in an RANAP Initial Message, which is also embedded in an SCCP Connection Request. The answer can be Attach Accept (ATAC) or Attach Reject (ATRJ).

Step 4: After sending the answer message the SCCP/RANAP connection on IuPS is released.

Step 5: Triggered by the release messages from the IuCS the RRC connection and its DCCH are also released.

3.15.2 Message Flow

Since all messages in this call flow have already been discussed, either in Iub-Iu LUP scenario or in Iub IMSI/GPRS Attach scenario, only the messages and their parameters will be listed in the following paragraphs (see Figures 3.107 to 3.109).

SCCP CR (source local reference=**a**, *RANAP Initial_UE_Message* (Id-CN-DomainIndicator=PS-Domain), NASmessage=**ATRQ** (IMSI or TMSI))

SCCP CC (source local reference=**b**, destination local reference=**a**)

SCCP DT1 (destination local reference=**a**, *RANAP initiatingMessage*, *SecurityModeControl*)

SCCP DT1 (destination local reference=**b**, *RANAP successfulOutcome*, *SecurityModeControl*)

SCCP DT1 (destination local reference=**a**, *RANAP DirectTransfer*, NASmessage=**ATAC (opt.: new P-TMSI)**)

Optional (if new P-TMSI in ATAC): **SCCP DT1** (destination local reference=**b**, *RANAP DirectTransfer*, NASmessage=**ACOM**)

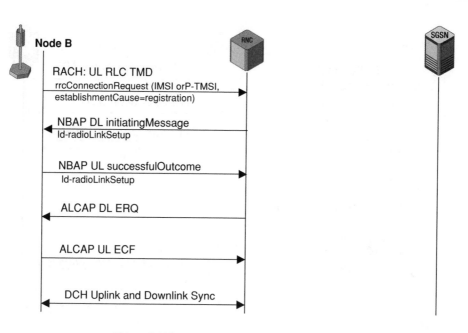

Figure 3.107   Iub-Iu GPRS Attach call flow 1/3.

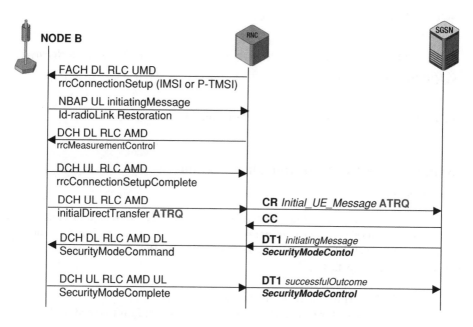

Figure 3.108   Iub-Iu GPRS Attach call flow 2/3.

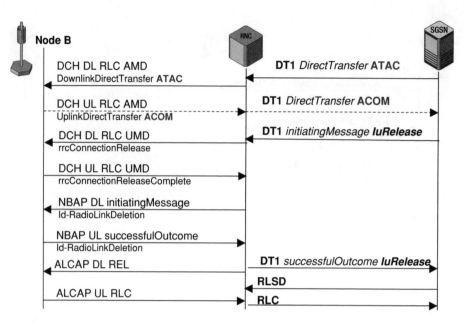

Figure 3.109   Iub-Iu GPRS Attach call flow 3/3.

SCCP DT1 (destination local reference=**a**, *RANAP initiatingMessage*, **IuRelease** (Id cause))

SCCP DT1 (destination local reference=**a**, *RANAP successfulOutcome*, **Iu Release**)

SCCP RLSD (source local reference=**b**, destination local reference=**a**, Release Cause)

SCCP RLC (source local reference=**a**, destination local reference=**b**)

3.16 Iub-Iu – PDPC Activation/Deactivation

3.16.1 Overview

The main difference between a mobile-originated voice call and a mobile-originated PDPC activation are the SGSN as peer core network element to the RNC and that the Iu Bearer is represented by a GTP Tunnel (GTP = GPRS Tunneling Protocol) running on an AAL5 connection (see Figure 3.110).

Step 1: Set up the DCCH for the RRC connection on Iub interface.

Step 2: NAS messages of MM/CC/SM protocol are transparently forwarded to the RNC on behalf of RRC direct transfer messages.

Step 3: The reception of the first MM/CC/SM message at RNC triggers the setup of an SCCP/RANAP connection on IuPS interface toward SCGS. The NAS message is embedded in an RANAP Initial Message, which is also embedded in an SCCP Connection Request.

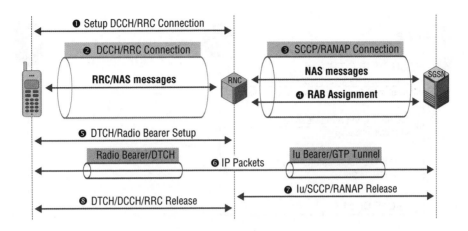

Figure 3.110 Iub-Iu PDPC activation/deactivation procedure overview.

The further NAS messages for PDPC activation and deactivation are exchanged in the same SCCP/RANAP connection.

Step 4: After PDPC activation process reaches a defined state, when it is necessary to have a traffic channel, RANAP RAB Assignment message is sent by the SGSN. RAB includes the Iu Bearer and the Radio Bearer. The Iu bearer will be realized on opening a GTP tunnel on an AAL5 connection.

Step 5: Triggered by reception of RAB Assignment message, RNC starts the setup of a DTCH, which follows the UMTS Bearer concept seen as the Radio Bearer.

Step 6: IP (Internet Protocol) packets are exchanged between the MS and the MSC using the RAB as the traffic channel.

Step 7: After release of the call in MM/CC/SM layer (messages are exchanged via SCCP/RANAP), the Iu Bearer and the SCCP/RANAP connection on IuPS are released as well. The deletion of the GTP tunnel is realized by another RAB Assignment procedure to ensure that other PDPC of the same MS will not be affected.

Step 8: Triggered by the release messages from the IuPS the RRC connection and its DCCH are also released.

3.16.2 Message Flow

Since most messages have been already discussed in other scenarios before, only those that are unique for PDPC activation/deactivation will be explained in detail. For description of NAS messages like PDPC Activation Request, see Iub PDPC activation/deactivation scenario (see Figures 3.111 and 3.113).

SCCP CR (source local reference=a, *RANAP Initial_UE_Message*
(Id-CN-DomainIndicator), NASmessage=**SREQ** (IMSI or TMSI, List of available
NSAPIs))

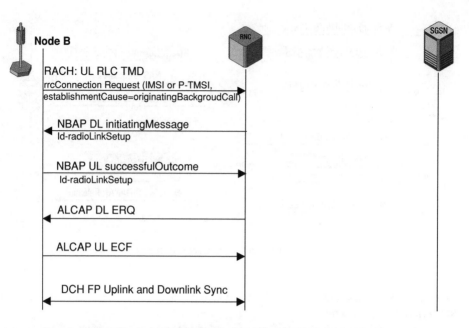

Figure 3.111   Iub-Iu PDPC activation/deactivation call flow 1/6.

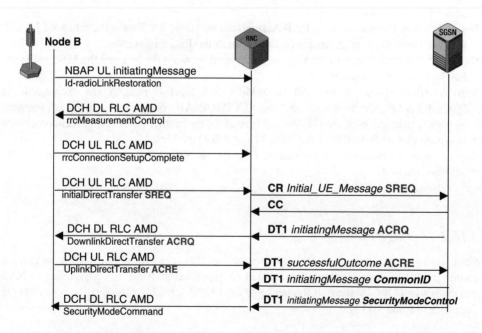

Figure 3.112   Iub-Iu PDPC activation/deactivation call flow 2/6.

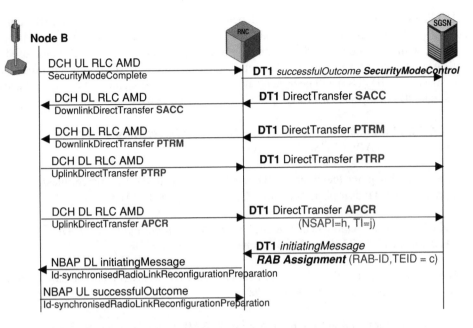

Figure 3.113   Iub-Iu PDPC activation/deactivation call flow 2/6.

SCCP CC (source local reference=**b**, destination local reference=**a**)

SCCP DT1 (destination local reference=**a**, *RANAP initiatingMessage*, NASmessage=**ACRQ**)

SCCP DT1 (destination local reference=**b**, *RANAP successfulOutcome*, NASmessage=**ACRE**)

SCCP DT1 (destination local reference=**a**, *RANAP initiatingMessage*, **Common ID** (IMSI))

SCCP DT1 (destination local reference=**a**, *RANAP initiatingMessage*, **SecurityModeControl**)

SCCP DT1 (destination local reference=b, *RANAP successfulOutcome*, **SecurityModeControl**)

SCCP DT1 (destination local reference=**a**, *RANAP DirectTransfer*, NASmessage=**SACC**)

SCCP DT1 (destination local reference=**a**, *RANAP DirectTransfer*, NASmessage=**PTRM**)

SCCP DT1 (destination local reference=**b**, *RANAP DirectTransfer*, NASmessage=**PTRP**)

The PDPC Activation procedure is used to activate the first PDPC for a given PDP address and APN, whereas all additional contexts associated to the same PDP address and APN are activated with the secondary PDP context activation procedure. The NSAPI value used in Activate PDPC Request message will be used as RAB-ID value in RAB Assignment procedure. The Transaction ID (TIO) will be used by all further session management messages related to this single PDPC, especially for PDPC modification and deactivation:

SCCP DT1 (destination local reference=**b**, *RANAP DirectTransfer*, NASmessage=**APCR** (NSAPI=**h**, TIO=**j**))

The Tunnel Endpoint Identifier (TEID) used in the first RAB Assignment message is the identifier of the GTP user plane entity on SGSN side (Figure 3.114).

SCCP DT1 (destination local reference=**a**, *RANAP initiatingMessage*, **RAB Assignment** (GTP-TEID=**c**, RAB ID=**h**))

The TEID used in the RANAP Successful Outcome RAB Assignment message identifies the GTP user plane entity on the RNC side:

SCCP DT1 (destination local reference=**b**, *RANAP successfulOutcome*, **RAB Assignment** (GTP-TEID=**d**, RAB ID=**h**)

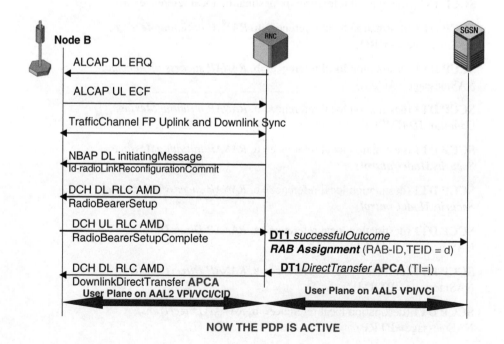

Figure 3.114 Iub-Iu PDP Context Activation/Deactivation call flow 4/6.

SCCP DT1 (destination local reference=**a**, *RANAP DirectTransfer*, NASmessage=**APCA** (TI=**j**))

Now the GTP tunnel is set up between the two endpoints (each named by its own TEID) and IP packets will be exchanged using an AAL5 connection with unique VPI/VCI parameter combination.

The release of the PDPC is started when the user sends Deactivate PDPC Request (DPCR) message (Figure 3.115). This message optionally includes a Tear Down Indicator that indicates wheteher only this single PDPC (identified by its Transaction ID [TIO]) shall be deleted or all PDPCs sharing the same PDP address with this single PDPC.

SCCP DT1 (destination local reference=**b**, *RANAP DirectTransfer*, NASmessage=**DPCR** (TIO=**j**, SM Cause, opt.: tear-down indicator))

SCCP DT1 (destination local reference=**a**, *RANAP DirectTransfer*, NASmessage=**DPCA** (TIO=**j**))

After the user requested deactivation of the PDPC and the network accepted this request the assigned RAB must be deleted as well (Figure 3.116). Since several PDPCs may be active for the same user, an RAB Release list is sent to specify which RABs shall be deleted and which not (or all).

SCCP DT1 (destination local reference=**a**, *RANAP initiatingMessage*, **RAB Assignment** (RAB Release List: RAB ID=**h**, cause))

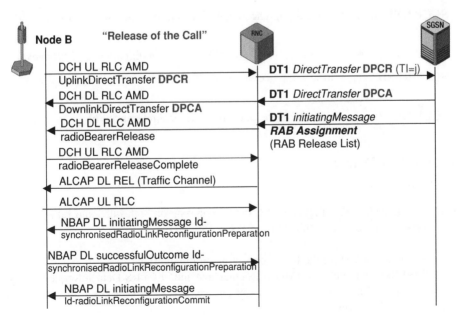

Figure 3.115   Iub-Iu PDPC activation/deactivation call flow 5/6.

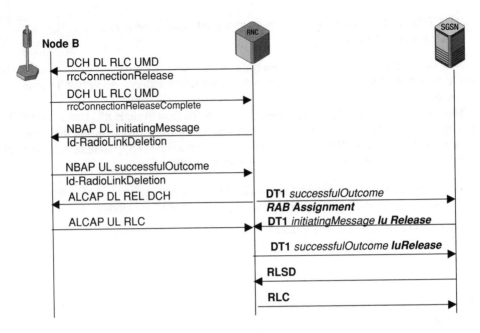

Figure 3.116 Iub-Iu PDPC activation/deactivation call flow 6/6.

SCCP DT1 (destination local reference=**b**, *RANAP successfulOutcome*, **RAB Assignment** (RAB ID=**h**)

SCCP DT1 (destination local reference=**a**, *RANAP initiatingMessage*, IuRelease (Id cause))

SCCP DT1 (destination local reference=**a**, *RANAP successfulOutcome*, **Iu Release**)

SCCP RLSD (source local reference=**b**, destination local reference=**a**, Release Cause)

SCCP RLC (source local reference=**a**, destination local reference=**b**)

3.17 Iub-Iu – Detach

3.17.1 Overview

The Detach procedure (see Figure 3.117) may be running on both IuCS and IuPS or just on IuPS interface as described in Iub IMSI/GPRS Detach scenario.

Step 1: Set up the DCCH for the RRC connection on Iub interface.
Step 2: IMSI Detach Indication (IMDETIN) message and/or GPRS Detach Request (DTRQ) is/are sent from MS to RNC. For different options in case of DTRQ see Iub IMSI/GPRS Detach scenario overview.

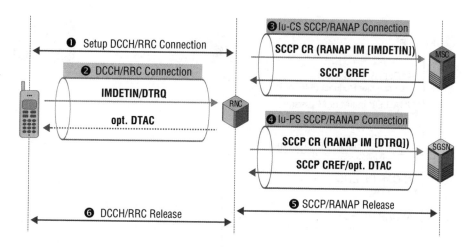

Figure 3.117   Iub-Iu IMSI/GPRS Detach procedure overview.

Step 3: IMDETIN is forwarded in an SCCP CR message to MSC/VLR. Since there is no response to this message, the SCCP connection request is refused by MSC to save resources for signaling traffic on IuCS interface.

Step 4: GPRS Detach Request is forwarded to the SGSN/SLR. In case the MS is switched off, the SCCP Connection Request is also rejected. Otherwise a Detach Accept (DTAC) message could be sent back to the MS.

Step 5: As far as IuPS SCCP connection was not refused, it is released now including the RANAP connection.

Step 5: Triggered by the release/refuse messages from the IuCS/IuPS interface the RRC connection and its DCCH are also released.

3.17.2 Message Flow

Compare the following call flow diagrams (Figures 3.118 to 3.120) with the descriptions of steps in the overview to get some more details. There are no comments because all SCCP and RANAP messages have already been described in previous scenarios.

>**SCCP CR** (source local reference=**a**, *RANAP Initial_UE_Message* (Id-CN-DomainIndicator), NASmessage=**IMDETIN** (IMSI or TMSI))
>
>**SCCP CREF** (destination local reference=**a**, RefusalCause)
>
>**SCCP CR** (source local reference=**a**, *RANAP Initial_UE_Message* (Id-CN-DomainIndicator), NASmessage=**DTRQ** (IMSI or P-TMSI))
>
>**SCCP CREF** (destination local reference=**a**, RefusalCause)

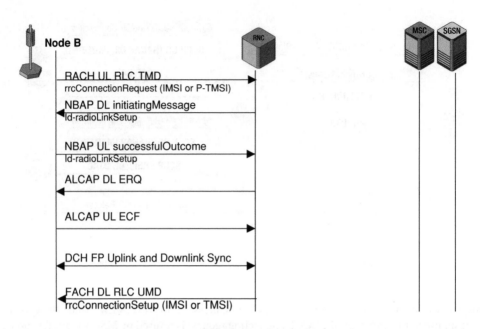

Figure 3.118   Iub-Iu IMSI/GPRS Detach call flow 1/3.

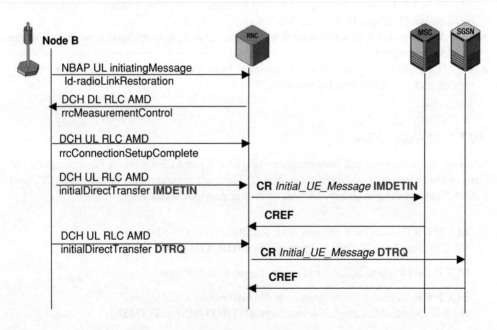

Figure 3.119   Iub-Iu IMSI/GPRS Detach call flow 2/3.

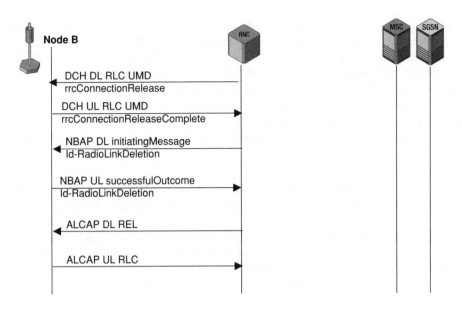

Figure 3.120   Iub-Iu IMSI/GPRS Detach call flow 3/3.

3.18 Iub-Iur – Soft Handover (Inter-Node B, Inter-RNC)

3.18.1 Overview

In comparison to the Inter-Node B/Intra-RNC Soft Handover procedure, this Inter-Node B/Inter-RNC handover introduces a new network element, the Drift RNC (DRNC), and a new interface, the Iur – used for interconnection of RNCs (see Figure 3.121).

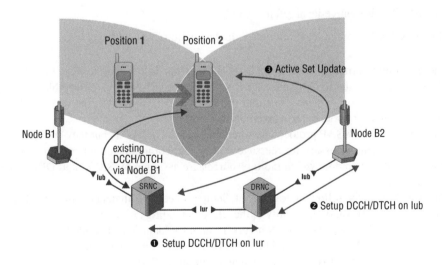

Figure 3.121   Iub-Iur Soft Handover procedure overview.

In all scenarios discussed before the RNC was always the serving RNC (SRNC): the RNC that controls the connections to the core network domains, to MSC and SGSN.

If the MS moves from Position 1 to Position 2 into the range of a cell that is actually controlled by a different RNC than the SRNC of the connection, the signaling and traffic channels from Node B2 must be set up not only on the Iub interface, but also on the Iur between SRNC and DRNC. The procedure is controlled by the SRNC and triggered by incoming measurement report. The old DTCH/DCCH via Node B1 is out of scope of the signaling example call flow and may stay active or be deleted depending on further moves of the UE.

Step 1: After successful radio link setup (also triggered by SRNC), the DCCH and DTCH on Iur interface are installed on behalf of RNSAP (Radio Network Subsystem Application Part) and ALCAP messages.

Step 2: The same DCCH and DTCH are set up on Iub interface between DRNC and Node B2. During this procedure, DRNC assigns downlink resources for dedicated physical channel in new cell and reports the assigned values back to SRNC.

Step 3: An Active Set Update procedure between SRNC and UE takes the new established links into service.

Note: If there is no Iur interface available or if the new RNC is in a different UTRAN (connected to a different MSC/SGSN pair), Soft Handover is not possible anymore. In such cases an Inter-3G_MSC as described in Chapter 4 will be performed.

3.18.2 Message Flow

The procedure starts with the setup of an SCCP connection on Iur interface (Figure 3.122):

Iur: **SCCP CR** (source local reference=**a**)

Iur: **SCCP CC** (source local reference=**b**, destination local reference=**a**)

The signaling information field of the first DT1 message is too small to carry the whole first RNSAP message. So the RNSAP message is segmented on SRNC and reassembled on DRNC side. The more data bit = 1 indicates that the RNSAP data content in the message is just a fragment and more fragments for reassembly will follow.

The messages in RNSAP protocol are often the same as in NBAP, but they contain specific information element necessary for message routing and communication between two RNCs. Furthermore, it will emerge which radio resources are assigned by SRNC and DRNC during the procedure.

The new thing in the following Radio Link Setup procedure is that it contains the identity of the sender (SRNC-ID) and the identity of the UE for signaling exchange over common control channels (S-RNTI = Serving Radio Network Temporary Identifier). On Iub during Call Setup procedure (see Iub MOC and Iub MTC scenarios), both elements together represent the U-RNTI. However, U-RNTI is not mentioned anymore in RNSAP.

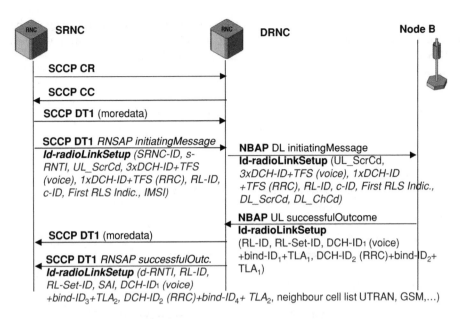

Figure 3.122   Iub-Iur Soft Handover call flow 1/3.

Included in the message is the uplink scrambling code of the dedicated physical channel proviced by SRNC during call setup scenario. It is the only physical radio identifier that will remain unchanged during the whole procedure.

The next section contains information about dedicated (transport) channels (DCHs) to be set up on both Iur and new Iub interfaces. It depends on active services of the UE about how many DCHs are necessary and how their TFSs have to be defined. Our next message example shows Radio link setup for an active voice call with one DCH for RRC signaling and three DCHs for AMR speech. However, it would be also possible that just one DCH is necessary (plain RRC signaling connection) or two DCHs have to be set up: one DCCH for RRC and one DTCH for IP payload.

Depending on included amount of information it is necessary in our call flow example (not always) to segment RNSAP Radio Link Setup Request message and transport it in multiple SCCP DT1 messages. If the message is relatively short it may be included already in SCCP Connection Request (CR).

Iur: **SCCP DT1** (destination local reference=**b**, more data=1)
Iur: **SCCP DT1** (destination local reference=**b**, more data=0 (*RNSAP
initiatingMessage* procedureCode=**id-radioLinkSetup** (longTID=**c**, SRNC-ID=**d**,
S-RNTI=**e**,
ULscramblingCode=**f**, DCH-ID=**g** + UL/DL Transport Format Set,
DCH-ID=**g′** + UL/DL
Transport Format Set, DCH-ID=**g″** + UL/DL Transport Format Set,
DCH-ID=**h** + UL/DL
Transport Format Set, rL-ID = **p**, c-ID= **mm**, IMSI = **nn**)))

The DCHs for AMR form a group of coordinated DCHs. In other words, they are bound together. That is the reason for using variables **g**, **g′** and **g″**. The following message example for RNSAP Radio Link Setup will make this fact better understandable:

```
|UMTS RNSAP acc. R99 TS 25.423 ver. 3.7.0 (RNSAP370)                |
|initiatingMessage (= initiatingMessage)                            |
|rnsapPDU                                                           |
|1 initiatingMessage                                               |
|1.1 procedureID                                                   |
|00010011 |1.1.1 procedureCode        |id-radioLinkSetup          |
|-01----- |1.1.2 ddMode               |fdd                        |
|---00--- |1.2 criticality            |reject                     |
|1.3 transactionID                                                 |
|***B2*** |1.3.1 longTransActionId    |25446                      |
|1.4 value                                                         |
|1.4.1 protocolIEs                                                 |
|1.4.1.1 sequence                                                  |
|***B2*** |1.4.1.1.1 id               |id-SRNC-ID                 |
|00------ |1.4.1.1.2 criticality      |reject                     |
|***B2*** |1.4.1.1.3 value            |742                        |
|1.4.1.2 sequence                                                  |
|***B2*** |1.4.1.2.1 id               |id-S-RNTI                  |
|00------ |1.4.1.2.2 criticality      |reject                     |
|***B2*** |1.4.1.2.3 value            |50504                      |
|1.4.1.3 sequence                                                  |
|***B2*** |1.4.1.3.1 id               |id-UL-DPCH-Information-RL-  |
|                                      SetupRq...                  |
|00------ |1.4.1.3.2 criticality      |reject                     |
|1.4.1.3.3 value                                                   |
|1.4.1.3.3.1 ul-ScramblingCode                                     |
|***B3*** |1.4.1.3.3.1.1                                           |
|ul-ScramblingCodeNumber               |1038948                    |
|1------- |1.4.1.3.3.1.2                                           |
|ul-ScramblingCodeLength               |long                       |
|-100---- |1.4.1.3.3.2                                             |
|minUL-ChannelizationCodeLength        |v64                        |
|00001000 |1.4.1.5.3.1.5.1.1 dCH-ID   |8                          |
|-00----- |1.4.1.5.3.1.5.1.2                                       |
|trCH-SrcStatisticsDescr               |speech                     |
|1.4.1.5.3.1.5.1.3 ul-transportFormatSet                           |
~~~~~~~~~~~~~~~~~~~~~~~~~~~~~~~~~~~~~~~~~~~~~~~~~~~~~~~~~~~~~~~~~~~~~~~~
|1.4.1.5.3.1.5.1.4 dl-transportFormatSet                           |
~~~~~~~~~~~~~~~~~~~~~~~~~~~~~~~~~~~~~~~~~~~~~~~~~~~~~~~~~~~~~~~~~~~~~~~~
|00001001 |1.4.1.5.3.1.5.2.1 dCH-ID   |9                          |
|-00----- |1.4.1.5.3.1.5.2.2                                       |
```

```
|trCH-SrcStatisticsDescr                                    |speech            |
|1.4.1.5.3.1.5.2.3 ul-transportFormatSet                                       |
~~~~~~~~~~~~~~~~~~~~~~~~~~~~~~~~~~~~~~~~~~~~~~~~~~~~~~~~~~~~~~~~~~~~~~~~~~~~~~~~~~~
|1.4.1.5.3.1.5.2.4 dl-transportFormatSet                                       |
|00001010 |1.4.1.5.3.1.5.3.1 dCH-ID                         |10                |
|-00----- |1.4.1.5.3.1.5.3.2 trCH-SrcStatisticsDescr        |speech            |
|1.4.1.5.3.1.5.3.3 ul-transportFormatSet                                       |
~~~~~~~~~~~~~~~~~~~~~~~~~~~~~~~~~~~~~~~~~~~~~~~~~~~~~~~~~~~~~~~~~~~~~~~~~~~~~~~~~~~
|1.4.1.5.3.1.5.3.4 dl-transportFormatSet                                       |
~~~~~~~~~~~~~~~~~~~~~~~~~~~~~~~~~~~~~~~~~~~~~~~~~~~~~~~~~~~~~~~~~~~~~~~~~~~~~~~~~~~
|00011111 |1.4.1.5.3.2.5.1.1 dCH-ID                         |31                |
|-01----- |1.4.1.5.3.2.5.1.2 trCH-SrcStatisticsDescr        |rRC               |
|1.4.1.5.3.2.5.1.3 ul-transportFormatSet                                       |
~~~~~~~~~~~~~~~~~~~~~~~~~~~~~~~~~~~~~~~~~~~~~~~~~~~~~~~~~~~~~~~~~~~~~~~~~~~~~~~~~~~
|1.4.1.5.3.2.5.1.4 dl-transportFormatSet                                       |
~~~~~~~~~~~~~~~~~~~~~~~~~~~~~~~~~~~~~~~~~~~~~~~~~~~~~~~~~~~~~~~~~~~~~~~~~~~~~~~~~~~
|00000--- |1.4.1.6.3.1.3.1 rL-ID                            |0                 |
|***B2*** |1.4.1.6.3.1.3.2 c-ID                             |45429             |
|1------- |1.4.1.6.3.1.3.3 firstRLS-indicator               |not-first-RLS|
~~~~~~~~~~~~~~~~~~~~~~~~~~~~~~~~~~~~~~~~~~~~~~~~~~~~~~~~~~~~~~~~~~~~~~~~~~~~~~~~~~~
|***B2*** |1.4.2.1.1 id                      |id-Permanent-NAS-UE-Identity|
|01------ |1.4.2.1.2 criticality                            |ignore            |
|1.4.2.1.3 extensionValue                                                      |
|***B8*** |1.4.2.1.3.1 imsi                  |92 09 90 01 91 86 75 f6|
```

Message Example 3.38   RNSAP Initiating Message Radio Link Setup.

In this message example we see procedure code, long transaction ID (same value in RNSAP Successful Outcome Radio Link Setup message example), and SRNC-ID plus S-RNTI. Four dedicated transport channels shall be setup: channels 8, 9, and 10 for speech, and channel 31 for carrying RRC signaling. The c-ID of the new cell is 45429 (the SRNC is able to translate the primary scrambling code of the cell received in RRC measurement report into a valid c-ID on behalf of a configuration table installed by O&M operation).

The NBAP Radio Link Setup Request message contains many parameters received from SRNC. However, especially the codes for the downlink dedicated physical resources on radio interface are assigned by CRNC, which is the RNC that controls the cell directly and communicates with the cell using NBAP protocol. In case of radio link setup on new Iub interface for Inter-RNC handover, CRNC is the same physical network element as DRNC. Since CRNC of Iub 2 is independent from SRNC, the downlink channelization code assigned for the new cell with NBAP Radio Link Setup procedure can be completely different from the one used in the old cell. However, the TFSs are the same on both links, because these settings are controlled by SRNC.

Iub: **NBAP DL** initiatingMessage, procedureCode=**id-radioLinkSetup** (longTID=**i**, id-CRNC-CommunicationContextID=**j**, ULscramblingCode/DLChannelizationCode =**f**/δ, DCH-ID=**g** + UL/DL Transport Format Set, DCH-ID=**g**′ + UL/DL Transport Format Set, DCH-ID=**g**″ + UL/DL Transport Format Set, DCH-ID=**h** +, UL/DL Transport Format Set, rL-ID = **p**, c-ID= **mm**)

NBAP Successful Outcome message of Radio Link Setup procedure contains two binding IDs, one for each DCH (one to carry DCCHs, the other one to carry DTCHs). Because of the coordinated set of DCHs for AMR speech (see explanation in Iub MOC scenario), only DCH-ID=**g** is included in the message to indicate which bindingID is related to speech channels. The other bindingID is related to RRC signaling channel.

> Iub: **NBAP** UL successfulOutcome, procedureCode=**id-radioLinkSetup**
> (longTID=**i**, id-CRNC-CommunicationContextID=**j**,
> id-NodeB-CommunicationContextID=**k**, DCH-ID=**g**, bindingID=**m**, DCH-ID=**h**,
> bindingID=**n**, rL-ID = **p**, rL-Set-ID=**q**)

Also on Iur a Successful Outcome message for the Radio Link Setup procedure is monitored. This is the appropriate RNSAP message forwarded to the SRNC. Owing to a long appendix containing neighboring cell information, this message could be one of the largest to be monitored on UTRAN interfaces. So it is segmented again:

> Iur: **SCCP DT1** (destination local reference=**a**, more data=1)
> Iur: **SCCP DT1** (destination local reference=**a**, more data=0, *RNSAP*
> (*successfulOutcome* procedureCode=***id-radioLinkSetup*** (longTID=**c**, d-RNTI=**o**,
> id-CN-PS-DomainIdentifier, id-CN-CS DomainIdentifier, rL-ID=**p**, rL-Set-ID=**q**,
> DCH-ID=**g**, bindingID=**r**, DCH-ID=**h**, bindingID=**s**,
> id-Neighboring-UMTS-CellInformation, id-Neighboring-GSM-CellInformation))

The following message example will show only fragments of Neighboring cell information part. As a rule there can be more than 30 neighbor cells with their specific parameters reported to SRNC.

```
|UMTS RNSAP acc. R99 TS 25.423 ver. 3.7.0 (RNSAP370)        |
|successfulOutcome (= successfulOutcome)                     |
|rnsapPDU                                                     |
|1 successfulOutcome                                         |
|1.1 procedureID                                             |
|00010011 |1.1.1 procedureCode        |id-radioLinkSetup    |
|-01----- |1.1.2 ddMode               |fdd                  |
|---00--- |1.2 criticality            |reject               |
|1.3 transactionID                                           |
|***B2*** |1.3.1 longTransActionId     |25446               |
|1.4 value                                                   |
|1.4.1 protocolIEs                                           |
|1.4.1.1 sequence                                            |
|***B2*** |1.4.1.1.1 id               |id-D-RNTI            |
|01------ |1.4.1.1.2 criticality      |ignore               |
|***B2*** |1.4.1.1.3 value            |51581                |
|1.4.1.2 sequence                                            |
|***B2*** |1.4.1.2.1 id               |id-CN-PS-DomainIdentifier |
```

```
|01------ |1.4.1.2.2 criticality             |ignore           |
|1.4.1.2.3 value                             |                 |
|***B3*** |1.4.1.2.3.1 pLMN-Identity         |299 00           |
|***B2*** |1.4.1.2.3.2 lAC                   |00 01            |
|11010001 |1.4.1.2.3.3 rAC                   |01               |
|1.4.1.3 sequence                            |                 |
|***B2*** |1.4.1.3.1 id                      |id-CN-CS-        |
|                                            |  DomainIdentifier |
|01------ |1.4.1.3.2 criticality             |ignore           |
|1.4.1.3.3 value                             |                 |
|***B3*** |1.4.1.3.3.1 pLMN-Identity         |299 00           |
|***B2*** |1.4.1.3.3.2 lAC                   |00 01            |
|1.4.1.4.3.1.3 value                         |                 |
|***b5*** |1.4.1.4.3.1.3.1 rL-ID             |0                |
|--00000- |1.4.1.4.3.1.3.2 rL-Set-ID         |0                |
|1.4.1.4.3.1.3.3 sAI                         |                 |
|***B3*** |1.4.1.4.3.1.3.3.1 pLMN-Identity   |299 00           |
|***B2*** |1.4.1.4.3.1.3.3.2 lAC             |00 01            |
|***B2*** |1.4.1.4.3.1.3.3.3 sAC             |00 11            |
|***B2*** |1.4.1.4.3.1.3.4 received-total-   |80               |
|wide-band-po..                              |                 |
|1.4.1.4.3.1.3.5 dl-CodeInformation          |                 |
|1.4.1.4.3.1.3.5.1 fDD-DL-CodeInformationItem|                 |
|***b4*** |1.4.1.4.3.1.3.5.1.1 dl-ScramblingCode |0           |
|***B2*** |1.4.1.4.3.1.3.5.1.2 fDD-DL-       |5                |
|ChannelizationCo..                          |                 |
|1.4.1.4.3.1.3.6 diversityIndication         |                 |
|1.4.1.4.3.1.3.6.1 nonCombiningOrFirstRL     |                 |
|1.4.1.4.3.1.3.6.1.1 dCH-InformationResponse |                 |
|1.4.1.4.3.1.3.6.1.1.1 dCH-InformationResponseItem |           |
|00001000 |1.4.1.4.3.1.3.6.1.1.1.1 dCH-ID    |8                |
|00011000 |1.4.1.4.3.1.3.6.1.1.1.2 bindingID |'18'H            |
|**B20*** |1.4.1.4.3.1.3.6.1.1.1.3           |XXXXXXXXXXXXXXXXXXXX|
|transportLayerAddress                       |                 |
~~~~~~~~~~~~~~~~~~~~~~~~~~~~~~~~~~~~~~~~~~~~~~~~~~~~~~~~~~~~~~~~~~~~~~~~~~~~~~~~~~~
|1.4.1.4.3.1.3.6.1.1.2 dCH-InformationResponseItem |           |
|00011111 |1.4.1.4.3.1.3.6.1.1.2.1 dCH-ID    |31               |
|00001111 |1.4.1.4.3.1.3.6.1.1.2.2 bindingID |0f               |
|**B20*** |1.4.1.4.3.1.3.6.1.1.2.3           |XXXXXXXXXXXXXXXXXXXX|
|transportLayerAddress                       |                 |
|1.4.1.4.3.1.3.14 Neighboring-UMTS-CellInformation |           |
~~~~~~~~~~~~~~~~~~~~~~~~~~~~~~~~~~~~~~~~~~~~~~~~~~~~~~~~~~~~~~~~~~~~~~~~~~~~~~~~~~~
|1.4.1.4.3.1.3.14.1 sequenceOf               |                 |
|***B2*** |1.4.1.4.3.1.3.14.1.1 id           |id-Neighboring-  |
|                                            |UMTS-CellInformat...|
|01------ |1.4.1.4.3.1.3.14.1.2 criticality  |ignore           |
```

```
|1.4.1.4.3.1.3.14.1.3 value                                                |
|***B2*** |1.4.1.4.3.1.3.14.1.3.1 rNC-ID              |142               |
|1.4.1.4.3.1.3.14.1.3.2 neighboring-FDD-CellInformation                    |
|1.4.1.4.3.1.3.14.1.3.2.1 neighboring-FDD-CellInformationItem              |
|***B2*** |1.4.1.4.3.1.3.14.1.3.2.1.1 c-ID            |37726             |
|***B2*** |1.4.1.4.3.1.3.14.1.3.2.1.2 UARFCNforNu     |9762              |
|***B2*** |1.4.1.4.3.1.3.14.1.3.2.1.3 UARFCNforNd     |10712             |
|***B2*** |1.4.1.4.3.1.3.14.1.3.2.1.4                 |315               |
|primaryScramblin..                                                        |
|1.4.1.4.3.1.3.14.1.3.2.2 neighboring-FDD-CellInformationItem              |
|***B2*** |1.4.1.4.3.1.3.14.1.3.2.2.1 c-ID            |37215             |
|***B2*** |1.4.1.4.3.1.3.14.1.3.2.2.2 UARFCNforNu     |97                |
|***B2*** |1.4.1.4.3.1.3.14.1.3.2.2.3 UARFCNforNd     |10712             |
|***B2*** |1.4.1.4.3.1.3.14.1.3.2.2.4                 |320               |
|primaryScramblin..                                                        |
|***B2*** |1.4.1.4.3.1.3.14.2.3.1 rNC-ID              |341               |
|1.4.1.4.3.1.3.14.2.3.2 Neighboring-FDD-CellInformation                    |
|1.4.1.4.3.1.3.14.2.3.2.1 Neighboring-FDD-CellInformationItem              |
|***B2*** |1.4.1.4.3.1.3.14.2.3.2.1.1 c-ID            |32737             |
/\/\/\/\/\/\/\/\/\/\/\/\/\/\/\/\/\/\/\/\/\/\/\/\/\/\/\/\/\/\/\/\/\/\/\/\/\/\
|1.4.1.4.3.1.3.15 Neighboring-GSM-CellInformation                          |
|***B2*** |1.4.1.4.3.1.3.15.1 id                      |id-Neighboring-GSM-|
|                                                    CellInformation    |
|01------ |1.4.1.4.3.1.3.15.2 criticality             |ignore            |
|1.4.1.4.3.1.3.15.3 value                                                  |
|1.4.1.4.3.1.3.15.3.1 Neighboring-GSM-CellInformationItem                  |
|1.4.1.4.3.1.3.15.3.1.1 cGI                                                |
|1.4.1.4.3.1.3.15.3.1.1.1 lAI                                              |
|***B3*** |1.4.1.4.3.1.3.15.3.1.1.1.1 pLMN-Identity   |299 00            |
|***B2*** |1.4.1.4.3.1.3.15.3.1.1.1.2 lAC             |00 08             |
|***B2*** |1.4.1.4.3.1.3.15.3.1.1.2 cI                |10 02             |
|010100-- |1.4.1.4.3.1.3.15.3.1.2 cellIndividualOffset |0                |
|1.4.1.4.3.1.3.15.3.1.3 bSIC                                               |
|***b3*** |1.4.1.4.3.1.3.15.3.1.3.1 nCC               |'111'B            |
|-011---- |1.4.1.4.3.1.3.15.3.1.3.2 bCC               |'011'B            |
|-----0-- |1.4.1.4.3.1.3.15.3.1.4 band-Indicator      |dcs1800Band       |
|***B2*** |1.4.1.4.3.1.3.15.3.1.5 bCCH-ARFCN          |575               |
|1.4.1.4.3.1.3.15.3.2 Neighboring-GSM-CellInformationItem                  |
|1.4.1.4.3.1.3.15.3.2.1 cGI                                                |
|1.4.1.4.3.1.3.15.3.2.1.1 lAI                                              |
|***B3*** |1.4.1.4.3.1.3.15.3.2.1.1.1 pLMN-Identity   |299 00            |
|***B2*** |1.4.1.4.3.1.3.15.3.2.1.1.2 lAC             |00 04             |
|***B2*** |1.4.1.4.3.1.3.15.3.2.1.2 cI                |09 88             |
----------------------------------------------------------------------------
```

Message Example 3.39   RNSAP Successful Outcome Radio Link Setup.

A new identifier is the drift RNTI (D-RNTI). D-RNTI is allocated by drift RNC (DRNC) when SRNC requests setup of DCHs in DRNC's RNS. It shall be unique within the DRNC. SRNC knows the mapping between S-RNTI and D-RNTI. DRNC shall know the S-RNTI and SRNC-ID related to existing D-RNTI within the DRNC.

Core Net Identities are self-explaining; rL-ID has the value assigned by SRNC before, but rL-Set-ID might have a different value than the one in the NBAP Radio Link Setup procedure on Iub 2. The reason is that rL-Set-ID in NBAP is related to an active Node B Communication Context. Hence, it is valid on CRNC level only, but not for communication between SRNC and DRNC.

The SAI identifies the service area to which the new cell in the link set belongs. DL codes for dedicated physical channels have been assigned by DRNC already with NBAP Radio Link Setup request. Now SRNC is informed about which parameters have been taken in use successfully to identify dedicated physical downlink channel of the new cell.

Once again we see two bindingIDs and two transport layer addresses bound to DCH-IDs of DCCH and DTCH. The bindingID values are completely different from the values used on Iub 2, because Iur is a different part of the transport network. Transport layer addresses may identify the peer RNC on ATM level, but if there is an ATM network used to interconnect UTRAN network elements, it might only be the end-user address of the next ATM switch. Transport layer address numbering follows ITU-T E.191 numbering plan (Figure 3.123). These addresses are also known as E.164 AESA addresses for broadband ISDN (B-ISDN). They contain an initial domain part (IDP) and a domain specific part (DSP). IDP is subdivided into authority and format identifier (AFI) and initial domain identifier (IDI). If AFI = 0x45 (45 hex) IDI consists of an ISDN telephone number following E.164 standard (see chapter on global title translation in Chapter 4 of this book to have an example of E.164 addressing). In the message example the address values are shown as "XXX..." to hide the true address identities.

After DCH information response items the neighborhood cell information lists are found. UTRAN cells are reported according to IDs of their CRNCs. So it becomes clear which RNC controls which neighbor cell and could act as a new DRNC in following soft handover situations. The cells are identified by their c-IDs, primary scrambling codes, and UTRA absolute radio frequency channel number (UARFCN). This UARFCN designates the carrier frequency. The value of the UARFCN in the IMT2000 band is defined as follows:

Uplink UARFCN (Nu) = 5 × Frequency (MHz) and Downlink UARFCN (Nd) = 5 × Frequency (MHz).

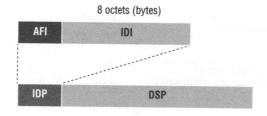

Figure 3.123   E.191 address format.

GSM cells in the next section are identified on behalf of their Cell Global Identity (CGI). This parameter is a concatenation of the LAI (Location Area Identity) and the CI (Cell Identity) and uniquely identifies a given GSM cell in the wire lined network.

To ensure that the same cell can be identified on radio interface as well, base station identity code (BSIC) is necessary. BSIC is broadcast on synchronization channel (SCH) of a GSM cell to inform MS about network color code (NCC) and base station color code (BCC) before registering on the network. Network Color Code (NCC) is used to differentiate between operators utilizing the same GSM frequency, for instance on an international border. Base Station Color Code (BCC) is used to discriminate cells using the same frequency during cell selection and camping on GSM process. BCC is also used to identify training sequence code (TSC) to be used when reading BCCH of a GSM cell.

Now on Iur and Iub the AAL2 SVCs for DCCH and DTCH are established (see Figure 3.124):

Iur: **ALCAP DL ERQ** (Originating Signal. Ass. ID=**l**, AAL2 Path=**m**, AAL2 ChannelId=**n**, served user gen reference=**r**)
Iur: **ALCAP UL ECF** (Originating Signal. Ass. ID=**q**, Destination Sign. Assoc. ID=**l**)

Iur: **ALCAP DL ERQ** (Originating Signal. Ass. ID=**t**, AAL2 Path=**u**, AAL2 ChannelId=**v**, served user gen reference=**s**)
Iur: **ALCAP UL ECF** (Originating Signal. Ass. ID=**w**, Destination Sign. Assoc. ID=**t**)

Also the RNSAP Dedicated Measurement Initiation will later be forwarded to Node B2 using NBAP on Iub. There might be one or more dedicated measurement initiated over Iur. A typical example is measurement of transmitted code power. It is also possible that RNSAP

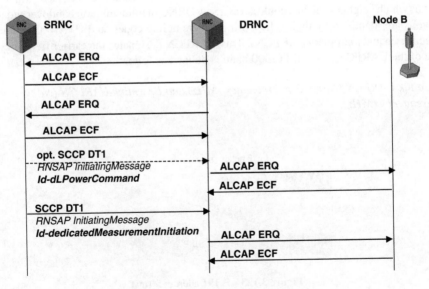

Figure 3.124   Iub-Iur Soft Handover call flow 2/3.

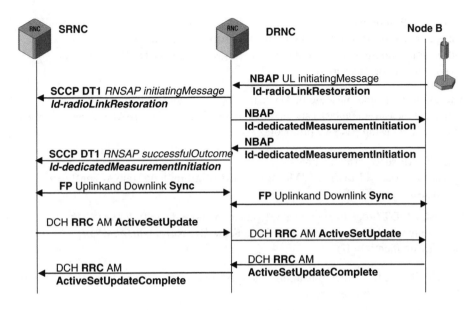

Figure 3.125 Iub-Iur Soft Handover call flow 3/3.

Downlink Power Commands are sent during this part of the procedure from SRNC to DRNC (Figure 3.125).

> Iur: **SCCP DT1** (destination local reference=**b**, more data=0, *RNSAP SuccessfulOutcome* procedureCode=***id-dedicatedMeasurementInitiation***, id-MeasurementID=**ff**, details and parameters to be measured, e.g. transmitted code power)

> Iub: **ALCAP DL ERQ** (Originating Signal. Ass. ID=**x**, AAL2 Path=**y**, AAL2 Channel id=**z**, served user gen reference=**m**)
> Iub: **ALCAP UL ECF** (Originating Signal. Ass. ID=**aa**, Destination Sign. Assoc. ID=**x**)

> Iub: **ALCAP DL ERQ** (Originating Signal. Ass. ID=**bb**, AAL2 Path=**cc**, AAL2 ChannelId=**dd**, served user gen reference=**n**)
> Iub: **ALCAP UL ECF** (Originating Signal. Ass. ID=**ee**, Destination Sign. Assoc. ID=**bb**)

The CRNC Communication Context is valid between the Node B2 and the DRNC, which acts as CRNC related to this Node B on NBAP layer.

> Iub: **NBAP** initiatingMessage procedureCode=**id-radioLinkRestoration** (shortTransActionID=**gg**, id-CRNC-CommunicationContextID=**j**)

> Iur: **SCCP DT1** destinationLocalReference=**a**, *RNSAP* (initiatingMessage, procedureCode=***id-radioLinkResoration***, longTransActionID=**hh**, rL-Set-ID=**q**)

Iub: **NBAP** initiatingMessage
procedureCode=**id-dedicatedMeasurementInitiation**, shortTransActionID=**ii**,
id-NodeB-CommunicationContextID=**k** , id-MeasurementID=**ff**, details and
parameters to be measured)

Successful Outcome message of Dedicated Measurement Initiation indicates that the UE
accepted the measurement tasks and parameters sent before. The NBAP message is forwarded
via Iur to SRNC using RNSAP protocol.

Iub: **NBAP** successfulOutcome
procedureCode=**id-dedicatedMeasurementInitiation**, longTransActionID=**ii**,
id-MeasurementID=**ff**)

Iur: **SCCP DT1** (destination local reference=**a**, more data=0, *RNSAP*
(initiatingMessage procedureCode=*id-dedicatedMeasurementInitiation*,
id-MeasurementID=**ff**)

The FP UL and DL Synchronization messages are used to align both channels, DCCH and
DTCH, on Iur and Iub interface.

Iur: DCH **FP** Up and Downlink Sync "for both channels"
Iub: DCH **FP** Up and Downlink Sync "for both channels"

With the Active Set Update procedure performed on both Iur and Iub interfaces, the new
links are taken into service and the traffic channel becomes available for the UE.

Iur: DCH **RRC** AM **ActiveSetUpdate**
Iub: DCH **RRC** AM **ActiveSetUpdate**
Iub: DCH **RRC** AM **ActiveSetUpdateComplete**
Iur: DCH **RRC** AM **ActiveSetUpdateComplete**

3.19 Iub-Iu – Forward Handover (Inter-Node B, Inter-RNC)

There are two general types of hard handovers defined for the UTRAN in 3GPP 25.931: Forward
Handover and Backward Handover. The MS initiates the Forward Handover by sending RRC
Cell Update message, while the Backward Handover is initiated by the network after SRNC
receives an RRC Measurement Report. The confusing thing is if one reads other 3GPP specs
related to handover scenarios, especially 3GPP 25.832 and 3GPP 23.009, the terms "Forward
Handover" and "Backward Handover" are not used again. In addition, Backward Handovers
using Iur resources have not been monitored yet by the authors. If network resources are taken
into account it makes more sense to perform interfrequency handovers between FDD cells using
SRNS relocation (UE involved) procedure, which is a hard handover plus relocation without
using Iur resources. On the other hand, intrafrequency as well as interfrequency hard handover
between TDD cells (there is no Soft Handover in TDD!) will be probably more efficient using
Iur procedures instead of RANAP relocation functions that cause a higher signaling load on
IuCS and IuPS. All in all this subject is interesting for further studies.

While Hard Handover can be both, Forward or Backward Handover, Soft Handover procedures are always controlled by the SRNC, so all Soft (and Softer) Handover belong to the
group of Backward Handover. To understand the relations, see Table 3.3.

Table 3.3   Handover types

	Soft Handover	Hard Handover
Forward Handover (UE sends RRC Cell Update)		x
Backward Handover (triggered by RRC Measurement Report)	x	x

When a Hard Handover is performed – in difference to a Soft Handover – the MS loses contact with the network (when it switches from old radio link to the new one) for a very short time period (short enough to be not noticed by the user who has an ongoing active voice call). Usually this time period shall be not more than 200 ms to avoid interruptions of voice call. However, there have been longer delays measured in case of a forward handover. This indicates that a forward handover is never a preferred solution from the point of view of network optimization, but only a recovery function in case that radio contact with UE is suddenly completely lost.

3.19.1 Overview

The case of forward handover described in this chapter can also be described as a kind of transport channel re-establishment. The authors are not even sure if the procedure described in this chapter meets the definition behind the term "forward handover" as given in 3GPP 25.931. Maybe a "forward handover" in case of 3GPP only happens if a UE with active PS call performs cell reselection/cell update in CELL_FACH state. Hence, there is probably no forward handover for CS calls defined by 3GPP, but the scenario described in this chapter is seen pretty often and hence it should have its own name. It happens in case radio links are suddenely lost. Possible reasons for losing radio contact are cell breathing effects (as shown in Figure 3.126), but also there can be technical problems on the SRNC side. For instance, the

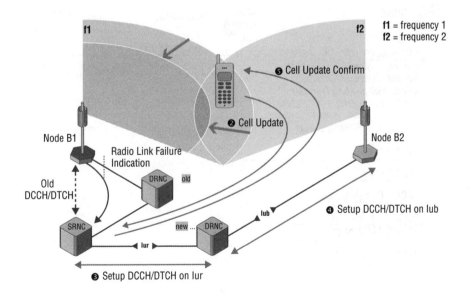

Figure 3.126   Iub-Iur Forward Hard Handover overview.

authors monitored a case when SRNC received RRC Measurement report including Event-ID *e1c* (*A nonactive primary CPICH becomes better than an active primary CPICH*). The correct reaction of SRNC would have been to delete the most weak radio link of the active set and subsecquently add the new strong radio link to the active set. Instead, SRNC deleted all radio links of the active set and UE lost contact with all dedicated radio links at the same time.

Step 1: After UE loses radio contact with the network during an active RRC connection, it changes its RRC state into CELL_FACH and performs cell reselection to evaluate the strongest available cell using the same frequency that was used before by radio links of the active set. In the example, this cell is controlled by Node B2 while the cell of Node B1 that was also involved in the previous active set is still out of range. UE sends RRC Cell Update message via Iub 2 and Iur interface to its SRNC.

Step 2: SRNC orders setup of physical transport bearers for DCCHs/DTCHs on Iur interface between itself and DRNC.

Step 3: Triggered by the messages coming from SRNC, an NBAP Radio Link Setup procedure on Iub 2 interface between DRNC and Node B2 is performed. Physical transport bearers for DCCHs and DTCHs are established as well.

Step 4: Then SRNC sends RRC Cell Update confirm to UE using downlink common transport channel (FACH) of cell on Node B2. Included in this message the UE finds all necessary parameters to have access to provided dedicated radio link again. The procedure is complete if communication between SRNC and UE uses finally dedicated channels again.

3.19.2 Message Flow

The procedure starts when the MS sends a Cell Update message via new Iub/Iur interface (Figure 3.127). The message contains the U-RNTI including SRNC-ID that enables the network

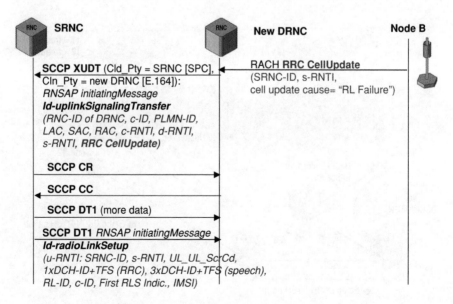

Figure 3.127   Iub-Iur Forward Hard Handover call flow 1/4.

to decide which RNC shall finally receive the message and also the S-RNTI, which is the identifier of the UE on the common transport channels. The cell update cause "radio link failure" indicates that the connection to the previously used UTRAN cell was lost.

Iub: RACH **RRC CellUpdate** (U-RNTI: SRNC-ID=**hh** + S-RNTI=**e**, CS and/or PS Domain Id, cellUpdateCause="radioLinkFailure")

The Layer 3 message Cell Update is included in RNSAP Uplink Signaling Transfer message and on behalf of this is forwarded from the new DRNC to the SRNC. Since there is no DCCH available on this Iur interface, SCCP uses the Extended Unitdata (XUDT) message format to transport the Uplink Signaling Transfer. XUDT has in comparison to UDT an extended header with a hop counter information element to prevent routing loops and a segmentation indicator ("more data" bit) to allow segmented transport and reassembly of large higher layer messages on the receiver side if necessary.

SRNC-ID received with Cell Update message is translated by DRNC into an SS#7 Signaling Point Code address that is used as called party number in SCCP XUDT, while DRNC identifies itself by using an E.164 Global Title address. (Refer to core network signaling part [Chapter 4] of this book to learn more about SCCP addressing.)

There are different kinds of location information included in RNSAP Uplink Signaling Transfer, especially information as to which Location Area (LA), Routing Area (RA), and Service Area (SA) the new cell (identified by its c-ID) belongs. For the different areas, appropriate area codes (LAC, RAC, SAC) are included together with PLMN Identity that contains Mobile Country Code and Mobile Network Code according to ITU-T E.212 numbering plan.

Since the UE is in CELL_FACH state, CRNC function of the new DRNC assigns a c-RNTI to identify the UE uniquely within the new cell. In addition there is a relation between this c-RNTI and the d-RNTI that is assigned and valid for time of RNSAP connection via Iur interface. D-RNTI will never be used on Iub/Uu interface! D-RNTI value is transmitted to SRNC together with the S-RNTI so that SRNC can store a fixed relation between the RNSAP context on Iur and the RRC context that is valid for the whole connection between SRNC and UE including links on Iub and Uu interfaces.

Iur: **SCCP XUDT** *RNSAP initiatingMessage*, ***id-uplinkSignalingTransfer***
(S-RNTI=**e**, c-RNTI=**kk**, d-RNTI=**o**, c-ID=**mm**, RNC-ID=**nn**, Layer3-Info: **RRC CellUpdate**)

After receiving the Cell Update message SRNC starts seting up an SCCP Class 2 connection on Iur toward the new DRNC.

Iur: **SCCP CR** (source local reference=**a**)
Iur: **SCCP CC** (source local reference=**b**, destination local reference=**a**)

Then SRNC starts Radio link setup procedure and, following this, the setup of DCCH and DTCH on Iur that triggers the same procedure on Iub (Figure 3.128). Depending on manufacturer's implementation, the procedure might be guided by short or long transaction ID. Besides this, the message is nearly identical with the one shown in message example 3.38. However, first the radio link set indicator this time is set to "first RLS," because there is no active radio link to this UE anymore.

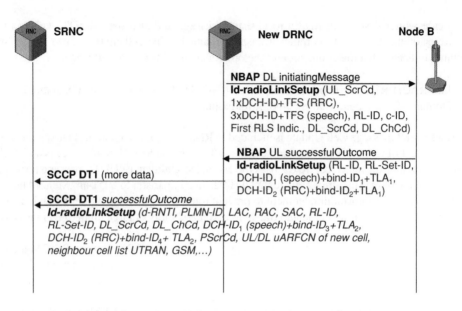

Figure 3.128 Iub-Iur Forward Hard Handover call flow 2/4.

Iur: **SCCP DT1** (destination local reference=**b**, more data=1)
Iur: **SCCP DT1** (destination local reference=**b**, more data=0, *RNSAP*
[*initiatingMessage* procedureCode=***Id-radioLinkSetup***, shortTID=**c**, u-RNTI:
SRNC-ID=**d** + S-RNTI=**e**, ULscramblingCode=**f**, DCH-ID=**g** + UL/DL Transport
Format Set, DCH-ID=**g**′ + UL/DL Transport Format Set, DCH-ID=**g**″ + UL/DL
Transport Format Set, DCH-ID=**h** + UL/DL Transport Format Set, rL-ID=**p**,
c-ID=**mm**, First Radio Link Set Indicator = "first RLS", IMSI= **oo**])

It shall be noticed again that SRNC only assigns the uplink radio resources (UL scrambling
code and UL channelization code length), but CRNC function on DRNC side assigns downlink
resources, especially DL channelization code for the radio link. This downlink channelization
code assignment is completely independent from the same procedure used during call setup.
The values may be equal by chance, but they do not need to be the same. NBAP transport channel
settings directly using the definitions received with RNSAP Radio Link Setup Request:

Iub: **NBAP** DL initiatingMessage, procedureCode=**Id-radioLinkSetup** (longTID=**i**,
id-CRNC-CommunicationContextID=**j**, ULscramblingCode/DLChannelizationCode
=**f**/δ, DCH-ID=**g** + UL/DL Transport Format Set, DCH-ID=**g**′ + UL/DL Transport
Format Set, DCH-ID=**g**″ + UL/DL Transport Format Set, DCH-ID=**h** + UL/DL
Transport Format Set, rL-ID=**p**, c-ID=**mm**, First Radio Link Set Indicator = "first
RLS")

NBAP Radio Link Setup Request is successfully answered with NBAP Radio Link Setup
Response:

Iub: **NBAP** UL successfulOutcome, procedureCode=**Id-radioLinkSetup**
(longTransActionID=**i**, id-CRNC-CommunicationContextID=**j**,
id-NodeB-CommunicationContextID=**k**, id-CommunicationControlPortID=i,
DCH-ID=**g** + bindingID=**m** + Transport Layer Address = **qq**, DCH-ID=**h** +
bindingID=**n** + Transport Layer Address = **qq**, rL-ID=**p**, rL-Set-ID=**q**)

After DRNC receives NBAP Radio Link Setup Response it sends RNSAP Radio Link
Setup Response to SRNC. Once again this message is quite the same as the one shown in the
appropriate message example of the previous call flow scenario. To leave no doubt that this
message is related to the Cell Update that was not transmitted as part of this SCCP Class 2
connection, d-RNTI is included again. In addition some information about the new radio link
that SRNC still does not know is included in the message: primary scrambling code of new
cell and uplink and downlink UTRA absolute radio frequency channel number (uARFCN),
which are both used to identify the cell uniquely on Uu.
Another very interesting fact is that rL-Set-ID in RNSAP Radio Link Setup procedure may
have a different value than in appropriate NBAP message. This happens because NBAP rL-
Set-ID identifies a radio link within one Node B context while RNSAP rL-Set-ID identifies a
radio link within a UE context (Figure 3.129):

Iur: **SCCP DT1** (destination local referece=**a**, more data=1)
Iur: **SCCP DT1** (destination local referece=**a**, more data=0, *RNSAP*
(successfulOutcome procedureCode=***Id-radioLinkSetup***, shortTransActionID=**c**,
d-RNTI=**o**, id-CN-PS-DomainIdentifier, id-CN-CS DomainIdentifier, rL-ID=**p**,
RL-Set-ID=**rr**, DCH-ID=**g** + bindingID=**r** + Transport Layer Address=**ss**,

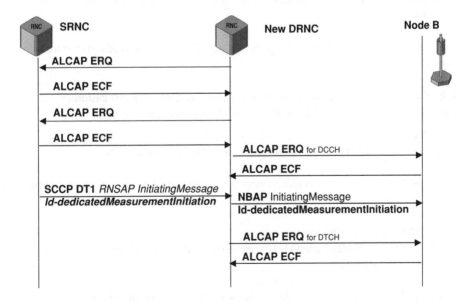

Figure 3.129   Iub-Iur Forward Hard Handover call flow 3/4.

DCH-ID=**h** + bindingID=**n** + Transport Layer Address =**ss**, PScrCd=**uu**,
Neighboring Cell Information: UMTS, GSM, . . .))

Iur: **ALCAP DL ERQ** (Originating Signal. Ass. ID=**l**, AAL2 Path=**m**, AAL2
ChannelId=**n**, served user gen reference=**r**)
Iur: **ALCAP UL ECF** (Originating Signal. Ass. ID=**q**, Destination Sign. Assoc.
ID=**l**)

Iur: **ALCAP DL ERQ** (Originating Signal. Ass. ID=**t**, AAL2 Path=**u**, AAL2
ChannelId=**v**, served user gen reference=**s**)
Iur: **ALCAP UL ECF** (Originating Signal. Ass. ID=**w**, Destination Sign. Assoc.
ID=**t**)

Iur: **SCCP DT1** (destination local reference=**b**, more data=0, *RNSAP*
initiatingMessage (procedureCode=*id-dedicatedMeasurementInitiation*,
shortTransactionID=**ii**, id-MeasurementID=**ff**, measurementThreshold)

Iub: **NBAP** initiatingMessage procedureCode=**id-dedicatedMeasurementInitiation**
(id-Node B-CommunicationContextID=**k**, id-MeasurementID=**ff**,
measurementThreshold)

Iub: **ALCAP DL ERQ** (Originating Signal. Ass. ID=**x**, AAL2 Path=**y**, AAL2
ChannelId=**z**, served user gen reference=**m**)
Iub: **ALCAP UL ECF** (Originating Signal. Ass. ID=**aa**, Destination Sign. Assoc.
ID=**x**)

Iub: **ALCAP DL ERQ** (Originating Signal. Ass. ID=**bb**, AAL2 Path=**cc**, AAL2
Channel id=**dd**, served user gen reference=**n**)
Iub: **ALCAP UL ECF** (Originating Signal. Ass. ID=**ee**, Destination Sign. Assoc.
ID=**bb**)

The Radio Link Restoration message indicates that the cell is now ready to start transmission
on radio (Uu) interface (Figure 3.130).

Iub: **NBAP** initiatingMessage procedureCode=**Id-radioLinkRestoration**
(shortTransActionID=**gg** , id-CRNC-CommunicationContextID=**j**)

Iur: **SCCP DT1** destinationLocalReference=**a**, *RNSAP* (*initiatingMessage*,
procedureCode=*id-radioLinkRestoration*, shortTransActionID=**tt**)

After signaling and traffic connection between SRNC and UE are set up, measurement is
initiated:

Iub: **NBAP** successfulOutcome
(procedureCode=**id-dedicatedMeasurementInitiation**, shortTransactionID=ii,
id-MeasurementID=**ff**)
Iur: **SCCP DT1** (destination local reference=**a**, more data=0, *RNSAP*
(initiatingMessage procedureCode=*id-dedicatedMeasurementInitiation*,
id-MeasurementID=**ff**))

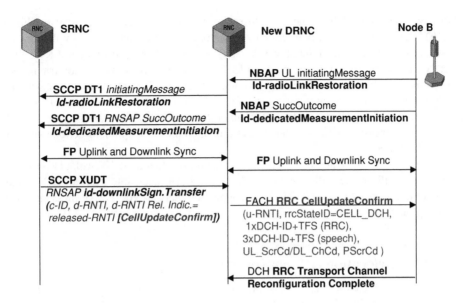

Figure 3.130   Iub-Iur Forward Hard Handover call flow 4/4.

FP Synchronization for initial alignment of DCCH and DTCH follows:

Iur: DCH **FP** Up and Downlink Sync for both Iur physical transport bearers (DCCH + DTCH)
Iub: DCH **FP** Up and Downlink Sync for both Iub physical transport bearers

Finally SRNC sends Cell Update Confirm still using the SCCP signaling transfer capabilities to the UE. This is necessary because UE still has not received any information about the provided dedicated resources.

Iur: **SCCP DT1** destinationLocalReference=**b**, *RNSAP (initiatingMessage*, procedureCode=*id-downlinkSignalingTransfer*, d-RNTI=**o**, Layer3-Info: **RRC CellUpdateConfirm**)

Since until now there was no information exchanged that informs the UE that the setup DCCH/DTCH is related to its Cell Update, the Cell Update Confirm message contains the dedicated physical parameters, especially uplink scrambling code and downlink channelization code, and also the TFSs of the different DCHs and primary scrambling code of the new cell we have already seen during radio link setup procedures. Cell Update Confirm message on Iub is sent in downlink direction using the FACH of the new cell. Using RRC state indicator the UE is requested to switch into CELL_DCH after receiving this message:

Iub: FACH **RRC CellUpdateConfirm** (u-RNTI: SRNC-ID=**hh** + S-RNTI=**e**, rrcStateID=**CELL_DCH**, ULscramblingCode/DLchannelizationCode=**f**/δ, DCH-ID=**g** + UL/DL Transport Format Set, DCH-ID=**g′** + UL/DL Transport Format Set, DCH-ID=**g″** + UL/DL Transport Format Set, DCH-ID=**h** + UL/DL Transport Format Set, PScrCd=**uu**)

After receiving the Cell Update Confirm the mobile switches to the new established DCCH/DTCH of the new cell belonging to new Node B. UE is now served by the new cell. How it confirms the successful handover depends on the implemented version of RRC protocol. In case that earlier versions are used, an UTRAN Mobility Information Confirm message might be sent on DCH to SRNC. RRC version 3.17 (2003–12) contains a clear statement if Cell Update Confirm message does not include radio bearer information elements, but does include transport channel information elements (as in our call flow example case) "the UE shall transmit a Transport Channel Reconfiguration Complete" in uplink direction using RLC acknowledged mode.

In any case there will be another confirm message transmitted to SRNC using DCH/DCCH. This message will be the indicator that DCHs have been taken into service by UE and it will trigger the release of the SCCP Class 2 connection that carried RNSAP message for this handover procedure.

3.20 SRNS Relocation (UE not Involved)

The purpose of an SRNS Relocation that does not involve the UE is to minimize traffic on Iur interface. Thus, SRNC is changed and Iu connections are reorganized. The decision to change the SRNC (the RNC that controls the connections to the core network domains) is triggered by a previous mobility management procedure like Inter-RNC Hard or Soft Handover. The UE is not involved if it is already located in the new cell, which is the case after finished soft handover and forward hard handover procedures.

An SRNS Relocation (UE not involved) can be performed in any state of the call: if only RRC signaling is exchanged between UE and network, but also if voice and/or data calls are active.

The signaling example in this chapter is not based on a real network trace, but constructed following different descriptions in different 3GPP "specs." One reason why the authors have not monitored an SRNS Relocation (UE not involved) so far may be that it only appears in RNS border areas. In addition, long-distance moves during active calls with long duration are required.

As the reader will see, there is a quite incredibly long list of parameters embedded in different RANAP and RRC containers. Most of these parameters have already been discussed in Iub and Iu signaling examples. Owing to huge number of parameters the authors decided not to assign variables to indicate parameter values for SRNS relocation scenarios. Message examples will be shown as far as available.

3.20.1 Overview

The following example shows relocation during an active PDPC (see Figure 3.131). In case that a voice call would be active, in addition the same RANAP procedures need to be performed on old and new IuCS interface. It is assumed that the UE sets up the PDP context in a cell that is controlled by RNC 1 (not shown in the figure). Then it performs soft handover in two steps. First a radio link is added that belongs to the cell controlled by RNC 2 (the cell that is shown in Figure 3.131). Later the first radio link is released. Now the UE has only radio contact with

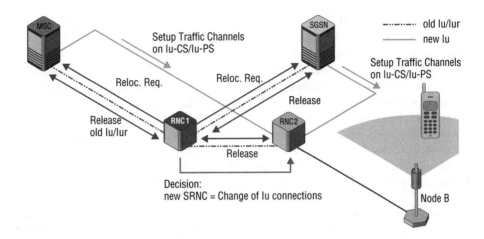

Figure 3.131 SRNS Relocation (UE not involved) procedure overview.

the cell controlled by RNC 2 and there are transport bearers for RRC signaling and IP payload established on Iur.

In this situation RNC 1 decides on behalf of an algorithm which is part of its SRNC function that it is necessary to perform a resource optimization, which means blocked transport capacities of Iur bearers can be freed if core network connection terminates directly at RNC 2. Since only a serving RNC (SRNC) can terminate the core network connections it is also necessary to hand over SRNC functionally from RNC 1 to RNC 2.

This procedure in steps is as follows:

Step 1: RNC 1 is SRNC, but has no direct connection via its RAN to the MS anymore. All signaling and traffic connections are running on Iub controlled by RNC 2, which after successful soft handover may still act as DRNC. Iur transport bearers are necessary to exchange traffic and signaling between SRNC and DRNC. To optimize the used network resources, SRNC (RNC 1) makes the decision to hand over its function to the DRNC (RNC 2). Since MS has direct contact with RNC 2 via Iub, the Iur connections will not be necessary any longer after RNC 2 becomes SRNC.

Step 2: RNC 2, the old SRNC, sends a Relocation Required message to SGSN, which then executes the next necessary steps.

Step 3: New IuPS signaling connection (SCCP Class 2) and new Iu bearer as part of RAB are set up toward RNC 2 and RNC 1 using RANAP Relocation Resource Allocation procedure. When the new connection to core network are ready to be used, RNC 1 commits to hand over SRNC function to RNC 2 sending an RNSAP message via Iur.

Step 4: After RNC 2 (former DRNC) became SRNC, Iur connection and old IuCS/IuPS connections between RNC 1 and SGSN are released.

3.20.2 *Message Flow*

The message flow shows in the first step the transport bearer situation before relocation trigger is received. There are transport bearers on Iur as well as on two Iub interfaces that "feed" two radio links of the active link set. UE is in soft handover using one cell of Node B1 and one cell of Node B2. The RANAP signaling connection as well as the Iu bearer for IP payload terminate at RNC 1 that acts as SRNC.

Now UE sends RRC Measurement Report including event-ID "e1b" and primary scrambling code of cell of Node B1 (Figure 3.132). It indicates that radio links of this cell became too weak to stay in active set. Hence, Active Set Update including Radio Link Deletion is performed. Successively transport bearers for DCHs on Iub 1 are deleted as well. Theoretically the measurement report can be monitored on both Iub interfaces depending on quality of received RLC frames (see discussion of quality estimate parameter and macrodiversity in an earlier section). In the presented call flow example it is assumed that radio link on Iub 1 is already too bad, so we see RRC Measurement report only on Iub 2 and Iur interfaces.

Now SRNC (RNC 1) decides to perform the relocation procedure. It is started by sending a Relocation Required message to the SGSN. "Relocation Required" is the message name that is used in 3GPP documents to describe the message flow. However, this message is embedded in a procedure and the procedure name/code is "Relocation Preparation." The association between procedure code and message name is defined in ASN.1 RANAP procedure description:

```
relocationPreparation RANAP-ELEMENTARY-PROCEDURE ::={
     INITIATING MESSAGE    RelocationRequired
     SUCCESSFUL OUTCOME    RelocationCommand
     UNSUCCESSFUL OUTCOME    RelocationPreparationFailure
     PROCEDURE CODE        id-RelocationPreparation
     CRITICALITY    reject
}
```

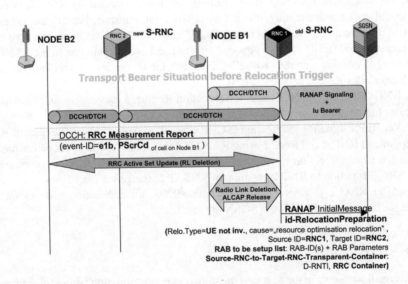

Figure 3.132   SRNS relocation (UE not involved) call flow 1/3.

Following this specification it emerges that RANAP Relocation Required message is defined as RANAP Initial Message that contains procedure code = "id-RelocationPreparation". In the following message descriptions we will use the procedure codes to identifiy messages, because this is what is shown on a protocol tester's monitor.

The SourceRNC-to-TargetRNC-Transparent-Container contains information that need to be forwarded by SGSN to RNC 2. Included D-RNTI that was assigned by DRNC (RNC 2) during radio link setup for soft handover shall later be used in RNSAP messages related to this required relocation on Iur interface. In RRC Container, RNC 2 finds all information that are necessary to take over SRNC function from RNC 1. The summary of these information elements and parameters is also known as RRC context. Message examples that show RRC Signaling Radio Bearer (SRB) Info List and Radio Bearer Info List as well as appropriate transport channel mapping can be found in Iub IMSI/GPRS Attach and MOC scenarios described earlier.

IuPS1: **RANAP** InitialMessage **id-RelocationPreparation**

- RelocationType = **"UE not involved"**
- cause = **"resource optimisation relocation"**
- Source ID=**RNC 1**
- Target ID=**RNC 2**
- **RAB to be setup list** (*If active PDP context*):
  – **RAB-ID(s)** + RAB Parameters
- **SourceRNC-to-TargetRNC-Transparent-Container**:
  – **D-RNTI**
  – **RAB-ID(s)** +Transport Channel Mapping, **DCH-ID(s)**,
  – **RRC Container**:
    ▪ StateofRRC = CELL_DCH
    ▪ StateofRRCConnection = "await no RRC message"
    ▪ CipheringStatus + Parameters
    ▪ IntegrityProt.Status + Parameters
    ▪ **U-RNTI**
    ▪ UE RadioAccessCapabilities
    ▪ RRC Measurement Info
    ▪ **SRB Info List** + DCH-Mapping
    ▪ **Radio Bearer Info List** + DCH-Mapping

Note: The RAB-to-be-setup-list in this message is an optional parameter that is only included if there are active PDP contexts on IuPS. On IuCS Relocation Required message a voice call needs to be active to fullfil the condition.

When SGSN receives Relocation Required message, it can identify the new IuPS inter-face (IuPS 2) on behalf of the target ID that contains the global RNC identity of RNC 2 (Figure 3.132). Now SGSN sends RANAP Relocation Request message to RNC 2, which is also the start of RAB setup on new IuPS interface. Encryption and integrity-specific information is added to the RAB-to-be-setup-list by SGSN to ensure that both security func-tions will be continued without problems after RNC 2 became SRNC. The SourceRNC-to-TragetRNC-Transparent-Container including RRC container is the same as in Relocation

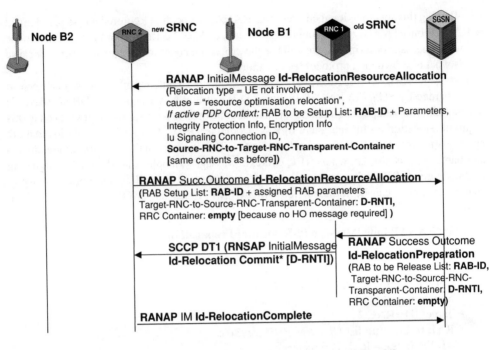

Figure 3.133   SRNS relocation (UE not involved) call flow 2/3.

Required. RANAP Relocation Request message belongs to Relocation Resource Allocation procedure.

IuPS2: **RANAP** InitialMessage **id-RelocationResourceAllocation**

- RelocationType = **"UE not involved"**
- cause = **"resource optimisation relocation"**
- **RAB to be setup list** (*If active PDP context*):
  - **RAB-ID(s)** + RAB Parameters
  - Integrity Protection Info
  - Encryption Info
  - Iu Signaling Connection ID
- **SourceRNC-to-TargetRNC-Transparent-Container:**
  - **D-RNTI**
  - **RAB-ID(s)** +Transport Channel Mapping, **DCH-ID(s)**,
  - **RRC Container**:
    - StateofRRC = CELL_DCH
    - StateofRRCConnection = "await no RRC message"
    - CipheringStatus + Parameters
    - IntegrityProt.Status + Parameters
    - **u-RNTI**
    - UE RadioAccessCapabilities

- RRC Measurement Info
- **SRB Info List** + DCH-Mapping
- **Radio Bearer Info List** + DCH-Mapping

In the next step RNC 2 acknowledges the Relocation Resource Allocation with a Successful Outcome message. Also in case one or more RABs cannot be set up, the Successful Outcome will be sent, but including a RABs-failed-to-setup-list. If all RABs can be set up successfully, the message has the following structure:

RANAP SuccessfulOutcome **id-RelocationResourceAllocation**

- **RAB Setup List: RAB-ID(s)** + assigned RAB parameters
- TargetRNC-to-SourceRNC-Transparent-Container:
  – **d-RNTI**
  – **RRC Container: empty**

The SGSN is informed about the parameters of successfully established RABs on behalf of the enclosed RAB-setup-list, and the TargetRNC-to-SourceRNC-Transparent-Container contains the D-RNTI and an empty RRC Container. D-RNTI will be used as a unique identifier within the following Relocation Commit procedure on Iur interface. The RRC container is empty in case of UE not involved in relocation, because there is no handover to be executed on radio interface. In the UE-involved case the RRC Container contains the handover message constructed by target RNC (see next signaling scenario to compare both relocation types).

After SGSN received Relocation Request Acknowledge message it sends a Relocation Command to RNC 1 that will trigger forwarding of SRNC function. Relocation Command is the Successful Outcome message of Relocation Preparation procedure. It contains a list of RABs to be released including their RAB-IDs that indicate which RABs have not been established successfully on new Iu interface. Based on internal rules, the source RNC will decide whether Relocation procedure is continued or aborted in case not all RABs have been established between core network and target RNC. In addition we see our friend, the TargetRNC-to-SourceRNC-Transparent-Container, with the same contents as in Relocation Request Acknowledge before.

RANAP SuccessfulOutcome **id-RelocationPreparation**

- **RAB to be Release List: RAB-ID(s)**
- TargetRNC-to-SourceRNC-Transparent-Container:
  – **d-RNTI**
  – **RRC Container:** *empty*

Now it is time to involve Iur interface. It is guessed that RNSAP Relocation Commit is sent embedded in DT1 message of an SCCP Class 2 connection that was set up during Inter-RNC soft handover procedure on Iur, because 3GPP 25.423 (RNSAP) specifies that "connection-oriented signaling transport service function." Relocation Commit is signed with a "*" in the message flow because in a Rel. 99 environment two RNSAP messages (Relocation Detect and Relocation Commit) are standardized to execute the SRNC forwarding via Iur. Rel. 4 standards have deleted Relocation Detect from RNSAP and owing to short life cycle time of protocol versions the authors preferred to show Rel. 4 signaling flow version.

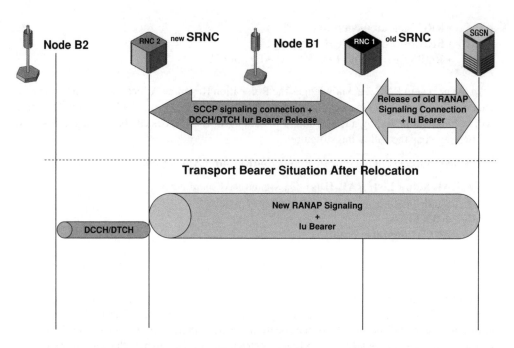

Figure 3.134   SRNS relocation (UE not involved) call flow 3/3.

The RNSAP Relocation Commit message contains the D-RNTI that was exchanged in RANAP Relocation Preparation and RANAP Resource Allocation procedures. It is the unique identifier that indicates to RNC 2 that starting with reception of RNSAP Relocation Commit it will be responsible to handle SRNC function for RRC connection that was specified on behalf of RRC context data in RRC container before.

Following Relocation Commit the SCCP Class 2 connection that carried RNSAP messages as well as Iur transport bearer for RRC signaling and IP payload exchange are deleted. The same happens at the SCCP Class 2 connection and GTP user plane tunnel on old IuPS interface.

After the relocation (UE not involved) is successfully finished a new RANAP signaling connection is active between RNC 2 and SGSN(Figure 3.134). In parallel there was/have been GTP user plane tunnel(s) (Iu bearer) for one/or more PDP contexts established. AAL2 SVCs for RRC signaling and IP payload remained active on Iub interface between RNC 2 and Node B2.

3.21 SRNS Relocation (UE Involved)

If the UE is involved in the relocation procedure it always means that a (backward) hard handover controlled by old SRNC (RNC 1) is performed. As shown in Figure 3.135 this relocation procedure may once again have impact on all ongoing signaling and user traffic exchanged between UE and CS/PS core network domains. When RNC 1 decides to perform hard handover and change of SRNC in one step (❶), a RANAP Relocation Required message will be sent to participating core network elements MSC and/or SGSN (❷), which then will set up new Iu signaling connections and Iu bearers toward RNC 2 (❸). After the handover was

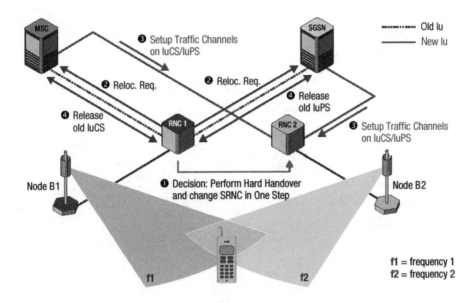

Figure 3.135   SRNS Relocation (UE involved) principle.

performed successfully signaling connections and user plane transport bearers on Iu interfaces between core network elements and RNC 1 can be released (❹).

Maybe the most significant difference to UE-not-involved SRNS relocation in the signaling flow is that neither signaling nor any other kind of data is exchanged via Iur interface. The procedure is used to perform SRNS relocation if no Iur interface is available between RNCs of the same UTRAN and to support interfrequency hard handover between UTRAN cells that use different UMTS frequency bands.

An UE-involved relocation is also executed in case of intersystem handover, which is a handover from a UTRAN cell into a neighbor cell that uses a different RAT like GSM, CDMA2000, etc. These kinds of handovers are also named inter-RAT handovers. Since in today's networks inter-RAT handovers cannot be executed without involving the core network elements and transport functions, these scenarios will be discussed in Chapter 4. However, the reader should keep in mind that especially with introduction of new interfaces and protocols, standard enhancements of Release 5 intersystem handovers become possible, which are directly executed between radio access networks using different radio technologies. An example is the Iurg interface between GSM BSC and UTRAN RNC where a new set of RNSAP messages can be used to perform intersystem handovers without involving the core network.

3.21.1 Overview

Also for this signaling example the authors have not been able to monitor a complete scenario in any network or testbed, but they have seen parts of this message flow on some interfaces. On behalf of this information a quite precise description of the overall procedure is possible despite that some uncertainties remain. For instance, it is proved that event-ID "e2a" ("change of best frequency") is used to trigger execution of the UE-involved relocation, but it might

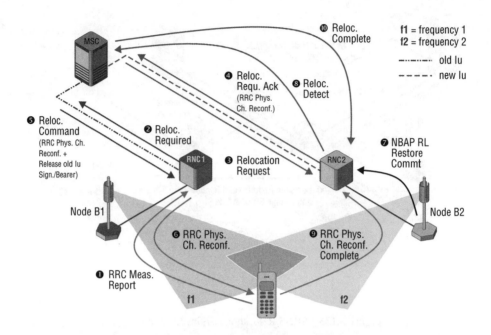

Figure 3.136   SRNS relocation (UE involved) procedure overview.

not be the only one. Following the understanding of the authors, event-ID "e2b" could also be used to define the trigger event. ("The estimated quality of the currently used frequency is below a certain threshold *and* the estimated quality of a non-used frequency is above a certain threshold.") Finally the real-network implementation of the procedure must be seen in all cases as a manufacturer-or operator-specific one that also might be changed with ongoing deployment of network structures driven by needs of network optimization.

The scenario of the call flow example shows a UE-involved SRNS relocation during an active voice call (Figure 3.136). MSC is participating, but SGSN is not involved, because there is neither a PDP context nor a present signaling connection between UE and PS domain active. The UE is still served by the cell with frequency f1 (cell 1) that belongs to Node B1, but the cell using frequency f2 (cell 2) that belongs to Node B2 is becoming stronger while received primary CPICH strength of cell 1 is fading away owing to UE's move.

Step 1: RNC 1 is SRNC and receives RRC Measurement Report from UE that cell 2 (with frequency f2) offers better conditions for the connection compared to situation on radio interface using frequency f1. Based on this measurement report the decision is made to perform hard handover and SRNS relocation in one step.

Step 2: RNC 1 sends RANAP Relocation Required message to serving MSC.

Step 3: Serving MSC sends RANAP Relocation Request to RNC 2. This message includes all information necessary to establish RANAP signaling connection and Iu bearer on new Iu interface between MSC and RNC 2.

Step 4: After RNC 2 receives Relocation Request message it builds the handover command message, in this case, RRC Physical Channel Reconfiguration Request. Physical Channel

Reconfiguration message is enclosed in RANAP Relocation Request Acknowledgment message sent from RNC 2 to MSC to confirm that necessary Iu signaling and user plane transport resources have been assigned.

Step 5: MSC sends RANAP Relocation Command to RNC 1, which is ordered to execute the handover now. The message contains the handover message as constructed and sent by RNC 2.

Step 6: On Iub and radio interface of cell 1, RRC Physical Channel Reconfiguration message is sent to UE. This message contains all parameters necessary to find the already provided dedicated physical channels of cell 2. Based on this information UE performs interfrequency hard handover.

Step 7: After UE synchronizes with cell 2, Node B2 sends an NBAP Radio Link Restore Commit message to indicate successful handover on physical radio layer.

Step 8: CRNC function of RNC 2 tiggers sending of RANAP Relocation Detect message to "tell" MSC that handover was executed on physical layer. Now MSC "knows" that UE is physically not connected anymore to cell 1. RNC 1 will be informed about this fact when it receives NBAP Radio Link Failure Indication from Node B1. This message is not shown in Figure 3.136 and could be sent at any time after Step 6. NBAP Radio Link Failure Indicaton is not a mandatory message in all cases of hard handover. It is not seen whether RNC triggered by Iu Release from IuCS deletes the assigned dedicated radio resources faster than Node B is able to report that UE lost contact. However, following reception of Radio Link Failure Indication, RNC 1 will release RRC context data and dedicated physical resources for connection with this UE as well.

Step 9: After UE has full access to dedicated physical channels in cell 2 it sends RRC Physical Channel Reconfiguration Complete message using the new radio link and hence the new Iub interface between Node B2 and RNC 2. Now RNC 2 takes over SRNC function of this connection and reactivates connection to the core network.

Step 10: Since the handover is now also completed on RRC level, RNC 2 (new SRNC) sends RANAP Relocation Complete message to serving MSC.

3.21.2 Message Flow

The message flow part starts with the triggering procedure. When RRC connection is established with UE, a number of RRC measurement tasks are defined. Measurement necessary for this scenario is interfrequency measurement and appropriate event-ID group is "e2...".

Sometime after connection is established and measurement is activated UE starts to move and reaches an area where radio conditions of used UTRAN frequency f1 become worse. This is indicated by sending one or several RRC Measurement Reports with event-ID "e2d" to SRNC. This shows that "estimated quality of the currently used frequency is below a certain threshold" and the so-to-say "value added" information of these measurement reports for SRNC is that as a consequence UE starts to monitor cells with other UMTS frequencies and/or other radio access technologies (Figure 3.137).

When UE found a cell that seems to offer required radio parameters, it sends another RRC Measurement Report including event-ID "e2a" (change of best frequency). In addition, primary scrambling code and downlink UMTS Absolute Radio Frequency Channel Number (UARFCN DL) are reported to SRNC. Both additional parameters allow unique identification of the new cell on radio interface.

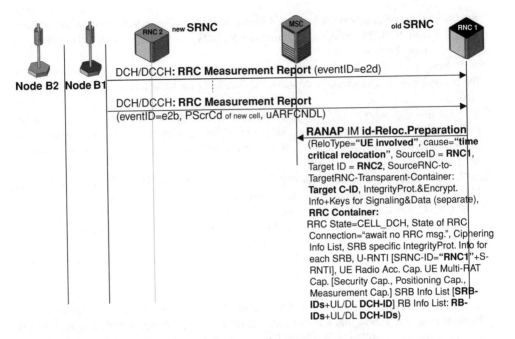

Figure 3.137   SRNS Relocation (UE involved) call flow 1/4.

The reception of the last mentioned RRC Measurement Report triggers the start of relocation procedure executed by SRNc, which starts RANAP Relocation Preparation procedure. For
relation of RANAP message names and procedure codes see previous section on SRNS Relocation (UE not involved). The RANAP message itself and also the embedded RRC container
are slightly different from the same message in the previous scenario. Target ID is derived
from primary scrambling code in RRC Measurement Report.

RANAP InitialMessage: ProcedureCode= **id-RelocationPreparation**

- RelocationType = "**UE involved**"
- Cause = "**time critical relocation**"
- Source ID = **RNC 1**
- Target ID = **RNC 2**
- **SourceRNC-to-TargetRNC-Transparent-Container**:
  – **Target Cell-ID**
  – Integrity Protection Info + Key for RRC Signaling
  – Integrity Protection Info + Key for user plane traffic
  – Encryption Info + Key for RRC Signaling
  – Encryption Info + Key for user plane traffic
  – **RRC Container**:
    ▪ RRC State=CELL_DCH
    ▪ State of RRC Connection = "await no RRC message"
    ▪ Ciphering Info List

- SRB specific IntegrityProt. Info for each SRB
- **U-RNTI** [SRNC-ID="**RNC1**"+S-RNTI]
- UE Radio Access Capabilities:
- UE Multi-RAT Cap.
  - Security Capabilities
  - Positioning Capabilities
  - Measurement Capabilities
- SRB Info List
  - **SRB-IDs**+UL/DL **DCH-ID**
- RB Info List
- **RB-IDs**+UL/DL **DCH-IDs**

```
|RANAP TS 25.413 V6.0.0 (2003-12) (RANAP)  initiatingMessage   |
|(= initiatingMessage)                                          |
|ranapPDU                                                       |
|1 initiatingMessage                                            |
|00000010 |1.1 procedureCode        |id-RelocationPreparation   |
~~~~~~~~~~~~~~~~~~~~~~~~~~~~~~~~~~~~~~~~~~~~~~~~~~~~~~~~~~~~~~~~~~~~
|***B2*** |1.3.1.1.1 id             |id-RelocationType          |
|00------ |1.3.1.1.2 criticality    |reject                     |
|-1------ |1.3.1.1.3 value          |ue-involved                |
|1.3.1.2 sequence                                               |
|***B2*** |1.3.1.2.1 id             |id-Cause                   |
|01------ |1.3.1.2.2 criticality    |ignore                     |
|1.3.1.2.3 value                                                |
|***b6*** |1.3.1.2.3.1 radioNetwork |time-critical-relocation   |
|1.3.1.3 sequence                                               |
|***B2*** |1.3.1.3.1 id             |id-SourceID                |
~~~~~~~~~~~~~~~~~~~~~~~~~~~~~~~~~~~~~~~~~~~~~~~~~~~~~~~~~~~~~~~~~~~~
|1.3.1.3.3.1 sourceRNC-ID                                       |
|***B3*** |1.3.1.3.3.1.1 pLMNidentity |92 02 f0                 |
|***B2*** |1.3.1.3.3.1.2 rNC-ID       |1001                     |
|1.3.1.4 sequence                                               |
|***B2*** |1.3.1.4.1 id             |id-TargetID                |
~~~~~~~~~~~~~~~~~~~~~~~~~~~~~~~~~~~~~~~~~~~~~~~~~~~~~~~~~~~~~~~~~~~~
|1.3.1.4.3.1 targetRNC-ID                                       |
|1.3.1.4.3.1.1 lAI                                              |
|***B3*** |1.3.1.4.3.1.1.1 pLMNidentity |92 02 f0               |
|***B2*** |1.3.1.4.3.1.1.2 lAC          |00 02                  |
|***B2*** |1.3.1.4.3.1.2 rNC-ID         |1002                   |
|1.3.1.5 sequence                                               |
|***B2*** |1.3.1.5.1 id             |id-SourceRNC-ToTargetRNC-  |
|                                   TransparentContainer        |
~~~~~~~~~~~~~~~~~~~~~~~~~~~~~~~~~~~~~~~~~~~~~~~~~~~~~~~~~~~~~~~~~~~~
|***B4*** |1.3.1.5.3.4 targetCellId |18196                      |
|toTarget-RRC-Container from 3GPP TS 25.331 V6.0.1 (2004-01) (TTRC)|
```

```
|toTargetRNC-Container                                                    |
|1 toTargetRNC-Container                                                  |
|1.1 srncRelocation                                                       |
|1.1.1 r3                                                                 |
|1.1.1.1 sRNC-RelocationInfo-r3                                           |
|----00-- |1.1.1.1.1 stateOfRRC                                 |cell-DCH  |
|***b4*** |1.1.1.1.2 stateOfRRC-Procedure                      |awaitNoRRC-|
|                                                              Message    |
|--1----- |1.1.1.1.3 cipheringStatus                           |notStarted|
|1.1.1.1.4 cipheringInfoPerRB-List                                        |
/\/\/\/\/\/\/\/\/\/\/\/\/\/\/\/\/\/\/\/\/\/\/\/\/\/\/\/\/\/\/\/\/\/\/\/\/\/\/\
|---1---- |1.1.1.1.6 integrityProtectionStatus                 |notStarted|
|1.1.1.1.7 srb-SpecificIntegrityProtInfo                                  |
/\/\/\/\/\/\/\/\/\/\/\/\/\/\/\/\/\/\/\/\/\/\/\/\/\/\/\/\/\/\/\/\/\/\/\/\/\/\/\
|1.1.1.1.8 U-RNTI                                                         |
|**b12*** |1.1.1.1.8.1 srnc-Identity                          |1001       |
|**b20*** |1.1.1.1.8.2 S-RNTI              |'00000000000000000010'B|
|1.1.1.1.9 ue-RadioAccessCapability                                       |
/\/\/\/\/\/\/\/\/\/\/\/\/\/\/\/\/\/\/\/\/\/\/\/\/\/\/\/\/\/\/\/\/\/\/\/\/\/\/\
|1.1.1.1.9.6 ue-MultiModeRAT-Capability                                   |
|1.1.1.1.9.6.1 multiRAT-CapabilityList                                    |
|-1------ |1.1.1.1.9.6.1.1 supportOfGSM                        |1         |
/\/\/\/\/\/\/\/\/\/\/\/\/\/\/\/\/\/\/\/\/\/\/\/\/\/\/\/\/\/\/\/\/\/\/\/\/\/\/\
|1.1.1.1.9.7 securityCapability                                           |
|**b16*** |1.1.1.1.9.7.1 cipheringAlgorithmCap                 |uea0      |
|**b16*** |1.1.1.1.9.7.2 integrityProtectionAlgorithmCap|uia1      |
|1.1.1.1.9.8 ue-positioning-Capability                                    |
/\/\/\/\/\/\/\/\/\/\/\/\/\/\/\/\/\/\/\/\/\/\/\/\/\/\/\/\/\/\/\/\/\/\/\/\/\/\/\
|1.1.1.1.9.9 measurementCapability                                        |
/\/\/\/\/\/\/\/\/\/\/\/\/\/\/\/\/\/\/\/\/\/\/\/\/\/\/\/\/\/\/\/\/\/\/\/\/\/\/\
|1.1.1.1.13 srb-InformationList                                           |
|1.1.1.1.13.1 sRB-InformationSetup                                        |
|***b5*** |1.1.1.1.13.1.1 rb-Identity                          |1         |
/\/\/\/\/\/\/\/\/\/\/\/\/\/\/\/\/\/\/\/\/\/\/\/\/\/\/\/\/\/\/\/\/\/\/\/\/\/\/\
|1.1.1.1.13.1.3 rb-MappingInfo                                            |
/\/\/\/\/\/\/\/\/\/\/\/\/\/\/\/\/\/\/\/\/\/\/\/\/\/\/\/\/\/\/\/\/\/\/\/\/\/\/\
|1.1.1.1.13.1.3.1.1.1.1 ul-TransportChannelType                           |
|--00100- |1.1.1.1.13.1.3.1.1.1.1.1 dch                        |5         |
|***b4*** |1.1.1.1.13.1.3.1.1.1.2 logicalChannelIdentity|1         |
|1.1.1.1.13.1.3.1.1.1.3 rlc-SizeList                                      |
|         |1.1.1.1.13.1.3.1.1.1.3.1 configured                 |0         |
|-----000 |1.1.1.1.13.1.3.1.1.1.4 mac-LogicalChannelPr..|1         |
|1.1.1.1.13.1.3.1.2 dl-LogicalChannelMappingList                          |
|1.1.1.1.13.1.3.1.2.1 dL-LogicalChannelMapping                            |
|1.1.1.1.13.1.3.1.2.1.1 dl-TransportChannelType                           |
|***b5*** |1.1.1.1.13.1.3.1.2.1.1.1 dch                        |5         |
|-0000--- |1.1.1.1.13.1.3.1.2.1.2 logicalChannelIdentity|1         |
|1.1.1.1.13.2 sRB-InformationSetup                                        |
|***b5*** |1.1.1.1.13.2.1 rb-Identity                          |2         |
```

```
|1.1.1.1.14 rab-InformationList                                        |
|1.1.1.1.14.1 rAB-InformationSetup                                     |
|1.1.1.1.14.1.1 rab-Info                                               |
|1.1.1.1.14.1.1.1 rab-Identity                                         |
|***b8*** |1.1.1.1.14.1.1.1.1 gsm-MAP-RAB-Identity       |1            |
|1.1.1.1.14.1.2 rb-InformationSetupList                                |
|1.1.1.1.14.1.2.1 rB-InformationSetup                                  |
|--00101- |1.1.1.1.14.1.2.1.1 rb-Identity                |6            |
|1.1.1.1.14.1.2.1.3 rb-MappingInfo                                     |
|1.1.1.1.14.1.2.1.3.1 rB-MappingOption                                 |
|1.1.1.1.14.1.2.1.3.1.1 ul-LogicalChannelMappings                      |
|1.1.1.1.14.1.2.1.3.1.1.1 oneLogicalChannel                            |
|1.1.1.1.14.1.2.1.3.1.1.1.1 ul-TransportChannelType                    |
|--00000- |1.1.1.1.14.1.2.1.3.1.1.1.1.1 dch              |1            |
|1.1.1.1.14.1.2.1.3.1.1.1.2 rlc-SizeList                               |
|         |1.1.1.1.14.1.2.1.3.1.1.1.2.1 configured       |0            |
|-000---- |1.1.1.1.14.1.2.1.3.1.1.1.3 mac-LogicalChann..|1            |
|1.1.1.1.14.1.2.1.3.1.2 dl-LogicalChannelMappingList                   |
|1.1.1.1.14.1.2.1.3.1.2.1 dL-LogicalChannelMapping                     |
|1.1.1.1.14.1.2.1.3.1.2.1.1 dl-TransportChannelType                    |
|00000--- |1.1.1.1.14.1.2.1.3.1.2.1.1.1 dch              |1            |
|1.1.1.1.14.1.2.2 rB-InformationSetup                                  |
|***b5*** |1.1.1.1.14.1.2.2.1 rb-Identity                |7            |
|1.1.1.1.14.2 rAB-InformationSetup                                     |
|1.1.1.1.14.2.1 rab-Info                                               |
|1.1.1.1.14.2.1.1 rab-Identity                                         |
|***b8*** |1.1.1.1.14.2.1.1.1 gsm-MAP-RAB-Identity       |5            |
|1.1.1.1.14.2.2 rb-InformationSetupList                                |
|1.1.1.1.14.2.2.1 rB-InformationSetup                                  |
|***b5*** |1.1.1.1.14.2.2.1.1 rb-Identity                |5            |
```

Message Example 3.40 RANAP Relocation Required incl. SourceRNC-to-TargetRNC-Transparent-Container and RRC Container

The RANAP Relocation Required message in the message example contains the discussed parameters, but RAB-ID values (1 and 5) in rb-InformationSetupList indicate that the UE in the example has a voice call and PDP contexts active simultaneously. Radio bearer mapping options and DCH parameters can be found more detailed in Iub scenarios "Mobile Originated Call (MOC)" and "PDP Context Activation/Deactivation."

On behalf of included target ID serving MSC is able to detect that RNC 2 shall become the new SRNC of the connection. So it starts RANAP Relocation Resource Allocation procedure with RNC 2 (Figure 3.138). In the Initial Message of this procedure, MSC defines the number of RABs and their parameters.

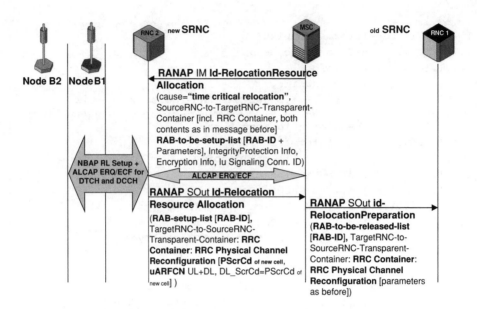

Figure 3.138   SRNS Relocation (UE involved) call flow 2/4.

RANAP InitialMessage: ProcedureCode= **id-RelocationResourceAllocation**

- RelocationType = **"UE involved"**
- Cause = **"time critical relocation"**
- **RAB-to-be-setup-list**
  - **RAB-IDs** + Parameters
- IntegrityProtection Info
- Encryption Info
- Iu Signaling Connection ID
- **SourceRNC-to-TargetRNC-Transparent-Container**:
  - **Target Cell-ID**
  - Integrity Protection Info + Key for RRC Signaling
  - Integrity Protection Info + Key for user plane traffic
  - Encryption Info + Key for RRC Signaling
  - Encryption Info + Key for user plane traffic
  - **RRC Container**:
    - RRC State=CELL_DCH
    - State of RRC Connection = "await no RRC message"
    - Ciphering Info List
    - SRB specific IntegrityProt. Info for each SRB
    - **U-RNTI** [SRNC-ID="**RNC1**"+S-RNTI]
    - UE Radio Access Capabilities:
      - UE Multi-RAT Cap.
      - Security Capabilities

      – Positioning Capabilities
      – Measurement Capabilities
      – SRB Info List
      ▪ **SRB-IDs**+UL/DL **DCH-ID**
      ▪ RB Info List
        – **RB-IDs**+UL/DL **DCH-IDs**

When RNC 2 receives RANAP Relocation Request message it provides necessary resources to establish dedicated physical channels on radio interface of new cell, dedicated transport channels for signaling and voice packets on Iub, and an AAL2 SVC that acts as physical transport bearer of Iu bearer on new IuCS interface.

When all these NBAP and ALCAP procedures are successfully finished, RNC 2 acknowledges RANAP Relocation Request. The appropriate signaling message contains the RRC handover message that is most likely an RRC Physical Channel Reconfiguration message. Depending on whether changes on transport channel level or QoS are necessary or required instead of Physical Channel Reconfiguration, an RRC Transport Channel Reconfiguration or RRC Radio Bearer Reconfiguration message could also be used as handover command that is always embedded in TargetRNC-to-SourceRNC-Transparent-Container. The RRC Physical Channel Reconfiguration message contains all parameters that allow UE to find provided dedicated physical channels in the new cell. The meaning of RAB Setup List is the same as in the UE-not-involved SRNS Relocation scenario.

RANAP SuccessfulOutcome: ProcedureCode = **id-Relocation Resource Allocation**

- **RAB-setup-list**
  – **RAB-IDs**
- TargetRNC-to-SourceRNC-Transparent-Container
  – **RRC Container**
    ▪ **RRC Physical Channel Reconfiguration**
    – **Primary Scrambling Code (PScrCd) new cell**
      ▪ **UARFCN** Uplink
      ▪ **UARFCN** Downlink
      ▪ Downlink Scrambling Code = PScrCd of new cell

MSC sends new Relocation Command to old SRNC. This message defines the IDs of RABs to be released in old RNS and it also contains the transparently forwarded RRC Physical Channel Reconfiguration message:

RANAP SuccessfulOutcome: ProcedureCode = **id-RelocationPreparation**

- **RAB-to-be-released-list**
  – **RAB-IDs**
- TargetRNC-to-SourceRNC-Transparent-Container
  – **RRC Container**:
    ▪ **RRC Physical Channel Reconfiguration**
    – **Primary Scrambling Code (PScrCd) new cell**

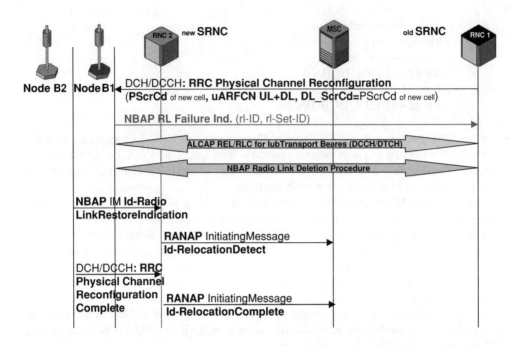

Figure 3.139 SRNS relocation (UE involved) call flow 3/4.

- **UARFCN** Uplink
- **UARFCN** Downlink
- Downlink Scrambling Code = PScrCd of new cell

RNC 1 extracts the Physical Channel Reconfiguration messages from the container and sends it to UE via Iub and Uu interfaces (Figure 3.139). Following the reception UE performs handover into the new cell.

Node B1 detects that radio contact with UE is lost and sends NBAP Radio Link Failure Indication. This triggers release of dedicated transport resources on old Iub executed by RNC 1.

Simultaneously Node B2 sends NBAP Radio Link Restore Indication to RNC 2 to inform that UE found provided DCHS on radio interface and synchronized with Node B. RNC 2 informs MSC that relocation was detected on physical level by sending RANAP Relocation Detect message.

Then UE sends RRC Physical Channel Reconfiguration Complete (in other cases: Transport Channel Reconfiguration Complete or Radio Bearer Reconfiguration Complete) to RNC 2, which is from now on SRNC of the active connection.

RNC 2 informs MSC that relocation is completed by sending RANAP Relocation Complete message.

Now MSC is sure that no further data or signaling regarding this single UE connection needs to be exchanged with RNC 1 anymore. Hence, RANAP Iu Release and SCCP Release procedures (see Figure 3.140) are triggered and executed by core network element, which is the last step of the successful relocation.

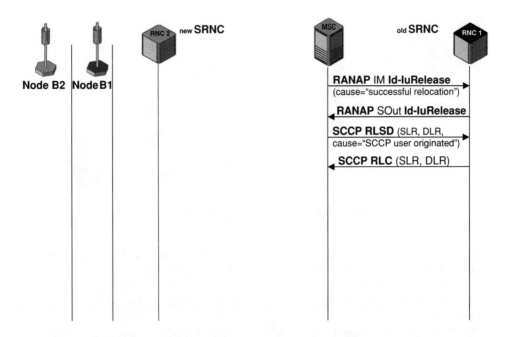

Figure 3.140   SRNS relocation (UE involved) call flow 4/4.

3.22 Short Message Service (SMS) In UMTS Networks

Also in UMTS the Short Message Service (SMS) – already well-known from GSM – is available and especially in the European region is one of the most important services. The following description gives an overview of SMS network architecture and signaling procedures in UTRAN.

3.22.1 SMS Network Architecture Overview

From the SMS point of view the network consists of the following network elements that are involved in short message (SM) exchange (see Figure 3.141).

The MS may submit SM to the network or receive SM that are delivered by the network.

Serving GPRS Support Node (SGSN) and serving Mobile Switching Center (MSC) provide alternative ways to transmit an SM from or to an MS. It is possible to perform a rerouting, e.g. in case that a paging from the circuit switched core network domain (MSC) is not successful, the same paging will be executed by the SGSN once again. The paging information is then forwarded using the Gs interface.

There is no general rule as to which core network domain is preferred for SM submission and delivery. 3GPP TS 23.040 recommends to use the packet switched CN domain (send/receive SM via SGSN) because of higher efficiency of resource allocation, but network operators as well as equipment manufacturers in the first stage of 3G deployment seem to prefer the way via MSC most likely because of the already proven high reliability of the SMS paths in this part of the network.

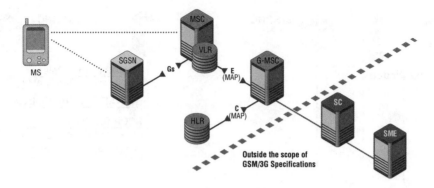

Figure 3.141   SMS network architecture overview.

The Visitor Location Register (VLR) is involved in case of paging an MS for SMS delivery and provides the MSISDN of the MS in case of mobile-originated SM.

On E-interface the SM is sent from/to a Gateway MSC (GMSC) to a Short Message Service Center (SC or SMSC). This is necessary to provide routing info in case of mobile-terminated SM (SMS delivery).

The SC is always connected to a GMSC, but not all GMSCs are connected to SCs. The interface between SC and GMSC is out of scope of GSM and 3G specifications, but often realized on behalf of an SS#7/MAP protocol stack.

Between Short Message Entity (SME) and SC as a rule IP-based protocol stacks run. Most SME belong to independent service providers and are not owned by GSM or 3G network operators. An SME can for instance be used to send SM from the Internet.

3.22.2 SMS Protocol Architecture

A look into the SMS protocol architecture (Figure 3.142) of the UTRAN shows that the SM protocols are users of the RRC and RANAP protocols. This means that SM are transported as low-priority NAS signaling messages transparently between the MS and the MSC (or SGSN, which is not shown in this figure).

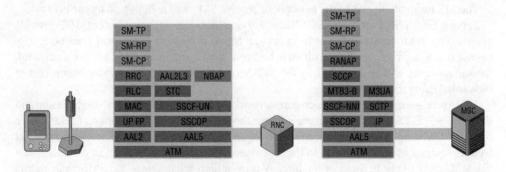

Figure 3.142   SMS protocol architecture.

The Short Message Control Protocol (SM-CP) provides SM transport functions between MS and MSC/SGSN as well as between following core network elements, which are involved in SM transport and routing, e.g. MSC/SGSN and GMSC. Messages belonging to the same SM-CP transaction, e.g. an SM MO, have the same transaction Id value.

Short Message Routing Protocol (SM-RP) provides addressing functions from MS to SMSC in case of short message mobile-originated (SM MO) services and from SMSC to MS in case of short message mobile-terminated (SM MT) services.

Short Message Transfer Layer is responsible for direct communication between MS and SM-SC and vice versa and contains the content of the message itself, e.g. the written text.

Short message transmission is specified as "low-priority NAS signaling" in 3GPP specifications. Hence, many messages regarding setup and release of RRC and RANAP connection are the same as in case of other already discussed signaling procedures like location update or GPRS attach. The focus in the following call flow diagrams will be on the messages belonging to new SM protocol layers and new parameter values in already known messages.

3.22.3 Mobile-Originated Short Message

Overview
If an MS sends a short message to the network, the international standard documents call this to submit a short message (Figure 3.143).

Step 1: Before an SM can be submitted an RRC connection needs to be established. The SM will be sent in a DCCH identified by the highest available SRB value, most likely logical channel = "4". This DCCH is only used for low-priority NAS signaling exchange and setup during RRC Connection Setup procedure (for details see description of Location Update procedure on Iub).

Step 2: Using the RRC direct transfer service an SMS-SUBMIT message is sent from the mobile to the network.

Step 3: When RNC receives this SMS-SUBMIT, it starts setting up an SCCP/RANAP connection and forwards this SMS-SUBMIT to the MSC (or SGSN) using RANAP direct transfer features.

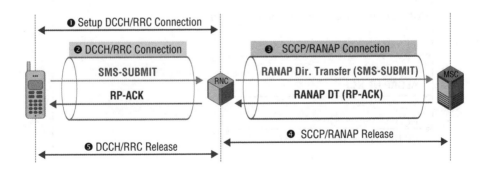

Figure 3.143   Short message mobile-originated (SM MO) procedure overview.

Step 4: The MS waits until SM-SC acknowledges reception of the mobile-originated SM. Then the SCCP/RANAP connection is released.

Step 5: Triggered by SCCP Release on IuCS (or IuPS), RRC connection may be released as well or RNC requests the MS to change into CELL_FACH and later CELL_PCH or CELL_URA state.

Message Flow

The setup of RRC connection and a dedicated transport channel that carries the DCCH for SM message exchange is already well known from other signaling procedures on Iub interface. However, a difference is found in the rrcConnectionRequest message since its establishment cause in case of SM MO is "originatingLowPrioritySignaling." Since the MS already needs to be attached either to CS or PS domain before it is allowed to send an SM, it uses either TMSI or P-TMSI for identification (see Figure 3.144):

> **RACH**: UL RLC TMD **rrcConnectionRequest** (TMSI or P-TMSI, establishmentCause=originatingLowPrioritySignaling)

It should be noted that all NAS messages including the short message will be transported in the DCH identified by a unique VPI/VCI/CID value on Iub interface that allows easy filtering of the NAS call flow sequence.

After RRC connection is established, MM/CC/SM messages can be exchanged embedded in RRC Direct Transfer messages. A Connection Management Service Request (CMSREQ)

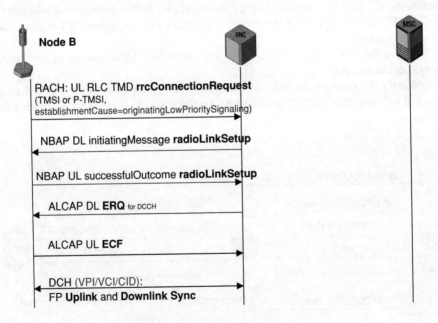

Figure 3.144   SM MO call flow 1/5.

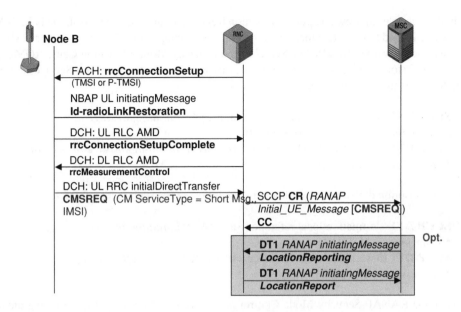

Figure 3.145   SM MO call flow 2/5

is sent by the mobile (Figure 3.145). The CM Service Type information element inside this message indicates that the MS wants to send a short message. In addition the IMSI is included as unique user identifier.

DCH: UL RRC initialDirectTransfer **CMSREQ (*CM ServiceType = "Short Message", IMSI*)**

When RNC receives this NAS message, it starts setting up SCCP connection on IuCS interface on behalf of SCCP Connection Request message. This CR message includes a RANAP Initial_UE_Message that carries the embedded CMSREQ message. The Source Local Reference Number in the CR message identifies the calling party of this SCCP connection. It will be used as destination local reference number in all messages sent by the other side (called party) of the SCCP connection:

SCCP CR (source local reference=**a**, *RANAP Initial_UE_Message*, NAS message = **CMSREQ [*CM ServiceType = "Short Message", IMSI*]**

When the RNC receives the SCCP Connection Confirm (CC) message from the MSC the SCCP connection is established successfully:

SCCP CC (source local reference=**b**, destination local reference=**a**)

Both values, source local reference and destination local reference, can be used as filter criteria for the SM call flow and to identify uplink and downlink messages on IuCS or IuPS interface.

In the example call flow an optional Location Report is requested by the MSC. This RANAP procedure is used to get actual location info from the Serving Mobile Location Center (SMLC) that will be forwarded by MSC or SGSN to the Gateway Mobile Location Center (GMLC). The GMLC stores all location relevant data of users subscribed to Location Services (LCS). The SMLC is usually colocated with the SRNC.

SCCP DT1 (destination local reference=**a**, *RANAP initiatingMessage* [*procedureCode = LocationReporting*])
SCCP DT1 (destination local reference=**b**, *RANAP initiatingMessage* [*procedure Code = LocationReport*])

As another option the already well-known Authentication procedure may follow:

SCCP DT1 (destination local reference=**a**, *RANAP initiatingMessage*, NASmessage=**AUTREQ**)
SCCP DT1 (destination local reference=**b**, *RANAP successfulOutcome*, NASmessage=**AUTREP**)

With the RANAP Security Mode Control procedure (see Figure 3.146), ciphering and/or integrity protection between RNC and UE are activated:

SCCP DT1 (destination local reference=**a**, *RANAP initiatingMessage* [*procedureCode = SecurityModeControl*])
Iub: **DCH** (VPI/VCI/CID): **RRC SecurityModeCommand**
Iub: **DCH** (VPI/VCI/CID): **RRC SecurityModeComplete**

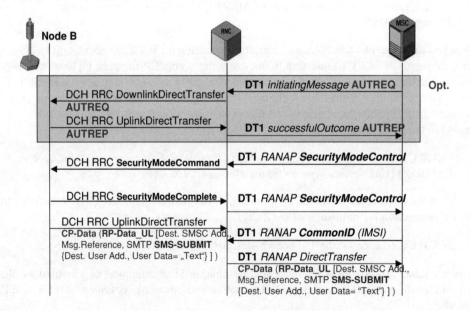

Figure 3.146   SM MO call flow 3/5.

SCCP DT1 (destination local reference=**b**, *RANAP successfulOutcome*
[*procedureCode = SecurityModeControl*])

Immediately after the security functions have been successfully activated the MS sends its short message embedded in an RRC UplinkDirectTransfer message:

Iub: **DCH** (VPI/VCI/CID): **RRC UplinkDirectTransfer**: CP-Data (RP-Data_UL
[***Destination SMSC Address***, MessageReference=**c**, **SMTP SMS-SUBMIT**
{*Destination User Address, User Data = "**Text**"*}])

The embedded NAS message contains Short Message Control Protocol (CP), Routing Protocol (RP), and Short Message Transport Protocol (SMTP) information. The CP is just a transport layer for SMS and provides services to the upper layer protocols that ensure end-to-end SM exchange.

The Short Message RP is responsible for the message exchange between MS (or any other short message entity [SME]) and SM-SC. The main parameter of the RP in uplink direction is the E.164 address of the SM-SC. This address is stored on the USIM inside the mobile and can be changed using remote operation of the SIM Application Toolkit by the network operator.

The SMTP layer finally provides the information entered by the subscriber: the B-Party Destination User Address for this SMS transaction and the contents of the short message, e.g. text, and also pictures, predefined animations, or e-mail are possible. There are also possibilities to concatenate several SM and perform SM compression as described in 3GPP TS 23.040. SMS Alphabet encoding is specified in 3GPP TS 23.038. In the standard alphabet letters and numbers are encoded in septets (each letter 7 bit).

While the SM arrives at SRNC via Iub interface on IuCS a RANAP Initiating Message that contains the Common ID procedure code is received by RNC to check the true identity of the subscriber (IMSI):

SCCP DT1 (destination local reference=**a**, *RANAP initiatingMessage*
[*procedureCode = Common ID {IMSI}*])

Then the SM is forwarded transparently on behalf of RANAP DirectTransfer from SRNC to MSC.

SCCP DT1 (destination local reference=**b**, *RANAP initiatingMessage*
DirectTransfer: CP-Data (RP-Data_UL [***Destination SMSC Address***,
MessageReference=**c**, **SMTP SMS-SUBMIT** {*Destination User Address, User*
*Data = "**Text**"*}])

The Short Message CP is designed in a way that every CD-Data block is acknowledged on each point-to-point-connection between the MS and the SM-SC to ensure that the under-laying transport layer (in this case RANAP and RRC) works error-free, because there is no

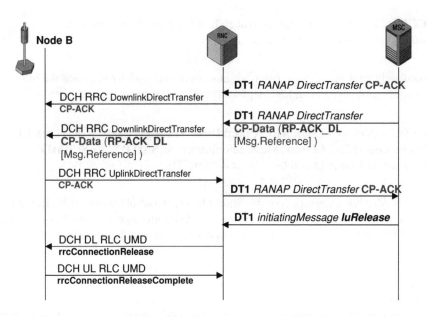

Figure 3.147   SM MO call flow 4/5.

explicit acknowledgment, e.g., to a RANAP DirectTransfer message. This is the reason why the following two messages are sent (see Figure 3.147):

SCCP DT1 (destination local reference=**a**, *RANAP initiatingMessage DirectTransfer: **CP-ACK***)

Iub: **DCH** (VPI/VCI/CID): RRC DownlinkDirectTransfer (**CP-ACK**)

After the SM-SC receives the submitted SM, it also sends an acknowledgment to the MS. However, this acknowledgment is on Short Message RP level. The RP-ACK message in downlink direction contains the same message reference value as the RP-Data block that carried the SM content in uplink direction before:

SCCP DT1 (destination local reference=**a**, *RANAP initiatingMessage DirectTransfer: CP-Data* [**RP-ACK** {MessageReference=**c**}])

Iub: **DCH** (VPI/VCI/CID): RRC DownlinkDirectTransfer (CP-Data [**RP-ACK** {MessageReference=**c**}])

Now error-free reception of these RP-ACK messages is also acknowledged on CP level on both interfaces:

Iub: **DCH** (VPI/VCI/CID): RRC UplinkDirectTransfer (**CP-ACK**)

SCCP DT1 (destination local reference=**b**, *RANAP initiatingMessage DirectTransfer: **CP-ACK***)

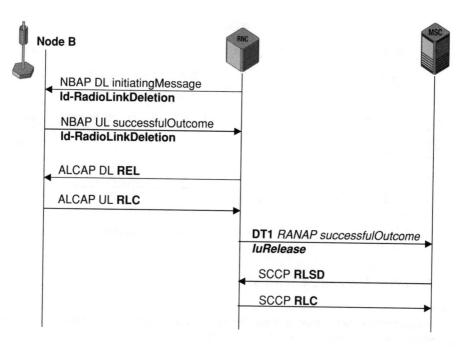

Figure 3.148   SM MO call flow 5/5.

Now the SM MO transaction procedure is complete and the SCCP/RANAP connection on IuCS can be released. The first IuRelease contains a release cause (see Figure 3.148):

SCCP DT1 (destination local reference=**a**, *RANAP initiatingMessage*, ***IuRelease*** (Id Cause))

This RANAP initial IuRelease message triggers the RRC Connection Release on Iub interface.

Following RRC Connection Release the radio resources (Scrambling Codes, Channelization codes, etc.) are deleted by CRNC, and then the AAL2 SVC of the DCH that carried the DCCHs is released as well.

On IuCS interface after successful release procedures on Iub the successfulOutcome of IuRelease is indicated by SRNC:

SCCP DT1 (destination local reference=**b**, *RANAP successfulOutcome*, ***IuRelease***)

Finally the SCCP connection is released:

SCCP RLSD (source local reference=**b**, destination local reference=**a, Release Cause**)
SCCP RLC (source local reference=**a**, destination local reference=**b**)

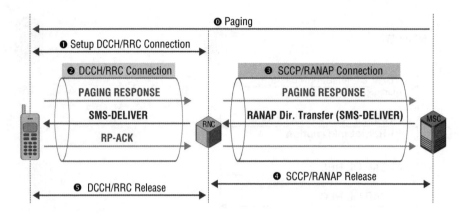

Figure 3.149   Short message mobile-terminated (SM MT) procedure overview.

3.22.4 Mobile-Terminated Short Message

Overview

If an MS receives a short message from the network the international standard documents name this to deliver a short message (Figure 3.149).

Step 0: The MS that shall receive the short message is paged in several cells belonging to the same LA, RA, or URA or in just one cell – depending from the present RRC state of the UE.

Step 1: If there is no active RRC connection then such a connection needs to be established. Once again the SM will be transmitted using the DCCH for low-priority NAS signaling.

Step 2: First the MS sends a Paging Response message to the network using the RRC direct transfer service.

Step 3: When RNC receives this paging response, it starts setting up an SCCP/RANAP connection to forward all NAS messages transparently to the MSC (or SGSN) using RANAP direct transfer features. After reception of the paging response the network sends the SM enclosed in an SMS-DELIVER message to the mobile and waits for positive acknowledgment of this transaction.

Step 4: After SMS RP Acknowledgment the SCCP/RANAP connection is released.

Step 5: Triggered by SCCP Release on IuCS (or IuPS) RRC Connection may be released as well or RNC requests the MS to change into CELL_FACH and later into CELL_PCH or CELL_URA state.

Message Flow

It is assumed that the Node B that is monitored on Iub interface has three cells. Hence, there are three PCHs that differ in the CID value of their AAL2 SVC (Figure 3.150). The (S)RNC receives the intial paging message from the MSC. It is a RANAP message embedded in an SCCP UDT message, which means connectionless SCCP message transfer. The Called and Calling Party Address in this UDT message that can be either signaling point codes (SPC) or E.164 (Global Title) format represent the addresses of RNC (called party) and MSC

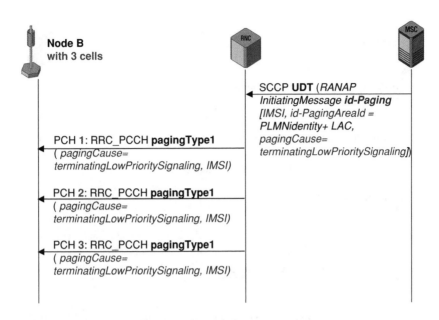

Figure 3.150 SM MT call flow 1/6.

(calling party). The RANAP paging message contains the IMSI as unique identifier of the MS to be paged, the paging area ID (in the example case, location area represented by mobile country code [PLMNidentity] + location area code), and the paging cause (in case of SM MT: "terminatingLowPrioritySignaling").

SCCP **UDT** (Called_Party_Address = **e** (**RNC**), Calling_Party_Address = **f** (**MSC**) [*RANAP InitiatingMessage* **id-Paging** {*IMSI, id-PagingAreaId = PLMNidentity + LAC, pagingCause= terminatingLowPrioritySignaling*}])

The RNC processes the received paging message and sends – depending on the RRC state of the UE – an RRC pagingType1 or pagingType2 message to all cells of all Node Bs within the paging area (in this example only one Node B is monitored):

PCH (VPI=**g**, VCI=**h**, CID=**i**): RRC_PCCH **pagingType1** (*pagingCause = terminatingLowPrioritySignaling, IMSI*)

PCH (VPI=**g**, VCI=**h**, CID=**k**): RRC_PCCH **pagingType1** (*pagingCause = terminatingLowPrioritySignaling, IMSI*)

PCH (VPI=**g**, VCI=**h**, CID=**l**): RRC_PCCH **pagingType1** (*pagingCause = terminatingLowPrioritySignaling, IMSI*)

The setup of RRC connection and dedicated transport channel is the same as in case of SM MO. Only the establishment cause in rrcConnectionRequest message is derived from the paging cause in pagingType1 message received by UE before. The UE is located in cell 2 of

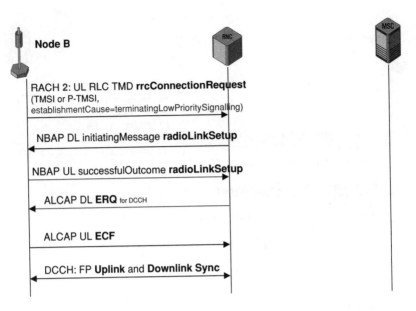

Figure 3.151   SM MT call flow 2/6.

the monitored Node B – hence it sends its rrcConnectionRequest on RACH 2, in which the
CID value of AAL2 SVC is different from those in RACH 1 and 3 (Figure 3.151).

RACH 2 (VPI=**g**, VCI=**h**, CID=**m**): UL RLC TMD **rrcConnectionRequest** (*TMSI
or P-TMSI*, establishmentCause=**terminatingLowPrioritySignaling**)

After radio link and RRC connection are established MM/CC/SM messages can be exchanged embedded in RRC Direct Transfer messages. This time a Paging Response (PRES)
instead of the Connection Management Service Request (CMSREQ) in case of SM MO is sent
by the mobile. The Paging Response message contains the IMSI.

Iub: **DCH** (VPI=**g** / VCI=**h** / CID=**o**): UL RRC initialDirectTransfer **PRES** (*IMSI*)

PRES is forwarded to the MSC.

SCCP CR (source local reference=**a**, *RANAP Initial_UE_Message*,
NASmessage=**PRES** [IMSI])

When the RNC receives the SCCP Connection Confirm message from the MSC the SCCP
connection is established successfully:

SCCP CC (source local reference=**b**, destination local reference=**a**)

Both values, source local reference and destination local reference, can be used once again
as filter criteria for the SM call flow and to identify uplink and downlink messages on IuCS or
IuPS interface.

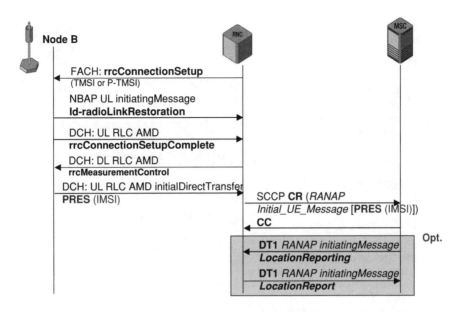

Figure 3.152   SM MT call flow 3/6.

Once again we also see the optional Location Report procedure (Figure 3.152) requested by the MSC.

SCCP DT1 (destination local reference=**a**, *RANAP initiatingMessage* [*procedureCode=LocationReporting*])

SCCP DT1 (destination local reference=**b**, *RANAP initiatingMessage* [*procedure Code=LocationReport*])

Authentication and Security Mode procedure (Figure 3.153) are exactly the same as in case of SM MO:

SCCP DT1 (destination local reference=**a**, *RANAP initiatingMessage*, NASmessage=**AUTREQ**)
SCCP DT1 (destination local reference=**b**, *RANAP successfulOutcome*, NASmessage=**AUTREP**)
SCCP DT1 (destination local reference=**a**, *RANAP initiatingMessage* [*procedureCode = SecurityModeControl*])
Iub: **DCH** (VPI=**g**/VCI=**h**/CID=**o**): **RRC SecurityModeCommand**
Iub: **DCH** (VPI=**g**/VCI=**h**/CID=**o**): **RRC SecurityModeComplete**
SCCP DT1 (destination local reference=**b**, *RANAP successfulOutcome* [*procedureCode = SecurityModeControl*])

Now the Common ID is sent on the IuCS interface in downlink direction:

SCCP DT1 (destination local reference=**a**, *RANAP initiatingMessage* [*procedureCode =Common ID*])

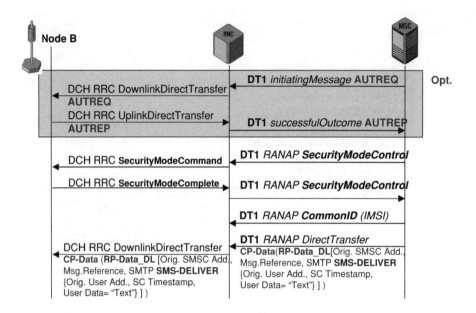

Figure 3.153   SM MT call flow 4/6.

In difference to the SM MO scenario now the short message content is delivered in downlink direction, but more or less with the same messages and similar parameters. There are differences besides the fact that the SM is sent in downlink direction.

Message Reference value is different and independent from the value used for same parameter in SM MO scenario.

The message type in SMTP is SMS-DELIVER. SMS-DELIVER contains the A-Party Originating User Address (this means the MSISDN of the SM sender) and the Service Center Timestamp. Both together, Originating User Address and SC Timestamp represent the unique identifier of each SM. Also if two SM from the same originating user arrive at SM-SC with only a very short time difference – the SC timestamp will always be different.

> **SCCP DT1** (destination local reference=**a**, *RANAP initiatingMessage*
> *DirectTransfer:* **CP-Data [RP-Data_DL** {Originating SM-SC Address,
> MessageReference=**n**, **SMTP SMS-DELIVER** {Originating User Address, Service
> Center Timestamp, User Data = "**Text**"}])

The SM is forwarded to the MS via Iub interface (Figure 3.154):

> Iub: **DCH** (VPI=**g**/VCI=**h**/CID=**o**): **RRC** DownlinkDirectTransfer: **CP-Data**
> (**RP-Data_DL** [Originating SM-SC Address, MessageReference=**n**, **SMTP**
> **SMS-DELIVER** {Originating User Address, Service Center Timestamp,
> User Data ="**Text**" }])

Now we will see CP-ACK on both Iub and Iu interface again as already commented in case of SM MO:

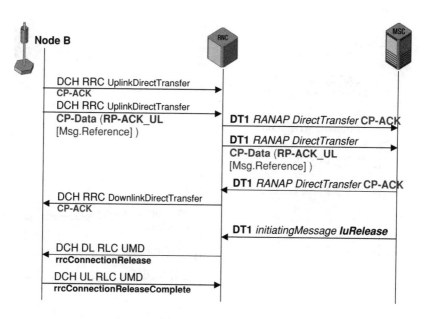

Figure 3.154 SM MT call flow 5/6.

Iub: **DCH** (VPI=**g**/VCI=**h**/CID=**o**): RRC UplinkDirectTransfer (**CP-ACK**)

In case of this example call trace, it is very obvious that CP-ACK and RP-ACK are completely independent from each other, because RP-ACK_UL on Iub is sent before CP-ACK on IuCS:

Iub: **DCH** (VPI=**g**/VCI=**h**/CID=**o**): RRC UplinkDirectTransfer (**CP-Data [RP-ACK** {MessageReference=**n**}])

SCCP DT1 (destination local reference=**b**, *RANAP initiatingMessage DirectTransfer: CP-ACK*)

SCCP DT1 (destination local reference=**b**, *RANAP initiatingMessage DirectTransfer: CP-Data [RP-ACK* {MessageReference=**n**}])

Now error-free reception of the RP-ACK messages is also acknowledged on CP level, but in different message order as in the case of SM MO:

SCCP DT1 (destination local reference=**a**, *RANAP initiatingMessage DirectTransfer: CP-ACK*)

Iub: **DCH** (VPI=**g**/VCI=**h**/CID=**o**): RRC UplinkDirectTransfer (**CP-ACK**)

Now the SM MO transaction procedure is complete and the SCCP/RANAP connection on IuCS can be released. The first IuRelease contains a release cause:

SCCP DT1 (destination local reference=**a**, *RANAP initiatingMessage, **IuRelease*** (Id Cause))

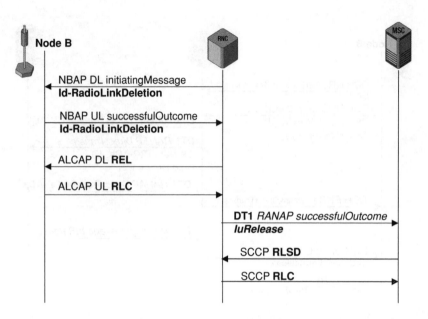

Figure 3.155   SM MT call flow 6/6.

This RANAP initial IuRelease message triggers the RRC Connection Release on Iub interface (Figure 3.154).

Following RRC Connection Release the radio resources (Scrambling Codes, Channelization codes, etc.) are deleted by CRNC, and then the AAL2 SVC of the DCH that carried the DCCHs is released as well (Figure 3.155).

On IuCS interface after successful release procedures on Iub, the successfulOutcome of IuRelease is indicated by SRNC:

SCCP DT1 (destination local reference=**b**, *RANAP successfulOutcome*, ***IuRelease***)

Finally the SCCP connection is released:

SCCP RLSD (source local reference=**b**, destination local reference=**a**, Release Cause)
SCCP RLC (source local reference=**a**, destination local reference=**b**)

4

Signaling Procedures in the 3G Core Network

The look at the UTRAN signaling procedures has already shown how Non-Access Stratum (NAS) messages are exchanged between the mobile station (MS) and the core network domains represented by serving MSC and SGSN. Now three aspects of core network signaling shall be analyzed more deeply:

1. Data exchange between core network switches and databases
2. Procedures for setup of circuit-switched and packet-switched calls between different core network nodes
3. How core network controls some handover procedures in Radio Access Network (RAN) and how it provides transport capabilities for handover-related RAN signaling messages

The following scenarios are not strictly related to the numbering of the previous enumeration. Their order is less formal and mostly dependent on growing complexity.

It is assumed that readers of this chapter have already understood the signaling procedures in the UTRAN and that they have at least a basic knowledge of Common Channel Signaling System #7 (CCS#7; also known as SS7 or SS#7).

4.1 ISUP/BICC Call Setup

On E interface between different MSCs, the SS#7 ISDN User Part (ISUP) is used for setup and release of calls through the circuit-switched (CS) core network domain. The same function has Bearer Independent Call Control (BICC) on Nc interface between different MSC Servers in a CS core network domain following 3GPP Rel. 4 specifications. BICC is an adaptation of ISUP, which means that in general many signaling messages in both protocols have the same name, but they are not peer-to-peer compatible with each other. The main difference is that ISUP can only assign time slots of PCM-30 or PCM-24 systems with a fixed data transmission rate (64 or 56 kbps) for traffic channels, while BICC is able to provide and control any necessary quality of service for an end-to-end-connection. The possible services offered to 3G subscribers with

UMTS Signaling Ralf Kreher and Torsten Rüdebusch
© 2005 Tektronix, Inc. ISBN: 0-470-01351-6.

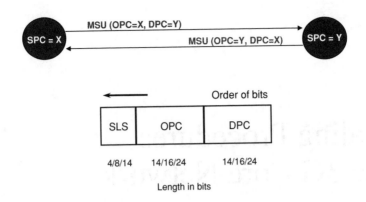

Figure 4.1 SS#7 MTP Routing Label.

introduction of Rel. 4 CS core network architecture range from plain speech to broadband real-time multimedia streaming.

4.1.1 Address Parameters for ISUP/BICC Messages

At least two protocols offer transport services for ISUP and BICC messages: SS#7 Message Transfer Part (MTP) and MTP Layer 3 User Adaptation Layer (M3UA). M3UA uses services of Stream Control Transmission Protocol (SCTP) and Internet Protocol (IP).

Figure 4.1 illustrates the addressing of MTP used for ISUP/BICC message routing.

The addresses are found in the so-called routing label. Each SS#7 exchange has its own address, the Signaling Point Code (SPC). The routing label is either part of the MTP Layer 2 Message Signal Unit (MSU) in case that physical layer is based on a PCM-30 or PCM-24 system or it is part of MTP3-B (Message Transfer Part Level 3 Broadband), which is used in case of ATM-based transport system.

The sender of an MSU or MTP3-B message is called the Originating Point Code (OPC), and the receiver is the Destination Point Code (DPC). The Signaling Link Selection (SLS) parameter gives information on which SS#7 signaling link that belongs to a bundle of links (Signaling Link Set) the message was sent.

The length of the SPC depends on the geographical region: In Europe 14-bit point codes are used, Japan 16 bit, and North America and China 24 bit. For MTP3-B the European standard applies.

In case of MTP3 signaling, the SLS length is 4 bit in European networks and 8 bit in North America. For MTP3-B a 14-bit SLS is used worldwide.

In case of M3UA there are no SPCs used for MSC addressing, but rather IP addresses of IP transport layer that identify the MSC Servers.

4.1.2 ISUP Call (Successful)

The call flow example in Figure 4.2 shows ISUP messages exchanged between two MSCs that are interconnected using a Signaling Transfer Point (STP). The only task of the STP is to

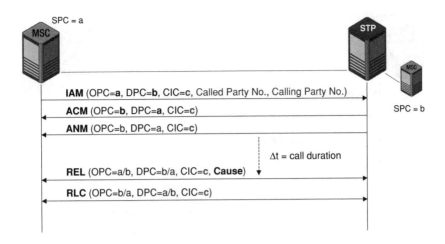

Figure 4.2   Successful call setup ISUP.

route the SS#7 signaling messages. It does not set up or release any calls. However, because of its central position in the network the STP is an excellent connection point for databases that enable the network to offer intelligent services like number portability or prepaid calling card.

In real life networks, as a rule a pair of mated STPs (some operators call this a Tandem STP) is installed for redundancy reasons. This ensures a higher reliability of the network, because it guarantees more possibilities of flexible message rerouting in case of error on single SS#7 signaling links.

Each ISUP call attempt starts with an Initial Address Message (IAM) containing the Called Party Number dialed by the originating user and Calling Party Number (MSISDN of mobile subscriber in case of mobile originated call). In case of a mobile terminated call, the Called Party Number contains the Mobile Station Roaming Number (MSRN).

The Address Complete Message (ACM) indicates that the SPC-B has received all dialing information that is necessary to reach the terminating exchange for this call. No additional dialing information can be sent by A-party after receiving this message.

Answer Message (ANM) indicates that B-party (called party) is now connected and the call is active until Release (REL) message from either A- or B-party of the call is received. This message includes a cause value that indicates, e.g., "normal call clearing."

The party that received the REL message confirms call release with a Release Complete (RLC) message.

4.1.3 ISUP Call Unsuccessful

In case of an unsuccessful call setup procedure, the call attempt is rejected by SPC-B that immediately sends a REL message including a cause value which indicates the reason why the call cannot be completed, e.g. (B-party) "user busy" (Figure 4.3).

When calls cannot complete, the cause value can provide useful hints as to the cause of the problems. Unfortunately, these cause values do not always tell the whole story by themselves. In many cases several different events can trigger the same cause value. To complicate things

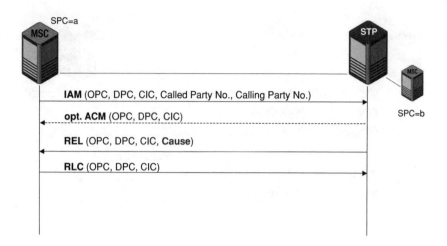

Figure 4.3 Unsuccessful call setup ISUP.

further different manufacturers may trigger the cause value but for different reasons. It is known that some SS#7 switches exist, which allow free configuration of cause values to be used in case of errors.

Figure 4.4 shows two ISUP call procedures between two SS#7 signaling points related to the same traffic channel that is marked by the same Circuit Identification Code (CIC). The first procedure is a successful call in which one of the B-party is obviously an analog telephone, because the call is suspended (SUS message) before it is released. In case of the second call attempt, the call cannot be completed for an unknown reason.

Figure 4.5 shows release cause values for different calls. It is often a hard discussion which cause values can be categorized as "good" or "bad."

Normal call clearing, normal–unspecified, user busy, and *user not responding* do mostly indicate a correct behavior of the network. *No circuit available* complains that there is no time slot for the traffic channel available, which is also a correct behavior from the technical point of view. It is a question of resource planning in the network if a certain amount is treated as normal, because it might not be very profitable for the network operator to buy additional expensive SS#7 exchanges just to ensure that enough traffic channels are available during

Long Time	2. Prot	2. MSG	OPC	DPC	3. Prot	3. MSG	CIC	SLS
13:45:34,506,381	MTP-L2	MSU	244-003-063	005-043-024	ISUP	IAM	331	139
13:45:34,662,164	MTP-L2	MSU	005-043-024	244-003-063	ISUP	ACM	331	24
13:45:47,673,082	MTP-L2	MSU	005-043-024	244-003-063	ISUP	ANM	331	24
13:46:05,405,302	MTP-L2	MSU	005-043-024	244-003-063	ISUP	SUS	331	24
13:46:07,968,350	MTP-L2	MSU	244-003-063	005-043-024	ISUP	REL	331	139
13:46:08,001,514	MTP-L2	MSU	005-043-024	244-003-063	ISUP	RLC	331	24
13:50:52,611,742	MTP-L2	MSU	244-003-063	005-043-024	ISUP	IAM	331	98
13:50:54,382,342	MTP-L2	MSU	005-043-024	244-003-063	ISUP	ACM	331	3
13:51:11,018,701	MTP-L2	MSU	244-003-063	005-043-024	ISUP	REL	331	98
13:51:11,053,067	MTP-L2	MSU	005-043-024	244-003-063	ISUP	RLC	331	3

Figure 4.4 Filtered ISUP call procedures .

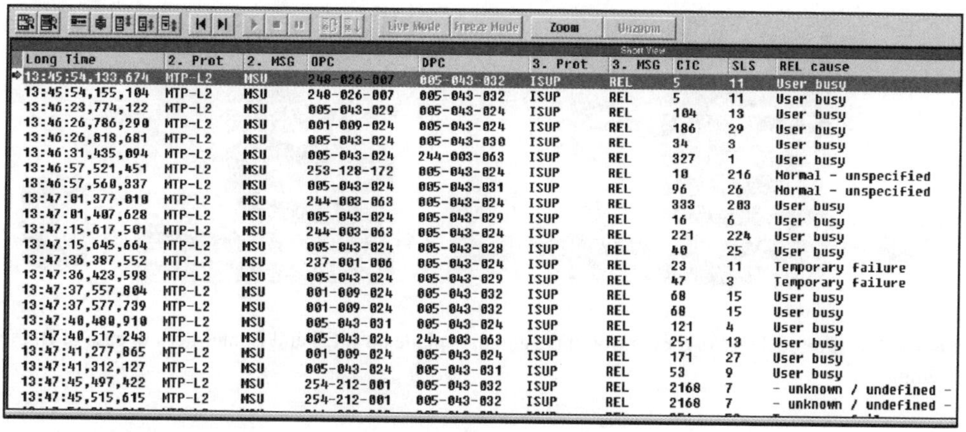

Long Time	2. Prot	2. MSG	OPC	DPC	3. Prot	3. MSG	CIC	SLS	REL cause
13:45:54,133,674	MTP-L2	MSU	248-026-007	005-043-032	ISUP	REL	5	11	User busy
13:45:54,155,104	MTP-L2	MSU	248-026-007	005-043-032	ISUP	REL	5	11	User busy
13:46:23,774,122	MTP-L2	MSU	005-043-029	005-043-024	ISUP	REL	104	13	User busy
13:46:26,786,290	MTP-L2	MSU	001-009-024	005-043-024	ISUP	REL	186	29	User busy
13:46:26,818,681	MTP-L2	MSU	005-043-024	005-043-030	ISUP	REL	34	3	User busy
13:46:31,435,094	MTP-L2	MSU	005-043-024	244-003-063	ISUP	REL	327	1	User busy
13:46:57,521,451	MTP-L2	MSU	253-128-172	005-043-024	ISUP	REL	10	216	Normal - unspecified
13:46:57,560,337	MTP-L2	MSU	005-043-024	005-043-031	ISUP	REL	96	26	Normal - unspecified
13:47:01,377,010	MTP-L2	MSU	244-003-063	005-043-024	ISUP	REL	333	203	User busy
13:47:01,407,628	MTP-L2	MSU	005-043-024	005-043-029	ISUP	REL	16	6	User busy
13:47:15,617,501	MTP-L2	MSU	244-003-063	005-043-024	ISUP	REL	221	224	User busy
13:47:15,645,664	MTP-L2	MSU	005-043-024	005-043-028	ISUP	REL	40	25	User busy
13:47:36,387,552	MTP-L2	MSU	237-001-006	005-043-024	ISUP	REL	23	11	Temporary failure
13:47:36,423,598	MTP-L2	MSU	005-043-024	005-043-029	ISUP	REL	47	3	Temporary failure
13:47:37,557,804	MTP-L2	MSU	001-009-024	005-043-032	ISUP	REL	68	15	User busy
13:47:37,577,739	MTP-L2	MSU	001-009-024	005-043-032	ISUP	REL	68	15	User busy
13:47:40,480,910	MTP-L2	MSU	005-043-031	005-043-024	ISUP	REL	121	4	User busy
13:47:40,517,243	MTP-L2	MSU	005-043-024	244-003-063	ISUP	REL	251	13	User busy
13:47:41,277,865	MTP-L2	MSU	001-009-024	005-043-024	ISUP	REL	171	27	User busy
13:47:41,312,127	MTP-L2	MSU	005-043-024	005-043-031	ISUP	REL	53	9	User busy
13:47:45,497,422	MTP-L2	MSU	254-212-001	005-043-032	ISUP	REL	2168	7	- unknown / undefined -
13:47:45,515,615	MTP-L2	MSU	254-212-001	005-043-032	ISUP	REL	2168	7	- unknown / undefined -

Figure 4.5   Cause values of ISUP Release messages.

extreme traffic peaks that happen once or twice a year, for instance in the early morning hours of New Year's Day.

In a similar differentiated way, *no route to destination* must be discussed. This cause value might indicate the misrouting of a call due to logical error in one of the network's routing tables or Global Title Translation databases. On the other hand, the same cause value is also returned if the calling party of the call is blacklisted, which means that the A-party subscriber is barred, e.g. because he or she did not paid an invoice.

Destination out of order indicates a hardware or software problem with one of the SS#7 switches on the way from A to B.

Finally, a *temporary failure* is always a tricky thing. Typically, it results from an IAM being sent to the network with no ACM or REL message to answer. The T_{iam} timer that guides the call attempt requires an answer to the sent IAM within a time value of, e.g., 10 s. There is a wide range of reasons why the ANM can be missed:

- The IAM was misrouted and sent to the wrong SS#7 signaling point; that signaling point will discard the IAM without returning any error indication.
- Similarly to misrouted IAMs the ANMs (ACM or REL) can also be misrouted.
- Glare may also cause temporary failures. This happens when two signaling points try to grab the same traffic channel (i.e. the same CIC) at the same time. A method to prevent such problems is not to allow that bidirectional trunk groups are specified in the routing tables. A pretty symptom that helps to identify glare is that temporary failures appear if the CIC value is either only odd or only even.

4.1.4 BICC Call Setup on E Interface Including IuCS Signaling

The example shown in Figure 4.6 is based on a version of BICC using MTP transport services for exchanging signaling messages over ATM links. The bearer service controlled by BICC is voice over ATM using an AAL2 SVC (Switched Virtual Connection) on E interface. To give an overview of a complete end-to-end scenario, IuCS procedures are shown as well.

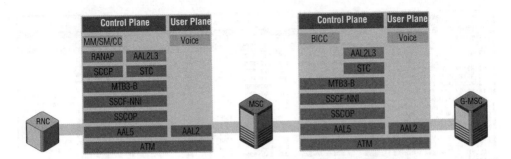

Figure 4.6   Protocol stacks for control plane and user plane on IuCS and E interfaces in case of BICC example.

The protocol stacks in the example on both interfaces look as shown in Figure 4.6.

The shown protocol stack configuration on E interface represents only one of three possibilities. For voice depending on quality requirements, a codec like AMR or G.711 can be used. BICC can also run over IP or on a PCM-24/30 (DS-1/E-1) SS#7 signaling link in case ISUP is simply replaced by BICC without changing the transport network.

4.1.4.1 Call Flow

In the call flow example (see Figure 4.7) each network node is identified by its SS#7 SPC, which is part of the MTP routing label.

The messages on IuCS interface can be filtered using SCCP SLR/DLR (Signaling Connection Control Part Source Local Reference/Destination Local Reference) parameter. On E

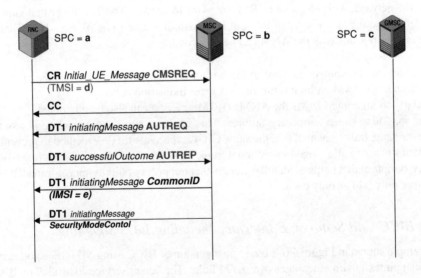

Figure 4.7   BICC Mobile Originated Call (MOC) call flow 1/5.

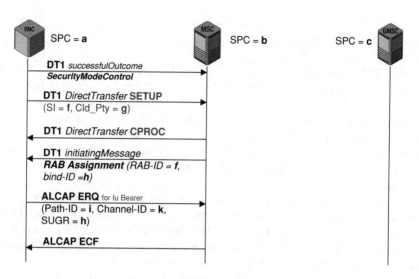

Figure 4.8 BICC Mobile Originated Call (MOC) call flow 2/5.

interface all BICC messages have the same OPC = "**b**" or "**c**" and DPC = "**c**" or "**b**" in the appropriate MTP routing label plus the same BICC CIC value if they belong to the same call.

First the already discussed exchange of NAS messages and RAB Assignment run on IuCS including authentication and security functions (Figure 4.8).

Then (as shown in Figure 4.9) after successful RAB setup the BICC IAM is sent on E interface toward the GMSC (Gateway MSC). However, it is also possible that BICC sends

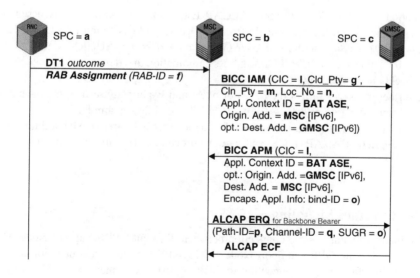

Figure 4.9 BICC Mobile Originated Call (MOC) call flow 3/5.

a so-called "early IAM" using continuity check procedure to withhold call completion until establishment of the RAB is complete. To check details in both cases, read *ITU-T Q.1901*, Annex E.4.1 Successful Call Setup.

The BICC IAM contains the Call Instance Code (CIC = **l**) that will be the same for all other BICC messages belonging to the same call. In addition, the called party number is included that might be slightly different from the one included in DMTAP SETUP message, because leading escape digits ("0" or "00") may have been deleted while Nature of Address parameter is changed into "national (significant) number" or "international number." This possible change is indicated by the Cld_Pty = **g**′ (compared to SETUP Cld_Pty = **g**). If Nature of Address is "unknown," all digits of the called party address signals are shown as they have been dialed by the A-party.

The included Location Number is an E.164 address that delivers information to identify the geographical area (e.g. region, country, city) of the origin of a call. The Application Context ID addressed the Bearer Association Transport (BAT) Application Service Element (ASE) of the peer BICC entity at GMSC. The BAT ASE will assign the necessary resources for establishing the backbone bearer, which is the "traffic channel" on E interface.

The Originating Address is the IP address (mostly IP version 6) of the MSC that sends the IAM. It is necessary to include this address, because in difference to SS#7-based transport networks where it is clear for the MSC which (physical) line leads to the adjacent GMSC in the ATM- or IP-based transport network, all MSC/GMSC can be connected to the same ATM- or IP-router and all logical signaling links can be running on the same physical lines. The Destination Address Information Element (IE; in the call flow example IP address of GMSC) can be included as well.

After the IAM is received, GMSC answers by sending an Application Transport Mechanism (APM) message back to the MSC. This message contains parameters of the backbone bearer to be established, especially a binding ID (bind-ID) if the bearer is represented by an AAL2 SVC.

Reception of BICC APM triggers ALCAP Establish procedures, as discussed in Section 3.1.2. Once again the binding ID from the BICC APM is found in ALCAP Establish Request (ERQ) message as Served User Generated Reference (SUGR) value. Path-ID and Channel-ID will lead to VPI/VCI/CID address combination that defines the logical connection for the backbone bearer (Figure 4.9).

The further messages reflect the behavior of A- and B-party subscriber and have the same name and same function as discussed before in ISUP call flow example.

BICC REL triggers the release of both the Radio Access Bearer (RAB) and the backbone bearer executed by RANAP (IuCS) and ALCAP procedures (Figures 4.10 and 4.11).

4.2 Gn Interface Signaling

The Gn interface identifies the connection between different GPRS Support Nodes (GSNs). They can be Serving GPRS Support Nodes (SGSNs) if they have a connection to UTRAN using IuPS interface and/or connection to GERAN using Gb interface, or Gateway GPRS Support Nodes (GGSNs) if they have a connection to a Packet Data Network (PDN; e.g. the

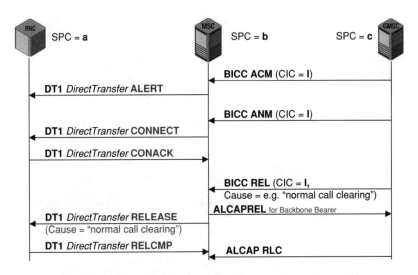

Figure 4.10   BICC Mobile Originated Call (MOC) call flow 4/5.

public Internet) using Gi interface or to other PLMN (Public Land Mobile Network) using Gp interface. The Gn interface is also used to connect all SGSNs to each other (Figure 4.12).

On both Gp and Gn interfaces the GPRS Tunneling Protocol (GTP) is used. The underlying transport network for GTP Control Plane (for GTP-C signaling messages) and GTP User Plane (for IP payload) is based on IP that runs on either Ethernet or ATM lines. To provide a fast transport service between peer GTP entities, User Datagram Protocol (UDP) is used. TCP, which is more reliable than UDP, is defined in the standard documents as an alternative, but not

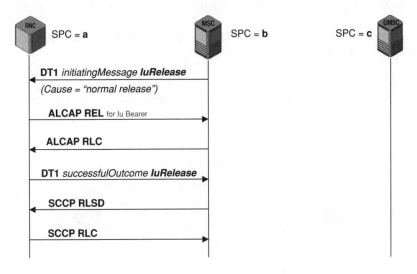

Figure 4.11   BICC Mobile Originated Call (MOC) call flow 5/5.

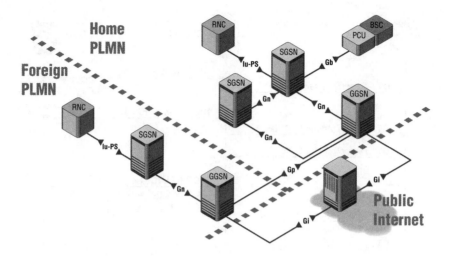

Figure 4.12   GPRS support nodes and interfaces in PS domain.

used by network operators and manufacturers because it would decrease the data throughput in the PS domain. To have an overview of Gn protocol stack, see Section 1.8.7.

As shown in Figure 4.13, the main purpose of the Gn interface is to encapsulate and tunnel IP packets. To tunnel data means to route it transparently through the core network. Between the GSNs a GTP-U (GTP User Plane) tunnel is created for each PDP context of a GPRS subscriber. Through this tunnel all IP packets in uplink and downlink directions are routed. A suite of GTP signaling messages are used to create, modify, and delete tunnels. These GTP-C messages are exchanged using a separate tunnel between the GSNs. Tunnel parameters like throughput rate, etc., are directly derived from the negotiated QoS (Quality of Service) of the PDP Context.

Because an IP transport layer carries GTP data packets that include IP user plane data, an IP-in-IP encapsulation can be monitored on Gn interface.

Note: Because of IP-in-IP encapsulation in the user plane 4 different IP addresses are monitored on Gn. The addresses of the lower (transport) IP layer are those of SGSN and GGSN and

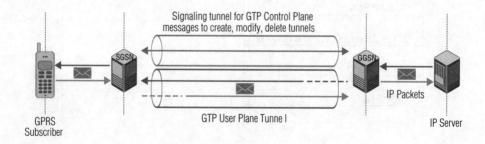

Figure 4.13   Gn interface IP tunneling.

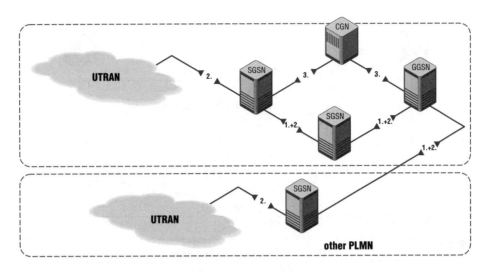

Figure 4.14   Three functions of GTP in relation to network architecture.

relevant only for the Gn interface. The IP addresses in the tunneled IP packets (transported by GTP T-PDUs) are the IP addresses of GPRS subscriber and IP Server and represent the packet-switched end-to-end- connection for exchange of payload. These latter IP addresses can be monitored on all other interfaces that carry PS data as well.

There are three parts of the GTP:

1. GTP-C – Control Plane
2. GTP-U – User Plane
3. GTP′ – GTP for Charging

Figure 4.14 shows between which node of the network architecture these functions can be found.

First, GTP-C establishes management and release of user-specific tunnels between GSNs for exchange of GTP signaling information. Second, it is also used to create, modify, and delete user plane tunnels (PDP contexts) between GSNs. The third task of GTP-C is to support the mobility management and optional location management.

The only task of GTP-U is to transport IP payload coming from or sent to PDN like the Internet. It is used on both IuPS and Gn interfaces. However, on IuPS the tunnels are controlled by RANAP signaling (RAB management).

GTP′ is used between the GSNs and the Charging Gateway Function (CGF) to transmit PDP-context-related Call Detail Records (CDRs).

4.2.1 PDP Context Creation on Gn (GTP-C and GTP-U)

The call flow in Figure 4.15 shows the activation (GTP term: creation) of a PDP context on Gn interface including both control plane and user plane.

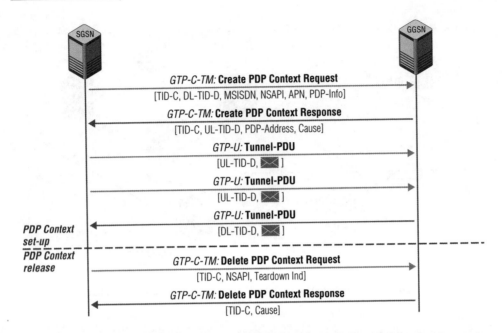

Figure 4.15 PDP context activation/deactivation on Gn interface.

Since this is a mobile-originated PDP context, the Create PDP Context Request message is sent by SGSN. It contains a TEID-C that identifies the signaling tunnel that is associated with the user plane tunnel, which is identified by a Downlink Tunnel Endpoint Identifier Data (DL-TEID-D) and an Uplink Tunnel Endpoint Identifier Data (UL-TEID-D). DL-TEID-D and UL-TEID-D are negotiated between peer GSNs during PDP context creation. The MSISDN is used as user identity for charging, NSAPI indicates the number of the PDP context for this specific user, and APN is the server that assigns the PDP address which is included in Create PDP Context Response message.

GTP T-PDUs (Packet Data Units) are used to transport IP payload within the user plane tunnel.

The REL messages for the previously created PDP context contain the signaling tunnel TEID-C. The appropriate user plane TEIDs have been stored by the GTP entities in relation to the TEID-C. Hence, the user plane tunnel will be deleted as well.

The purpose of the Teardown Indicator is to indicate if all PDP contexts that share the same PDP address as the deleted PDP context shall be deleted (Teardown Ind. = "1") or if only the PDP context with the NSAPI shown in Delete PDP Context Request shall be deleted (Teardown Ind. = "0").

A cause value gives information about reasons for PDP context deactivation as it was described in case of voice calls before.

4.2.2 GTP-C Location Management

The optional GTP-C Location Management messages are defined to support the case when Network-Requested PDP Context Activation procedures are used and a GGSN does not have an SS7 MAP (Mobile Application Part) interface, i.e. Gc interface. GTP-C is then used to transfer signalling messages between the GGSN and a GTP-MAP protocol-converting GSN in the GPRS backbone network. The function and software on this GTP-MAP-converting GSN is different from those of other GSNs in the network.

To obtain the IP address of the MS, the GGSN may send a Send Routing Information for GPRS Request message to the HLR (Home Location Register). This message contains the IMSI of the MS.

The appropriated Send Routing Information Response contains a Cause IE that indicates whether the request was accepted or not. In addition, the message may also contain a MAP Cause IE, an MS Not Reachable Reason IE, a GSN Address IE, and operator-specific information in the Private Extension IE.

If the MS cannot be reached by the GGSN, it may send a Failure Report Request message to the HLR. If the HLR receives this message, the MS not Reachable for GPRS (MNRG) flag for this IMSI is set in the HLR and a Failure Report Response message is sent to the peer entity. When MNRG flag is set, the MS need to perform a new attach to the PS domain.

If an MS becomes reachable for GPRS, again a Note MS GPRS Present Request message is sent to the HLR and the MNRG flag shall be cleared. The HLR answers with a Note MS GPRS Present Response message that indicates whether the request was accepted or not.

4.2.3 GTP-C Mobility Management

The GTP-C Mobility Management messages are the signaling messages that are exchanged between SGSNs during GPRS Attach and Inter-SGSN Routing Area Update (RAU) procedures. Generally, the purpose of this kind of signaling is to transfer data associated with the MS from the old SGSN to the new SGSN.

Note: The new SGSN derives the address of the old SGSN from the old Routing Area Identity.

If the MS, at GPRS Attach, identifies itself with P-TMSI and it has changed SGSN since detach, the new SGSN shall send an Identification Request message to the old SGSN to request the IMSI. This Identification Request message is answered with an Identification Response. If the cause value in this Identification Response is "request accepted," the IMSI will be included in the message, otherwise not.

An interesting signaling example is the call flow of an Inter-SGSN RAU. During an active PDP context the MS changes its location and is now served by a new SGSN. However, this new SGSN has still no idea about the special requirements of the ongoing PDP context. So it sends an SGSN Context Request message to get all the important parameters about active RABs, mobility management information (e.g. entries from SGSN register function), and PDP contexts (e.g. PDP address of active connection) from the old SGSN, as shown in Figure 4.16.

The SGSN Context Request message contains the following mandatory IEs:

- Old Routing Area Identity (RAI)
- TEID-C for identification of the existing signaling tunnel related to the user data
- Old SGSN Addr. {IPv4} for control plane to establish a signaling connection between old and new SGSN
- P-TMSI as user identity

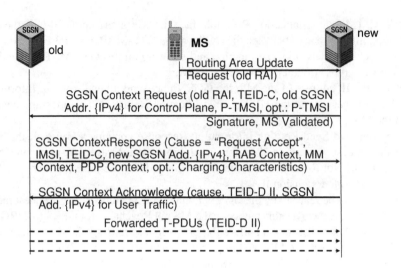

Figure 4.16   GPRS Routing Area Update with forwarded PDP context.

An optional MS Validated IE indicates that the new SGSN has successfully authenticated the MS. The IMSI shall be included if MS Validated indicates "Yes." Another optional IE is the P-TMSI signature, which is also used for security reasons.

With the appropriate SGSN Context Response message the new SGSN receives the RAB Context (important IuPS parameters), MM context (important parameters regarding location and subscribed services of the user), and PDP context (information regarding the current connection, especially the PDP Address of the user) if the cause value is "Request Accepted." For unique identification, the IMSI of the MS is included in the message as well.

If the RAU procedure was successful, the new SGSN completes the procedure with an SGSN Context Acknowledge message. This message contains a Tunnel Endpoint Identifier II for Data (TEID-D II), which is used to establish a temporary unidirectional tunnel between old and new SGSN to forward IP packets that have been queued while the RAU procedure was executed. Together with the TEID-D II, the SGSN data address of the new SGSN (IP address, most likely IP version 4) is included in the message.

After the SGSN Context Acknowledge is received, the old SGSN starts to forward user data packets (T-PDUs) to the new SGSN. The T-PDUs are identified by the previously negotiated TEID-D II.

4.2.4 SGSN Relocation

With the introduction of UTRAN the mobility management functions of the GTP protocol known from Rel. 98 have been enhanced with an additional one: the SGSN Relocation.

The SGSN Relocation becomes necessary if there is an SRNS Relocation in UTRAN and the new SRNC (former DRNC) is connected to a different SGSN. In this case not only the SRNC is changed, but also the SGSN, as shown in Figure 4.17.

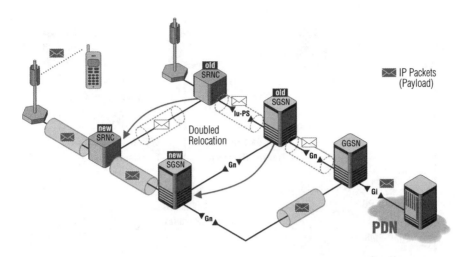

Figure 4.17   SGSN Relocation overview.

The messages for this operation on Gn interface are easy to understand and self-explaining:

- Forward Relocation Request
- Forward Relocation Response
- Forward Relocation Complete

On Iur interface the relocation procedure is executed as described in Section 3.20 and 3.21 of this book.

4.2.5 Example GTP′

Charging information in the GPRS network is collected for each UE by the SGSNs and GGSNs, which are serving that MS. The information that the operator uses to generate an invoice to the subscriber is operator-specific (Figure 4.18).

The Charging Gateway Function (CGF) provides the mechanism to transfer charging information from the SGSN and GGSN nodes to the network operator's chosen Billing System(s) (BSs). The main functions of the CGF are:

- Collection of GPRS CDRs from the GPRS nodes generating CDRs
- Intermediate CDR storage buffering
- Transfer of the CDR data to the BSs

The SGSN collects charging information for each UE related to the radio network usage, while the GGSN collects charging information for each UE related to the external data network usage. Both GSNs also collect charging information on usage of the GPRS network resources.

A Serving GPRS Support Node – Call Detail Record (S-CDR) is used to collect charging information related to the PDP context data information for a GPRS mobile in the SGSN.

A Mobility Management – Call Detail Record (M-CDR) is used to collect charging information related to the mobility management of a GPRS mobile in the SGSN.

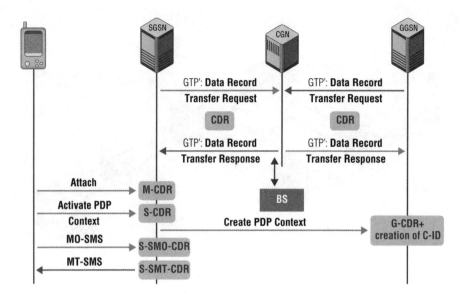

Figure 4.18 GTP message flow and events that trigger CDR creation.

A Gateway GPRS Support Node – Call Detail Record (G-CDR) is used to collect charging information related to the packet data information for a GPRS mobile in the GGSN.

SMS transmission (MO or MT) can be provided over GPRS via the SGSN. The SGSN should provide an SGSN delivered Short message Mobile Originated – Call Detail Record (S-SMO-CDR) when short message is mobile-originated and an SGSN delivered Short message Mobile Terminated – Call Detail Record (S-SMT-CDR) when it is mobile-terminated. In addition, SMS-IWMSC (MO-SMS) and SMS-GMSC (MT-SMS) may provide SMS-related CDRs. No active PDP context is required when sending or receiving short messages. If the subscriber has an active PDP context, volume counters of S-CDR are not updated because of short message delivery.

The CDRs will be transmitted to the CGF by using the GTP′ protocol. The Data Record Transfer Request message will transport the CDR and will be acknowledged by CGF.

When a PDP Context is created, it is necessary to define a Charging ID (C-ID). This is because two instances will now deliver charging information for one UE, and CGF will need to combine the CDRs from the different GSNs.

4.3 Procedures on Gs Interface

The optional Gs interface is used to exchange data between VLR and SGSN register function. Using the Gs interface signaling it is, for instance, possible to page a subscriber for a voice call via PS domain or to page a mobile-terminated PDP context via CS domain. Also combined attach procedures and location/routing area updates are possible.

SCCP UDT (CLD_PTY = VLR (E.164), SSN=**192**, CLN_PTY= SGSN
(E.164), SSN=**192** [**BSSAP+ LURQ** {IMSI=**a**, SGSN No.= SGSN (E.164),
GPRS Location Update Type = **IMSI Attach**, Cell Global Identity:
MCC+MNC+LAC=**b**+RAC+CI, MS Classmark, Location Area Identifier:
MCC+MNC+LAC=**b**}])

SCCP UDT (CLD_PTY = SGSN (E.164), SSN=**192**, CLN_PTY= VLR
(E.164), SSN=**192** [**BSSAP+ LUAC** {IMSI=**a**, Location Area Identifier:
MCC+MNC+LAC=**b**}])

Figure 4.19   Gs interface: IMSI Attach/Location Update Procedure call flow.

The protocol stack on Gs is the same as on GSM A interface, but a set of enhanced BSSAP+ procedures is used. All BSSAP+ messages are transported on behalf of SCCP Unitdata (UDT) messages using connectionless SCCP transport services.

4.3.1 Location Update via Gs

With this procedure the SGSN informs the VLR about a CS location area update that was combined with RAU procedure. IMSI is used in both messages to identify the subscriber uniquely. The SCCP Subsystem Number (SSN) for this service is often defined by network operators, but in some cases SSN=192 was seen as value in real network traces. So it is used in this example (Figure 4.19).

4.3.2 Detach Indication via Gs

Both IMSI and GPRS Detach can be indicated using Gs signaling. In the example call trace, a GPRS Detach procedure is shown (Figure 4.20). An additional identifier indicates that this detach is for GPRS services only and it is requested by the network.

4.3.3 Paging via Gs

The last example for Gs signaling shows a paging request message sent from VLR to SGSN. So this is a CS paging sent via PS domain (Figure 4.21).

The appropriate paging response to this request is expected to arrive embedded in a RANAP message via IuCS.

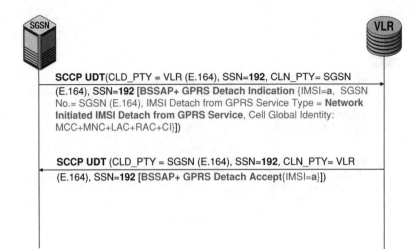

Figure 4.20   Gs interface: IMSI/GPRS Detach procedure call flow.

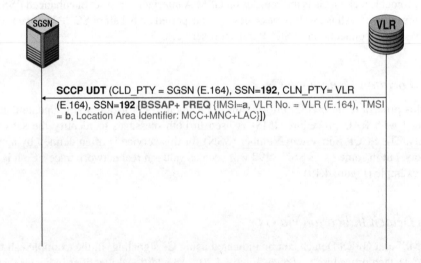

Figure 4.21   Gs interface: CS Paging Request call flow.

4.4 Signaling on Interfaces Toward HLR

The HLR is a main database of the PLMN that stores subscriber data. Here we find information about user identity (IMSI, currently used TMSI/P-TMSI), location of the subscriber (present location area, routing area, etc.), user rights, and information about subscribed services (e.g. the subscribed QoS profile for PDP context activation). The HLR also knows whether the user is attached to the network or to defined services of the network or not.

The HLR is especially important in case of mobile-terminated voice or data calls. The GMSC or GGSN retrieves the necessary information for routing of mobile-terminated calls from the HLR. If the user is not reachable, the HLR may have further information for alternative routing targets. For instance, a voice call can be routed to a voice mail system or a special announcement is played that informs the calling party that the subscriber is temporarily not available. In this case interaction with other Intelligent Network (IN) or North American Advanced Intelligent Network (AIN) components is necessary, as described in Section 4.7

As a rule, an HLR is colocated with a GMSC, but not every GMSC has a colocated HLR. It depends on the number of subscribers if there is more than one HLR in the network. In case of a so-called "greenfielder," the HLR could be the only database in the network. Greenfielders are service providers that build up a minimal own network structure only, e.g. one HLR plus one GMSC and/or one GGSN while they rent usage of network resources (like whole RAN including MSC/VLR and SGSN) from a full service network operator. It is important to know that one general concept of 3G standards is to provide total flexibility regarding ownership of network parts or single network components.

Figure 4.22 shows some Mobile Application Part (MAP) interfaces in the core network that are mandatory to offer basic circuit- and packet-switched services.

In today's networks the Authentication Center (AuC) is often integrated in the HLR device so that the H interface runs on a device internal bus system. Most PLMNs also do not have an Equipment Identity Register (EIR), but a number of companies or government institutions that operate mobile networks (like GSM-R networks of railway companies) use EIR to ensure that

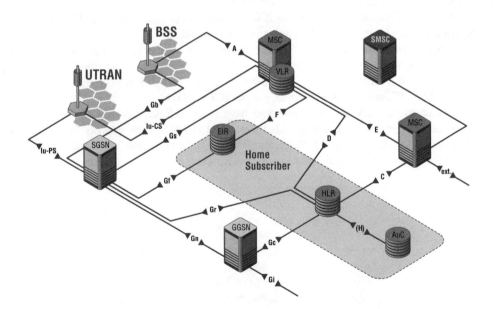

Figure 4.22   MAP interfaces in core network environment.

only registered phones can be used in their networks. Nevertheless, 3GPP Release 5 defines a database that combines EIR, HLR, and AuC functions. This database is the so-called Home Subscriber Server (HSS).

On D interface between VLR and HLR, it can be monitored how the location update procedure is continued in the core network after the MS send Location Update Request to the MSC/VLR.

On C interface between GMSC and HLR, it is observed how the GMSC retrieves routing information for mobile-terminated procedures.

The counterpart of the D interface in the PS domain is the Gr interface between SGSN and HLR and the Gc interface ensures the connections between GGSN and HLR.

As it was described in Section 3.22.1 of this book, MAP is also used for communication between GMSC and external Short Message Service Center (SMSC), but this interface is outside the scope of 3GPP international standards.

4.4.1 Addressing on MAP Interfaces

A typical example shall be given on how signaling information using MAP is exchanged via Gr interface. In the communication with the HLR, the CS Location Update and GPRS Attach procedures are the most complex ones, because a lot of IEs are included and different kinds of addressing are used.

Figure 4.23 shows the protocol stack used on the MAP interface between SGSN and HLR (Gr interface). The protocol layers are the same as for all core network interfaces using MAP, but SCCP subservice numbers (SSN) are different.

The SCCP SSN indicates the user of an SCCP connection. In case of current MAP signaling, only SCCP classes for connectionless data transfer are used.

The SSN either represents a higher layer protocol (as it was shown for SSN=192 that indicates BSSAP+ protocol on Gs interface) or it may also stand for a network element like SGSN (SSN=149), HLR (SSN=6), VLR (SSN=7), etc. Table 1-7 in Section 1.18 gives an overview about different SSNs as defined, e.g., in *3GPP 23.003*.

In addition to the SSN, the network nodes have other addresses as well. As a rule, each node in the CS core network domain has its own E.164 address, an address format that is well known from international telephone numbers. To be truly unique worldwide, it consists of a country code, a national destination code (also known as area code), and a subscriber number.

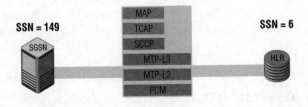

Figure 4.23   Protocol stack on Gr interface.

To give an example:

Country Code (CC) = 54 ⇒ *Argentina*

National Destination Code (NDC) = 11 ⇒ *Cap. Fed. Buenos Aires*

Subscriber Number = 43xxxxxx ⇒ *Silvina A.*

A various number of escape digits (e.g. 0054...) may have to be dialed before the country code to reach a B-party in a foreign country. The character "+" replaces the escape digits and indicates that the dialed number is an international ISDN number. (On protocol level in ISUP this will be reflected by the Numbering Plan IE of the called party number that will show "international number" when the "+" is dialed.)

However, in case of network element addressing the address part called "subscriber number" does not identify a telephone, but a switch or database in the network. So the E.164 number +54-11-43001000 could be the central office switch to which Silvina's phone is connected to (but indeed, we do not know the detailed numbering plan of *Telefónica de Argentina*).

In addition to their E.164 number, which is necessary for SCCP addressing purposes, the packet switches SGSN and GGSN do also have IP addresses that are used by GTP entity on Gn interface. By the way they are also identified by SS#7 SPCs on the MTP level (Gc, Gr, and IuPS interfaces).

4.4.2 MAP Architecture

MAP uses the functions provided by Transaction Capabilities Application Part (TCAP). Transaction Capabilities are necessary to exchange information that is not directly related to calls between network nodes like exchanges and databases.

To explain it in a popular way: It was necessary to define a way how data records in, e.g., VLR and HLR can be continuously updated. One solution would have been to install a separate data network between these databases. The other one would have been to connect the databases to the existing SS#7 signaling network and enable the SS#7 network to perform the necessary transactions. This second and more efficient solution was the reason why TCAP was defined by CCITT/ITU-T.

Figure 4.24 explains how MAP and TCAP are related to each other.

A single MAP User ASE represents a complete MAP function like Location Update or GRPS Attach function with all the necessary operations, parameters, results, and errors. Hence, for instance a VLR communication software entity (not the database itself!) consists of several ASEs to offer the necessary mobility management functionality. These operations, parameters, etc., are sent either to a peer entity or received from a peer entity.

All MAP operations are embedded in TCAP messages. The TACP ASE consists of two sublayers: the component sublayer and the transaction sublayer. A component is a request to execute an operation or the answer to such a request. In TCAP "language" the requests are called INVOKE while the answers are called RETURN. In case of successful INVOKE, a RETURN RESULT is sent by peer entity. Otherwise the entity that sent the INVOKE will receive RETURN ERROR. The peer entity is also able to REJECT an invoked operation.

TCAP transaction sublayer provides all necessary functions to exchange components between two different TCAP users. Especially the five generic TCAP message formats are built by

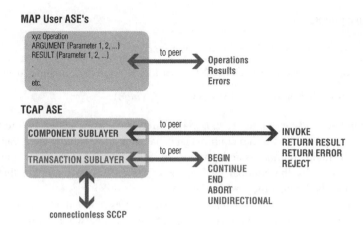

Figure 4.24   MAP architecture.

transaction sublayer using ASN.1 Basic Encoding Rules (BERs). These messages are named BEGIN, CONTINUE, END, and ABORT for a structured dialog (connection-oriented). For unstructured dialog (connectionless), the UNIDIRECTIONAL TCAP message is used.

All MAP operations and results are exchanged using structured TCAP dialogs only.

A very interesting IE is the TCAP Application Context (AC). The AC reflects the present software evolution stage of the MAP or any other user part on top of TCAP. If, for instance, a look at the HLR is taken, it emerges that the structure of such a database changes dramatically if a new level of network evolution is introduced. The most exiting evolution step for HLR in the past was the start of GPRS. There is a large number of specific HLR entries like flags, subscribed QoS parameter for PDP contexts, Routing Area Information, etc., that were not known in plain GSM networks. To "learn" about GPRS for the HLR was like to learn a new language, and the kind – or better version – of this "language" is reflected by the AC value. Indeed, not only the databases have grown with introduction of new services, but also the MAP protocol itself is growing from version to version by adding new operation codes and new parameters.

The purpose of the AC is to prevent communication problems if, e.g., one HLR that already knows GPRS-specific information elements "talks" with another HLR that still knows only GSM-related data.

In the example call flow in Figure 4.25, it is an external SMSC that wants to retrieve routing information for a mobile-terminated short message (SM) from an HLR in the target PLMN.

TCAP AC negotiation rules require that the proposed AC, if acceptable, is reflected in the first backward message.

If the AC is not acceptable, and the TCAP user does not wish to continue the dialog, it may provide an alternate AC to the initiator which can be used to start a new dialog.

The example in the figure shows the case when first Send Routing Info for Short Message, which is a local MAP operation, is sent embedded in a TCAP BEGIN message using an AC with version 3. Since the HLR does not support version 3, it rejects the TCAP dialog by sending ABORT including a dialog service user error value "Application context name not supported" and the alternate AC value 2. Now the sending TCAP entity knows that the receiver

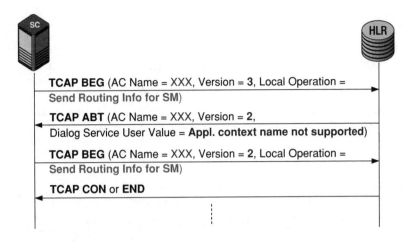

Figure 4.25   TCAP application context call flow example.

only supports AC version 2 and sends a new TCAP BEGIN message including the same local operation as before, but this time with AC version 2, which means that the local operation is still the same, but the number and structure of included parameters will be changed according to AC version 2 standards. The further TCAP dialog will be continued successfully with TCAP CONTINUE and/or END message(s).

The lesson a network troubleshooter can learn from this example is that not every TCAP ABORT message indicates an error in the network.

4.4.3 MAP Signaling Example

An Update GPRS Location procedure is shown as example of MAP signaling (Figure 4.26). This call flow can be monitored on Gr interface. The procedure uses connectionless SCCP data transfer with end-to-end signaling.

The first interesting fact in this example call flow is that the HLR address in the first message is derived from the IMSI. This is the so-called Mobile Global Title (Mobile GT) following ITU-T E.214 specification. The Mobile GT includes all IMSI digits or at least the MSIN digits if MCC and MNC have been already translated by SGSN Global Title Translation table into Country Code plus National Destination Code.

Example Global Title Translation table:

E.212 (IMSI)		E.164/E.214	SS#7 SPC	Comment
MCC = 262	⇒	CC = 49		Germany
MNC = 02	⇒	NDC = 172		D2 Vodafone
MSIN = 38...	⇒	⇒	4551	HLR of subscriber identified by IMSI

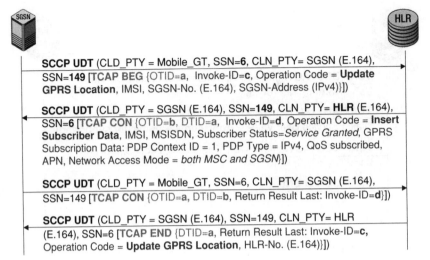

SCCP UDT (CLD_PTY = Mobile_GT, SSN=**6**, CLN_PTY= SGSN (E.164),
SSN=**149** [TCAP BEG {OTID=a, Invoke-ID=**c**, Operation Code = **Update GPRS Location**, IMSI, SGSN-No. (E.164), SGSN-Address (IPv4)}])

SCCP UDT (CLD_PTY = SGSN (E.164), SSN=**149**, CLN_PTY= **HLR** (E.164),
SSN=**6** [TCAP CON {OTID=**b**, DTID=a, Invoke-ID=**d**, Operation Code = **Insert Subscriber Data**, IMSI, MSISDN, Subscriber Status=*Service Granted*, GPRS Subscription Data: PDP Context ID = 1, PDP Type = IPv4, QoS subscribed, APN, Network Access Mode = *both MSC and SGSN*}])

SCCP UDT (CLD_PTY = Mobile_GT, SSN=6, CLN_PTY= SGSN (E.164),
SSN=149 [TCAP CON {OTID=a, DTID=b, Return Result Last: Invoke-ID=**d**}])

SCCP UDT (CLD_PTY = SGSN (E.164), SSN=149, CLN_PTY= HLR
(E.164), SSN=6 [TCAP END {DTID=a, Return Result Last: Invoke-ID=**c**, Operation Code = **Update GPRS Location**, HLR-No. (E.164)}])

Figure 4.26   MAP Update GPRS Location/Insert Subscriber Data call flow.

As a rule, a foreign PLMN shall not translate more than the Mobile Country Code (MCC) into a Country Code (CC) and – if necessary – the Mobile Network Code (MNC) into a National Destination Code (NDC). Then the message is routed based on this Mobile GT to a gateway of the network indicated by NDC. This gateway (most likely a GMSC) is finally able to translate the first two digits of the MSIN, which is still an unchanged part of the Mobile GT with its hybrid E.214 address format, into the SS#7 SPC or the E.164 number of the HLR.

The SSN indicates that the first SCCP UDT of the example call flow is sent by SGSN (SSN=149) and shall be received by HLR (SSN=6).

The first SCCP UDT transports a TCAP BEGIN message with Originating Transaction ID (OTID) = **a**. There is an Update GRPS Location operation invoked (using Invoke-ID = **c**) by the SGSN that is identified by both its E.164 and its IP address. The IMSI that identifies the subscriber who wants to be attached to the PS domain is also included. Such an Update GPRS Location procedure is performed if SGSN received a GPRS Attach or Routing Area Update Request message on IuPS.

HLR answers with a TCAP CONTINUE message containing Originating Transaction ID (OTID) = **b** and Destination Transaction ID (DTID) = **a**. These transaction ID values are the same in all messages belonging to this procedures. They can be used as filter conditions when monitoring MAP interfaces, because they link MAP messages belonging to a single subscriber together.

Before the HLR can proceed the GPRS Location Update procedure, it invokes to Insert Subscriber Data at the location register function of the SGSN. To this subscriber data belongs IMSI and MSISDN of the subscriber as well as subscriber status and GPRS subscription data including the QoS parameters as described in the contract between network operator and subscriber. There is a new Invoke-ID = **d** related to this new operation code.

After subscriber data was inserted successfully, SGSN answers with another TCAP CONTINUE message that contains a Return Result Last sequence including the Invoke-ID = **d** that is related to the Insert Subscriber Data operation.

Now in the last message, which is TCAP END, the HLR approves that Update GPRS Location was successful. Once again a Return Result Last sequence includes Operation Code and appropriate Invoke-ID plus the E.164 HLR address that will be used for SCCP addressing in further MAP transactions between SGSN and HLR.

4.5 Inter-3G_MSC Handover Procedure

MAP is not only used for communication between databases. There is also a number of MAP operations for information exchange between MSCs and last but not least MAP offers transport functions for different Layer 3 RAN protocols. For instance, MAP can carry RANAP messages that are exchanged between RNCs connected to different MSCs. Since RANAP for its part offers transport functions for RRC protocol information, it is also possible that on E interface between two 3G MSCs RRC Information that is carried by RANAP messages, which are embedded in MAP operations, can be monitored. Hence, the protocol stacks used by protocol monitors need to look very different from what is described in the international standards. In this and in the following two sections, the "real world" protocol stacks as necessary for complete decoding of captured signaling messages and the complete messages including their piggybacked parts will be described.

First, it shall be explained why MAP carries RANAP and RRC information. For this reason it is necessary to extend the view at the UTRAN, as given in Section 1.5.

Usually, network overview pictures show only one MSC/SGSN to which the RNCs are connected. This exactly reflects the UTRAN definition: A UTRAN is a RAN connected to one MSC of the CS domain and one SGSN of the PS domain (if CS and/or PS domain are available in the network). Thus, a 3G network consists of more than just one UTRAN and as a rule also more than just one MSC, SGSN, etc.

RNCs of the same UTRAN may be interconnected via Iur interface, but there is no Iur between RNCs of different UTRANs. Figure 4.27 shows how two different UTRANS are linked via E interface of the CS domain and Gn interface of the PS domain. If there is a UE in a handover situation between cells belonging to different UTRANs, the CRNCs of these cells need to communicate with each other to ensure a error-free Hard Handover. Soft Handovers are impossible in such a situation, also if the cells have the same frequency.

As one may remember from Chapter 3 of this book, all communication between UE and network starts with setup of an RRC connection. The RNC that controls this RRC connection and terminates the Iu interfaces of the connection is called the Serving RNC (SRNC). In Figure 4.28, this initial SRNC is RNC 1.

If the UE moves it will get in contact with a cell of Node B2 that is controlled by RNC 2. RNC 1 and RNC 2 belong to the same UTRAN and are interconnected via Iur interface. If the two cells work on the same frequency, a Soft Handover is possible with RNC 1 as SRNC and RNC 2 as DRNC. The RRC connection is still active between UE and RNC 1, and RNC 2 routes all RRC messages transparently in uplink and downlink directions.

If the UE loses contact with all cells directly controlled by RNC 1 and RNC 2 is the only one that provides radio resources for the connection, RNC 1 will make the decision to perform an SRNS relocation procedure as described in part 2 of this book. At the end of this SRNS relocation, RNC 2 will be the new SRNC.

However, the UE may continue to move while a call is still active. A completely new and much more difficult situation is given in case that a cell of Node B3 becomes better than cells

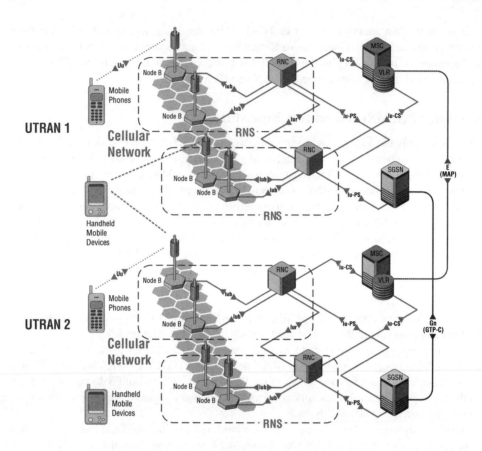

Figure 4.27   UMTS interfaces between two UTRANs.

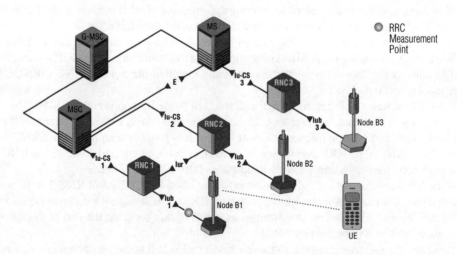

Figure 4.28   Initial RRC connection setup between UE and SRNC (RNC 1).

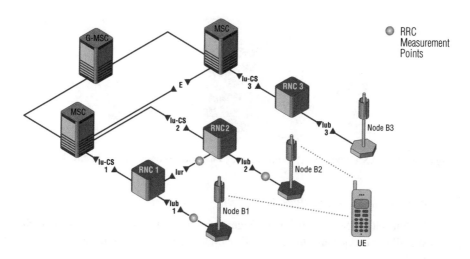

Figure 4.29   UE in Soft Handover situation, RRC connection controlled by RNC 1.

of Node B2, as shown in Figure 4.29. This event may trigger the decision to perform a Hard Handover between RNC 2 and RNC 3 (Figure 4.30).

Now in case of a Hard Handover the parameters of the RRC connection have to be forwarded to the new SRNC (RNC 3) and the only connection between RNC 2 and RNC 3 goes through E interface of CS core network domain.

As a result, a protocol tester must be equipped with the protocol stack shown in Figure 4.31 to be able to decode all higher layer protocol information on E interface.

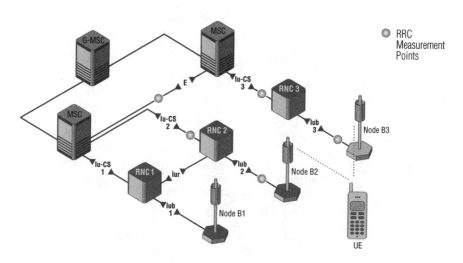

Figure 4.30   Hard Handover to cell of Node B3 performed by RNC 2.

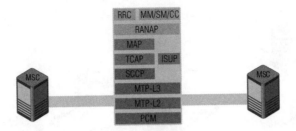

Figure 4.31   RANAP-over-MAP protocol stack on E interface for Inter-3G_MSC Handover.

4.5.1 Inter-3G_MSC Handover Overview

Step 1: The Inter-3G_MSC Handover is triggered by an RRC Measurement Report that reports event-ID=e1a, because the new cell operates at the same frequency as the old one (Figure 4.32).

Step 2: Following the reception of the RRC Measurement Report message the SRNC (RNC 1) decides to perform a handover to RNC 2. Since there is no Iur interface between these two RNCs, the handover must be a Hard Handover plus SRNS Relocation at the same time. It is completely controlled by old SRNC, which sends a RANAP Relocation Required message to its serving MSC.

Step 3: Based on a general routing table the serving MSC detects that RNC 2 is connected to a different MSC as shown in Figure 4.32. Hence, it is necessary to send the RANAP Relocation Request message through the other MSC to RNC 2. Because the traffic channel of the call also needs to be forwarded to the new MSC, the serving MSC sends a MAP Prepare Handover invoke to its peer MSC; the MAP message contains the RANAP messages sent to RNC 2.

Step 4: The new MSC forwards the RANAP Relocation Request message to RNC 2.

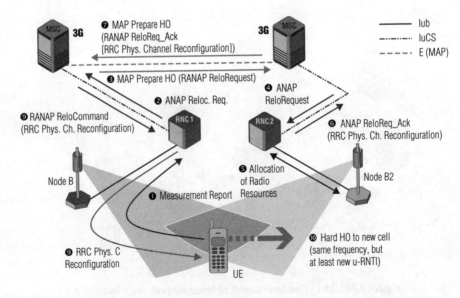

Figure 4.32   Inter-3G_MSC Handover/Relocation overview.

Step 5: RNC 2 allocates all necessary radio resources to take over the UE connection. Especially admission control and packet scheduler function are checking if the connection with its present QoS can be continued and if the same transport combination format set can be used as calculated by RNC 1. It depends on the result of this calculation which RRC message will be constructed and sent by RNC 2 to perform the handover. If QoS needs to be changed, a Radio Bearer Reconfiguration message will be sent. If changes in the transport format set are required, a Transport Channel Reconfiguration message is sent. If only typical identifiers like Spreading Code, Scrambling Code, or U-RNTI are changed, a Physical Channel Reconfiguration message will be sent. In this example Physical Channel Reconfiguration is assumed.

Step 6: RNC 2 sends a RANAP Relocation Acknowledge message to RNC 2 that piggybacks the RRC Physical Channel Reconfiguration message.

Step 7: The answer to the MAP Prepare Handover message is sent by new MSC to old serving MSC. This message contains the RANAP Relocation Acknowledge including the embedded RRC Physical Channel Reconfiguration.

Step 8: Old MSC sends RANAP Relocation Command to RNC 1. Once again the RRC Physical Channel Reconfiguration constructed by RNC 2 is embedded in this message. The Relocation Command message orders the RNC 1 to give up its rule as SRNC of the connection.

Step 9: The Physical Channel Reconfiguration message is forwarded by RNC 2 via old Iub and Uu (radio) interfaces to the UE. On behalf of this message the UE is informed into which new cell the handover shall be performed and which parameters like new U-RNTI become valid after handover.

Step 10: Based on the information found in RRC Physical Channel Reconfiguration Request message, the handover is performed and RRC Physical Channel Reconfiguration Complete message is sent on new Iub interface to RNC 2.

4.5.2 Inter-3G_MSC Handover Call Flow

The handover call flow examples show messages only on E interface. In difference to other call flow examples the main parameters are presented in a real number format to allow a better understanding of the procedures (Figure 4.33).

With this message the left MSC orders the right one to prepare a handover with invoke ID=1. Already the MAP message contains information on which RNC is the target RNC for this handover procedure. The unique target RNC ID consists of MCC and MNC (both parameters known from E.212 numbering plan used for IMSI) plus an RNC-ID (which is unique within the operator's network).

Inside the MAP message there is a container, the so-called an-APDU. "an" stands for access network and inside the an-APDU there is some more information about the access network protocol: 25.413 is the 3GPP specification for RANAP. IMSI is embedded as user identity and the radio resource = "speech" indicates that this is a voice call.

Now the complete RANAP Initial Message with Procedure Code = "Relocation Resource Allocation" follows. The message is forwarded transparently by the MSC. Once again IMSI can be found inside this message and a cause value tells that this procedure is part of a relocation owing to changing conditions on air interface.

Next embedded item in the RANAP message is the SourceRNC-to-TargetRNC-Transparent-Container including an RRC container with all the necessary information about the exiting

TCAP BEG ⇨ MAP Prepare Handover: Invoke-ID=**1**,
Target RNC: MCC=262, MNC=99, RNC-ID= 00 0a,

an-APDU:
- ▶ AccessNetworkProtocolID= ts3G-25413, IMSI=26299..., RadioResource=Speech
- ▶ **RANAP** *InitiatingMessage*: procedureCode=**RelocationResourceAllocation**,
  - – IMSI=26299..., Cause=Relocation-desireable-for-radio-reasons
  - – **SourceRNC-to-TargetRNC-TransparentContainer:**
    - ▶ **RRC Container** (rrcState ID = CELL_DCH, Ciphering Status, Integrity Prot. Status, **Cell ID** of target cell, **U-RNTI**, Radio Bearer Info List, Meas. Command + List)
    - ▶ No. Of Iu Instances=2, RelocationType=**UE not involved**
    - ▶ ChoosenIntegrityProt.Algorithm, IntegrityProt.Key
  - – RAB-SetupList:
    - ▶ RAB-ID=1, RAB Parameters, SDU Parameters
    - ▶ Iu-TransportAssociation: bindingID=**00 80 00 00** (hex)
  - – IntegrityProtectionInfo: ChoosenInt.Prot.Alg., IK
  - – EncryptionInformation: EncryptionAlgorithm, CK
  - – Iu-SigConID=**80 00 00 00** (hex)

Figure 4.33   Inter-3G_MSC Handover/Relocation call flow 1/6.

RRC connection between the source RNC and the UE. There will be information whether ciphering and integrity protection is used, which algorithms are used, which was the last ciphered/integrity-protected RLC frame on old radio interface (identified by their RLC sequence number). Cell-ID of the target cell was extracted from RRC Measurement Report and U-RNTI assigned during RRC connection setup by old SRNC. A radio bearer information list and all settings of RRC measurement complete the mandatory part of the RRC container. Optional information may be added as well.

The Relocation Type = "UE not involved" indicates that there is no change of the currently used frequency, so this a intrafrequency Hard Handover.

RAB parameters including Ciphering and Integrity Protection Info are repeated again in RANAP part of the message, because RRC and RANAP parts terminate in different entities. While the RANAP message is already read by target MSC the RRC containter can only be read by new RNC.

Iu Transport Association binding ID will be used by BICC (or ISUP) as CIC value for traffic channel setup between the two MSCs. (*Note*: In the PS domain this binding ID is used to define the TEIDs for data transport between two SGSNs.)

The Iu Signaling Connection Identifier is allocated by the first MSC, and target RNC is required to store and remember this identifier as long as the call is active on IuCS interface between second MSC and target RNC.

In next step of the example call flow the new MSC sends MAP Prepare Handover acknowledge with a Return Result Last for Invoke-ID = **1** (Figure 4.34). A handover number is allocated by second MSC, which is used like an MSRN for routing of BICC/ISUP signaling from first to second MSC. The reception of this handover number starts call setup in BICC/ISUP, as described in Section 4.1.

TCAP CON ⇨ MAP Prepare Handover, ReturnResultLast: Invoke-ID=1,
ReturnResultSequence: LocalOper.=PrepareHandover, HandoverNumber(E.164)=123451100,
an-APDU:
  ▸ AccessNetworkProtocolID= ts3G-25413
  ▸ **RANAP** *SuccessfulOutcome*: procedureCode=**RelocationResourceAllocation**,
    – **TargetRNC-to-SourceRNC-TransparentContainer:**
      ▸ **RRC Container (RRC PhysicalChannel Reconfig. (new U-RNTI))**
    – RAB-SetupList-RelocationRequestAcknowledge:
      ▸ RAB-Setup-Item-Relocation-Request-Acknowledge
        ↳ RAB-ID=1
    – ChoosenIntegrityAlgorithm
    – ChoosenEncryptionAlgorithm
  BICC IAM (Cld_Pty= 123451100, CIC = 0080000 (hex))

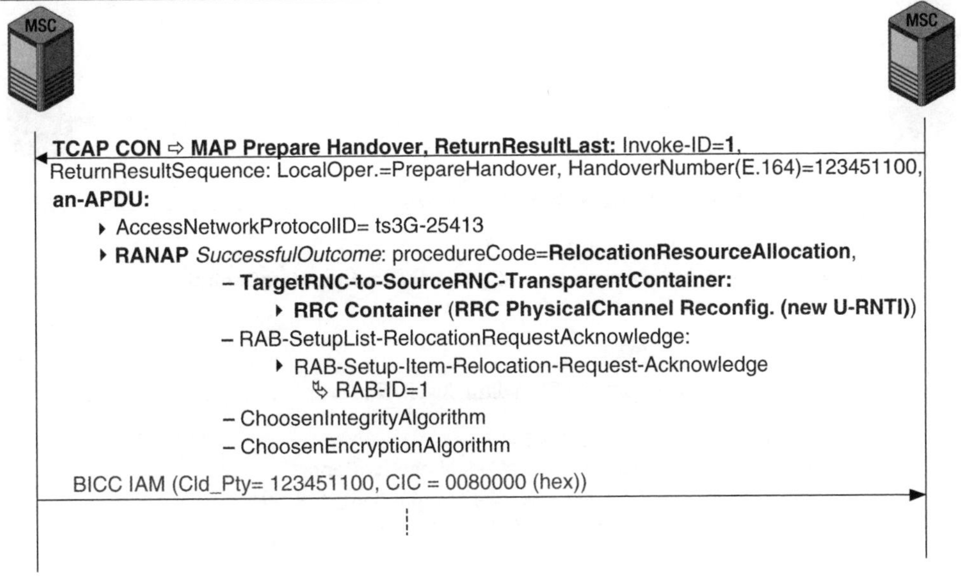

Figure 4.34   Inter-3G_MSC Handover/Relocation call flow 2/6.

Once again an an-APDU is embedded in the MAP message that contains the RANAP Successful Outcome message for Relocation Resource Allocation (in RANAP specifications also named Relocatioin Request Acknowledge message).

In this message a TargetRNC-to-SourceRNC-Transparent-Container is found that contains the RRC Physical Channel Reconfiguration message including all new assigned radio resources, at least a new U-RNTI is assigned when the cell is changed.

The RAB setup list informs that the new user plane bearer is already installed and waits to be detected by UE. Chosen Integrity Protection and Ciphering algorithms are confirmed by target RNC.

Now this RRC Physical Channel Reconfiguration message is sent to UE via the old radio link. After reception of the message the UE switches into the new cell.

While this procedure is ongoing, the RANAP entities of both RNCs update the stored location information for this UE. This procedure is mandatory, also if there is no location database colocated with the RNCs.

It clearly emerges that TCAP continues the dialog started with Relocation Preparation and MAP uses Forward Access Signaling operation to send RANAP messages to the target RNC while responses from target RNC sends back source RNC using MAP Process Access Signaling (Figure 4.35).

When the UE found the new radio link and synchronized with the Node B, target RNC receives NBAP Radio Link Restoration on the appropriate Iub interface. This triggers sending of RANAP Relocation Detect message. The detection of the traffic channel on target MSC side leads to sending BICC Answer message to the source MSC.

Reception of RCC Physical Channel Reconfiguration Complete message on new radio link (in new cell) is reported sending RANAP Relocation Complete to the source RNC. Since second MSC knows that this means successful finishing of the handover procedure it includes

TCAP CON ⇨MAP Forward Access Signaling: Invoke-ID=2,
an-APDU:
  ▸ AccessNetworkProtocolID= ts3G-25413
  ▸ **RANAP** *InitiatingMessage*: procedureCode=**LocationReportingControl**
      – EventID= **change of service area**

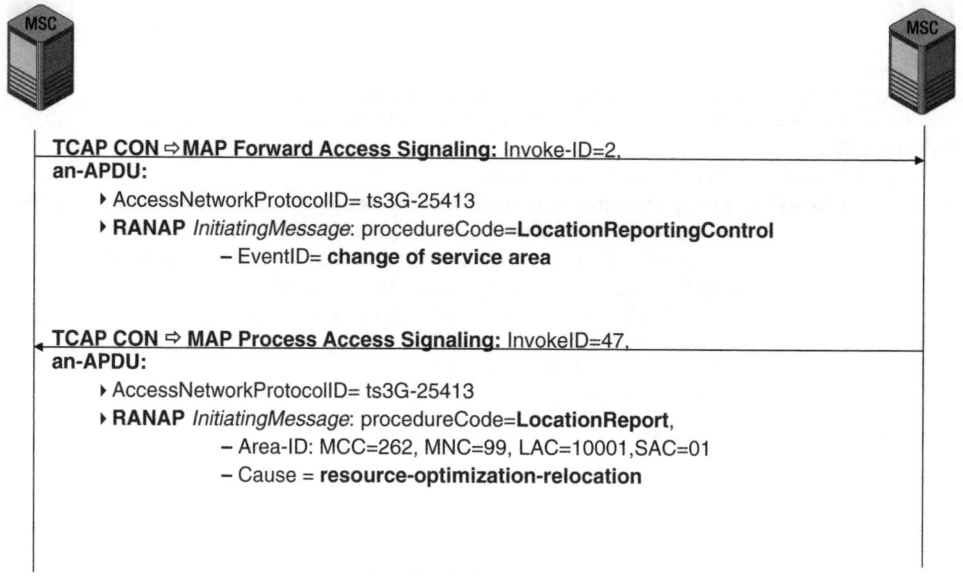

TCAP CON ⇨ MAP Process Access Signaling: InvokeID=47,
an-APDU:
  ▸ AccessNetworkProtocolID= ts3G-25413
  ▸ **RANAP** *InitiatingMessage*: procedureCode=**LocationReport**,
      – Area-ID: MCC=262, MNC=99, LAC=10001,SAC=01
      – Cause = **resource-optimization-relocation**

Figure 4.35   Inter-3G_MSC Handover/Relocation call flow 3/6.

this message in a MAP Send End Signal. Now the MAP entity of the first MSC is also, informed about successful handover procedure (Figure 4.36).

By sending MAP End Signal the handover is successfully completed and the UE is now served in the new cell. Former target RNC is now SRNC. However, the first MSC is still what is called the anchor MSC of a mobile call. There is a traffic channel active between first and second MSC, but the call is still controlled by anchor MSC (first MSC) while second MSC only lends its transmission resources, which are necessary to reach the UE. In other words, the Layer 3 signaling between UE and MSC using DMTAP protocol is running between UE and anchor MSC; second MSC forwards DMTAP messages transparently in uplink and downlink directions. To allow this DMTAP message forwarding, the TCAP dialog is not finished after successful handover. It is continued when the call is released or another handover (called a subsequent handover) needs to be performed.

In the example given in Figure 4.37, the call is released after approximately 12 s with a normal call clearing procedure.

A TCAP Continue message with MAP Process Access signaling operation is sent from second MSC to anchor MSC when call release starts. It includes a RANAP Direct Transfer carrying DMTAP Disconnect message. The Transaction ID (TIO) links all DMTAP messages of this call (as discussed in Section 3.3.2); a send sequence number is used to ensure flow control of the call. The release cause indicates normal call clearing.

Simultaneously, with this DMTAP Disconnect a BICC/ISUP Release message is sent on the same E interface to release the traffic channel on E interface as well.

The answer of the anchor MSC DMTAP entity is also embedded in TCAP/MAP/RANAP. UE confirms reception of DMTAP REL with DMTAP Release Complete. On BICC/ISUP level another RLC message is sent to finish the call procedure identified by CIC = 0080000 (hex).

TCAP CON ⇨ **MAP Process Access Signaling:** Invoke-ID=129,
an-APDU:
▸ AccessNetworkProtocolID= ts3G-25413
▸ **RANAP** *InitiatingMessage*: procedureCode=**RelocationDetect**

BICC ANM

TCAP CON ⇨ **MAP Send End Signal:** Invoke-ID=130,
an-APDU:
▸ AccessNetworkProtocolID= ts3G-25413
▸ **RANAP** *InitiatingMessage*: procedureCode=**RelocationComplete**

Figure 4.36 Inter-3G_MSC Handover/Relocation call flow 4/6.

Normal Call Release after 12 seconds:

TCAP CON ⇨ **MAP Process Access Signaling:** Invoke-ID=131,
an-APDU:
▸ AccessNetworkProtocolID= ts3G-25413
▸ **RANAP** *InitiatingMessage*: procedureCode=**DirectTransfer**
– **DMTAP DISCONNECT**: TIO=0, SendSequenceNumber=1,
Cause=normalcallclearing

TCAP CON ⇨ **MAP Forward Access Signaling:** Invoke-ID=3,
an-APDU:
▸ AccessNetworkProtocolID= ts3G-25413
▸ **RANAP** *InitiatingMessage*: procedureCode=**DirectTransfer**
– **DMTAP RELEASE**: TIO=0,
SendSequenceNumber=Message sent from the network

Figure 4.37 Inter-3G_MSC Handover/Relocation call flow 5/6.

TCAP CON ⇨ **MAP Process Access Signaling:** Invoke-ID=132,
an-APDU:
   ‣ AccessNetworkProtocolID= ts3G-25413
   ‣ **RANAP InitiatingMessage**: procedureCode=DirectTransfer
         – **DMTAP RELEASE COMPLETE**: TIO=0, SendSequenceNumber=2

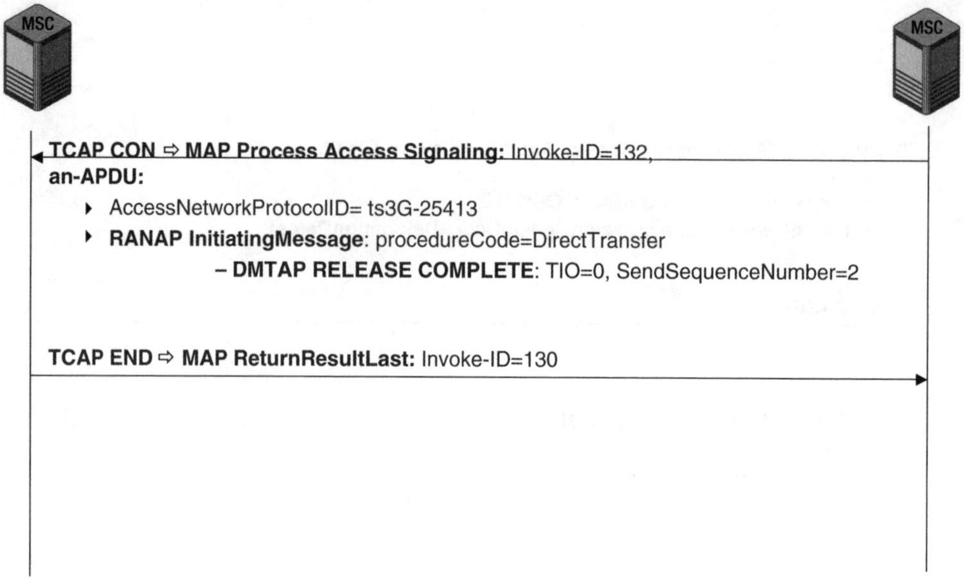

TCAP END ⇨ **MAP ReturnResultLast:** Invoke-ID=130

Figure 4.38   Inter-3G_MSC Handover/Relocation call flow 6/6.

Then the call is cleared from point of view of UE and anchor MSC. The last step is now to release the TCAP and MAP transport resources between the MSCs. This is done by sending a final MAP Return Result Last for all invokes directly followed to the one with Invoke-ID=130 (the successful handover). The ReturnResultLast is embedded in an End message that finishes the structured TCAP dialog (Figure 4.38).

4.6 Inter-3G-2G-3G_MSC Handover Procedure

Because UMTS cells will cover only urban areas, but here again the coverage cannot be guaranteed, it will be often necessary to hand over especially voice calls to cells with Radio Access Technology (RAT) different from UMTS. The examples in this book will always refer to GSM as alternative RAT, but other technologies are possible as well, for instance CDMA 2000. These procedures are also often called intersystem handovers.

The main difference to the Inter-3G_MSC Handover/Relocation procedure is that there is no target RNC for the handover, but a target BSC. And a BSC does not "speak" RANAP, it "understands" only BSSAP. This leads to the changes in the interface overview as shown in Figure 4.39.

However, as it will emerge in the call flow example the BSC also needs to deal with some RRC parameters. So it is not the BSC we used to know from plain GSM networks.

Because of the special needs of intersystem handovers the MAP transport capabilities have been adapted, and in the first step the protocol stack for full decoding of a handover procedure is introduced as shown in Figure 4.40. Alternatively to ISUP of course BICC can also be used.

Another important point to understand the intersystem handover procedure completely is to know that the BSC is not able to perform handover by its own. All handover procedures in

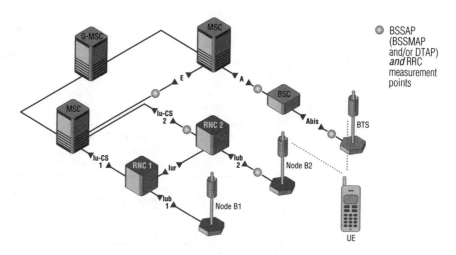

Figure 4.39 Interfaces involved in intersystem handovers between UMTS and GSM.

GSM are controlled by an MSC. This is also true for Intra-MSC Handovers – when the new cell is controlled by a new BSC that is connected to the same MSC as the first one.

As shown in Figure 4.41, the GSM Inter-MSC handover is triggered by measurement reports frequently (i.e. periodically) sent by the UE to the old BSC. On behalf of these measurement reports the BSC decides to perform a handover procedure and sends (in case of Inter-Cell Inter-BSC Handover) a BSSMAP Handover Required message to the serving MSC. On behalf of existing SCCP Class 2 connection (established during call setup, unique identifiers SLR/DLR) it can be recognized by the MSC to which user this Handover Required belongs. The Handover Required message contains the identifier of the new desired cell (new CID) and a cause value, e.g. "better cell."

MSC starts to prepare handover to the new BSC by sending Handover Request message including – besides other parameters related to radio resources – old and new Cell Identifiers and IMSI.

New BSC answers with Handover Request Acknowledge including a DTAP Handover Command with a Handover Reference Number (HO Ref). This DTAP Handover Command is sent from new BSC to UE, but since the only active radio link to UE is controlled by old BSC the Handover Command is sent to old BSC first. The message can be identified on the old A

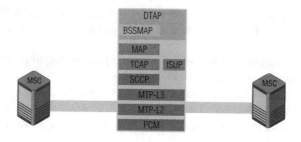

Figure 4.40 BSSAP-over-MAP protocol stack for 3G-2G Handover on E interface.

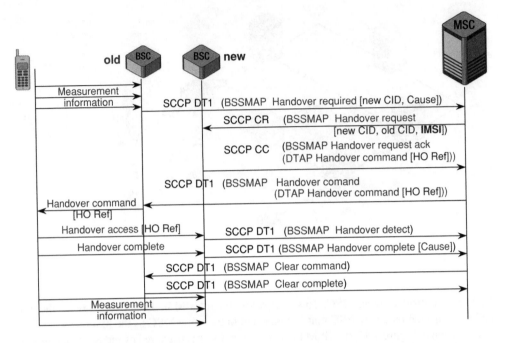

Figure 4.41   Intra-2G_MSC Handover procedure.

interface between MSC and old BSC, on Abis interface, and on radio interface on behalf of its HO Ref.

After switching to the new cell the new BSC sends Handover Detect and Handover Complete messages triggered by messages coming from the UE on radio interface and Abis. The Handover Ref is included in the handover access burst sent by MS via radio interface to the new BTS and reported to BSC with a Handover Detect message as response to successful channel activation procedure on Abis interface (the complete Abis call flow procedure is not shown in Figure 4.41, but in 3G-2G Handover call flow example).

After the successful handover procedure the frequently measurement reports are sent to the new BSC.

4.6.1 Inter-3G-2G_MSC Handover/Relocation Overview

Step 1: As in case of Inter-3G_MSC Handover the procedure is triggerd by RRC Measurement Report coming from UE. However, this time an event from event-ID group e3 (inter-RAT measurement) will be sent.

Step 2: RANAP Relocation Required message is sent from SRNC to its 3G MSC, including information about source and target of the handover.

Step 3: From the target information in the RANAP Relocation Required message the 3G MSC detects that the new desired cell is a GSM cell. Hence, it is necessary to send a BSSMAP Handover Request message to the target BSC. This BSSMAP Handover Request is part of a MAP Prepare Handover operation.

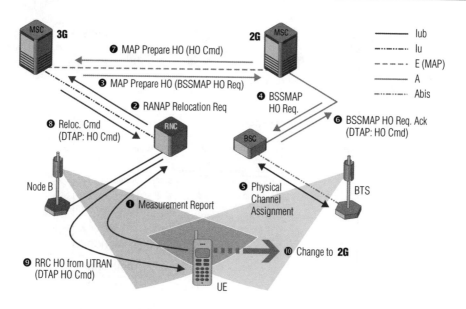

Figure 4.42   Inter-3G-2G_MSC Handover/Relocation overview.

Step 4: The BSSMAP Handover Request message is forwarded transparently by 2G MSC to target BSC.

Step 5: Target BSC allocates radio resources (especially a time slot for the connection) in target cell. The signaling of this procedure can be monitored on Abis interface.

Step 6: After successful resource allocation via Abis the BSC constructs a DTAP Handover Command that is sent to UE via E interface and later UTRAN. 2G MSC receives a BSSMAP Handover Request Acknowledge message that contains the DTAP Handover Command.

Step 7: MAP Prepare Handover Acknowledge message is used to transfer DTAP Handover Command via E interface.

Step 8: 3G MSC orders relocation of SRNC and forwards DTAP Handover Command via IuCS.

Step 9: RRC entity of SRNC sends RRC Handover Command including DTAP Handover Command to UE.

Step 10: Based on the information found in DTAP Handover Command the UE enters the GSM cell.

4.6.2 Inter-3G-2G_MSC Handover Call Flow

First a complete end-to-end call flow diagram of the procedure is introduced (Figure 4.43 and following) and then important messages and parameters as monitored on E interface will be presented as detailed as possible.

The trigger to start an Inter-RAT Handover procedure is defined when RRC measurement is set up as described in part 2 of this book. The appropriate trigger event is identified by ID "e3a" = "*The estimated quality of the currently used UTRAN frequency is below a certain threshold and the estimated quality of the other system is above a certain threshold.*" Including

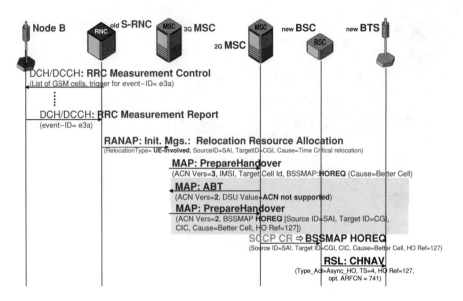

Figure 4.43   Inter-3G-2G_MSC Handover/Relocation call flow 1/4.

full parameterization (hysteresis, time-to-trigger value, etc.) the setup procedure of this trigger event is done by sending RRC Measurement Control message from SRNC to UE.

Anytime during an active call the UE moves into a position where the trigger conditions are fulfilled. This is when UE sends RRC Measurement Report with event-ID = "e3a."

Now the SRNC starts the handover procedure by sending RANAP Initiating Message with procedure code = "Relocation Resource Allocation" (in *3GPP 25.413* this combination is named *RANAP Relocation Preparation message*) including a source ID and a target ID to its serving MSC. The relocation type = "UE involved" indicates that UE will perform handover and relocation simultaneously. The cause value = "Time critical relocation" shall set a higher priority in comparision to Intra-3G Relocation/Handover procedures. The source ID is encoded using the UMTS Service Area Identifier (SAI). The target ID uses the Cell Group Identifier CGI to identify the target RNS. Since CGI is typical for Base Station Subsystems (BSSs), it already emerges at this point of the call that the handover target will be a GSM cell.

It is the task of the serving MSC to take the important parameters from the RANAP message and construct a BSSMAP Handover Request message. To highlight this point: It is the 3G MSC that constructs a BSSMAP message, which is actually used for communication between 2G MSC and BSC on GSM A interface. In TCAP transaction portion, the AC version 3 is used in the first step.

However, as it will happen often in early days of 3G networks and intersystem handovers, the target 2G MSC is not able to "understand" this version 3. Hence, it sends TCAP Abort message and requests a repetition of the dialog using AC version 2.

After successful reception of MAP Prepare Handover operation on 2G MSC side, an SCCP Class 2 connection is set up between 2G MSC and target BSC. It starts with an SCCP Connection Request (CR) message that carries the BSSMAP Handover Request as constructed by 3G MSC before.

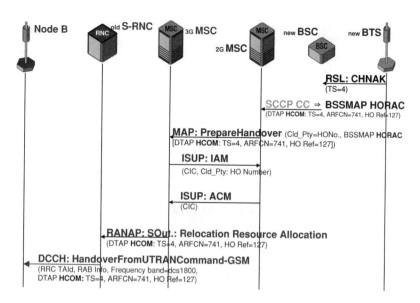

Figure 4.44   Inter-3G-2G_MSC Handover/Relocation call flow 2/4.

The BSC that receives the BSSMAP Handover Request proceeds as in case of normal Intra-2G Handover: it creates an RSL (Radio Signaling Link) Channel Activation message (CHNAV) that is sent to the GSM Base Transceiver Station (BTS). In the Channel Activation message the time slot number for the call after handover to GSM and the HO Ref value are found. It should be noticed that especially the channel activation messages of the RSL protocol often contain proprietary parameters and/or sequences, and so it mostly depends on the equipment manufacturer how these messages look in detail.

In the next step the BTS confirms the channel activation for time slot 4 with an RSL Channel Activation Acknowledge message (CHNAK) (Figure 4.44).

The setup of SCCP connection on A interface is confirmed with an SCCP Call Confirm (CC) sent by BSC. This CC message includes a BSSMAP Handover Request Accept as information to the MSC that requested the handover (the 3G MSC!) that radio resources have been allocated successfully. The BSSMAP message contains a DTAP Handover Command (HCOM), which is to order the UE to enter the GSM cell and use time slot 4 of this cell to continue the call. The Absolute Radio Frequency Code Number (ARFCN) is used to distinguish this cell from its neighbor GSM cells on radio interface. It indicates the unique frequency of this cell's broadcast channel, because in GSM each neighbor cell has its own unique BCH frequency. So the function of the ARFCN as identifier in GSM can be compared with the identification function of the primary scrambling code in UTRAN. The DTAP Handover Command also contains the HO Ref that will be used later by BSC to detect the successful handover.

With MAP Prepare Handover acknowledge the 3G MSC receives the handover number (HO No.) for routing of ISUP IAM related to this call. Now the traffic channel on E interface is set up using the well-known ISUP procedure while the embedded DTAP Handover Complete is forwarded on IuCS interface to SRNC using RANAP Successful Outcome message for Relocation Resource Allocation.

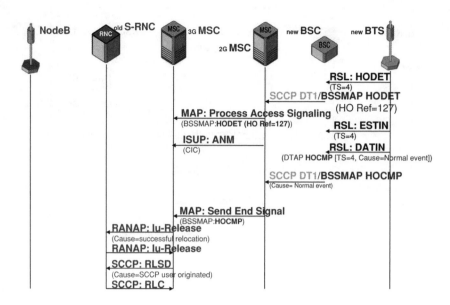

Figure 4.45   Inter-3G-2G_MSC Handover/Relocation call flow 3/4.

Now the SRNC sends an RRC Handover from UTRAN Command-GSM to the UE. This RRC message contains information about the new RAT (frequency band = dcs1800) as well as information of how the RAB shall be defined in the new cell. In addition, the message transports the DTAP Handover Command coming from target BSC to the UE.

After the UE switched into the new cell, the UE sends a handover access burst on radio interface containing the same HO Ref value that has been assigned with RSL Channel Activation procedure to the BTS. When the handover access burst is received and the HO Ref value is the same as expected (which is the normal case) an RSL Handover Detect (HODET) message is sent by BTS to BSC. The time slot number (TS=4) links this message to the channel activation procedure. So the BSC is enabled to send a BSSMAP Handover Detect (HODET) message to the 3G MSC using the active SCCP Class 2 connection. This BSSMAP HODET contains the HO Ref assigned to the procedure before (Figure 4.45).

To forward the BSSMAP HODET to the 3G MSC, 2G MSC needs to send this message enclosed in a MAP Process Access Signaling operation.

Simultaneously, 2G MSC activates the forwarded voice traffic channel on E interface by sending ISUP ANM for the defined CIC.

Meanwhile, radio link establishment on Abis interface is complete by RSL Establish Indication (ESTIN), and a Data Indication (DATIN) is used to transport the first Layer 3 DTAP message from UE to the BSC using GSM radio channels. This message is DTAP Handover Complete (HCOMP) with cause value = "normal event." So it is a regular and successful handover.

BSSMAP sends Handover Complete to anchor MSC of the call (3G MSC) embedded in MAP Send End Signal, which means that from the point of view of MAP the handover procedure is finished.

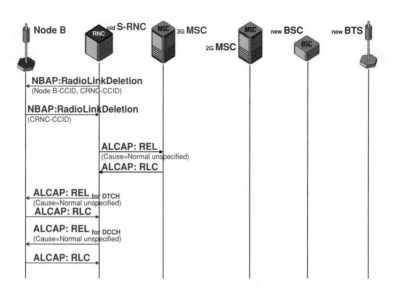

Figure 4.46   Inter-3G-2G_MSC Handover/Relocation call flow 4/4.

Reception of BSSMAP Handover Complete triggers release of the RANAP and SCCP connection on IuCS interface by 3G MSC. With this procedure also the RAB will be released as described in MOC/MTC scenarios in part 2 of this book.

The further messages in the UTRAN are used to release the assigned UMTS radio resources [RNTI(s), scrambling codes, spreading codes, etc.] with NBAP Radio Link Deletion (Figure 4.46). Also the AAL2 SVC established for transport of dedicated traffic channels and dedicated control channels are released on both Iub and IuCs interfaces.

4.6.3 Inter-3G-2G_MSC Handover Messages on E Interface

Now the focus is on the messages of the previously described handover procedure that can be monitored on E interface between 3G MSC and 2G MSC. Addressing aspects and parameter description shall be highlighted.

The end-to-end addressing between both MSCs is a task of SCCP. Since the application part is MAP, SCCP UDT messages will be exchanged between MSCs and each UDT contains a called party address as well as a calling party address, which are true E.164 addresses of the switches. There is no Global Title Translation in this case of MAP signaling (Figure 4.47).

TCAP OTID will change with the direction of TCAP message. Having a look at OTID together with SCCP address will make clear which MSC is the sender or which one is the receiver of a message.

The MAP Prepare Handover message already contains the target ID in CGI format as it is used for GSM cells only. In addition, an an-APDU is enclosed that contains a message as described in *3GPP 48.006*. Indeed, this is not completely true, because *3GPP 48.006* describes only the basic procedures on GERAN A interface while the encoding of messages and parameters used in these procedures is described in *3GPP 48.008*.

 SCCP Addr. (E.164):   SCCP Addr. (E.164):
                    4910040001   4910040002
          TCAP OTID: 134218239   **TCAP** OTID: 2768240847

TCAP BEG ⇨ **MAP Prepare Handover:** Invoke-ID=127, Target Cell ID: MCC=262, MNC=99,
LAC=120, Cell Identity=133
an-APDU:
> ▶ AccessNetworkProtocolID= ts3G-48006, IMSI=262996189990424
> ▶ **BSSMAP HOREQ:** ChannelType = Full Rate TCH/Speech
>> – EncryptionInfo: Encrypt. Algorithm, Encrypt. Key, Classmark Info 2
>> – Source ID (SAI): PLMN ID = 262 99, LAC = 301, RNC ID = 30541
>> – Target ID (CGI): MCC=262, MNC=99, LAC=120, Cell Identity=133
>> – Cause: Better Cell, Classmark Info 3
>> – **Old BSS to New BSS Information**:
>>> ▶ **RRC Inter RATHO Info:**
>>> UE Security Info (start CS)
>>> UE Radio Access Capability Info

Figure 4.47   Inter-3G-2G_MSC Handover/Relocation on E interface call flow 1/3.

```
|Old BSS to New BSS Information                                              |
|00111010 |IE Name                 |Old BSS to New BSS Information           |
|00011111 |IE Length               |31                                       |
|**B31*** |Information elemets |07 1d 40 00 0a cc 46 83 44 12 13... |
|Old BSS to New BSS Parameters (3GPP TS 08.08 V 8.12.0) (OLD2NEW_BSS)
BSS-PRM (= Old BSS to New BSS Parameters)  |
|Old BSS to New BSS Parameters                                               |
|Inter RAT Handover Info                                                     |
|00000111 |IE Name                 |Inter RAT Handover Info                  |
|00011101 |IE Length               |29                                       |
|**B29*** |Inter RAT Handover Info |40 00 0a cc 46 83 44 12 13 19
a1...|
|INTER RAT HANDOVER INFO from 3GPP TS 25.331 V3.12.0 (2002-09)
(IRHI) interRATHandoverInfo (= interRATHandoverInfo)  |
|interRATHandoverInfo-Message                                                |
|1 interRATHandoverInfo                                                      |
|1.1 predefinedConfigStatusList                                              |
|         |1.1.1 absent            |0                                        |
|1.2 uE-SecurityInformation                                                  |
|1.2.1 present                                                               |
|**b20*** |1.2.1.1 start-CS    |'00000000000000000010'B                    |
|1.3 ue-CapabilityContainer                                                  |
|**b200** |1.3.1 present           |88 d0 68 82 42 63 34 23 16 49 81... |
|1.4 v390NonCriticalExtensions                                               |
|         |1.4.1 absent            |0                                        |
```

```
|UE-RadioAccessCapabilityInfo from 3GPP TS 25.331 V3.12.0 (2002-09)
(UE-RACI)  ue-RadioAccessCapabilityInfo (= ue-RadioAccess
CapabilityInfo)  |
|ue-RadioAccessCapabilityInfo-Message                              |
|ue-RadioAccessCapabilityInfo                                      |
|ue-RadioAccessCapability                                          |
|-------- |accessStratumReleaseIndicator      |r99               |

|rlc-Capability                                                    |
|-------- |totalRLC-AM-BufferSize             |kb50              |
|-------- |maximumRLC-WindowSize              |mws2047           |
|-------- |maximumAM-EntityNumber             |am6               |
|transportChannelCapability                                        |
|dl-TransChCapability                                              |
|-------- |maxNoBitsReceived                  |b5120             |
|-------- |maxConvCodeBitsReceived            |b1280             |

|turboDecodingSupport                                              |
|-------- |supported                          |b5120             |
|-------- |maxSimultaneousTransChs            |e8                |
|-------- |maxSimultaneousCCTrCH-Count        |1                 |
|-------- |maxReceivedTransportBlocks         |tb16              |
|-------- |maxNumberOfTFC                     |tfc96             |
|-------- |maxNumberOfTF                      |tf64              |
|ul-TransChCapability                                              |

|-------- |maxNoBitsTransmitted               |b3840             |
|-------- |maxConvCodeBitsTransmitted         |b1280             |
|turboEncodingSupport                                              |
|-------- |supported                          |b3840             |
|-------- |maxSimultaneousTransChs            |e8                |
|modeSpecificInfo                                                  |
|-------- |fdd                                |0                 |
|-------- |maxTransmittedBlocks               |tb8               |

|-------- |maxNumberOfTFC                     |tfc32             |
|-------- |maxNumberOfTF                      |tf32              |
|rf-Capability                                                     |
|fddRF-Capability                                                  |
|-------- |ue-PowerClass                      |4                 |
|-------- |txRxFrequencySeparation            |mhz190            |
|physicalChannelCapability                                         |
|fddPhysChCapability                                               |

|downlinkPhysChCapability                                          |
|-------- |maxNoDPCH-PDSCH-Codes              |3                 |
```

```
|-------- |maxNoPhysChBitsReceived              |b9600            |
|-------- |supportForSF-512                     |0               |
|-------- |supportOfPDSCH                       |0               |
|simultaneousSCCPCH-DPCH-Reception             |               |
|-------- |notSupported                         |0               |
|uplinkPhysChCapability                        |               |

|-------- |maxNoDPDCH-BitsTransmitted           |b2400           |
|-------- |supportOfPCPCH                       |0               |
|ue-MultiModeRAT-Capability                    |               |
|multiRAT-CapabilityList                       |               |
|-------- |supportOfGSM                         |1               |
|-------- |supportOfMulticarrier                |0               |
|-------- |multiModeCapability                  |fdd             |
|securityCapability                            |               |

|-------- |cipheringAlgorithmCap                |uea1, uea0      |

|ue-positioning-Capability                     |               |
|-------- |standaloneLocMethodsSupported        |1               |
|-------- |ue-BasedOTDOA-Supported              |1               |
|-------- |networkAssistedGPS-Supported  |noNetworkAssistedGPS |
|-------- |supportForUE-GPS-TimingOfCellFrames  |0               |
|-------- |supportForIPDL                       |0               |

|measurementCapability                         |               |
|downlinkCompressedMode                        |               |
|-------- |fdd-Measurements                     |1               |
|gsm-Measurements                              |               |
|-------- |gsm900                               |1               |
|-------- |dcs1800                              |1               |
|-------- |gsm1900                              |0               |
|-------- |multiCarrierMeasurements             |0               |

|uplinkCompressedMode                          |               |
|-------- |fdd-Measurements                     |1               |
-------------------------------------------------------------------
```

Message Example 4.1   Old BSS to New BSS Information.

The UE identifier that can be used as a search parameter to find a single handover procedure is the subscriber's IMSI.

The main part of the an-APDU is the BSSMAP Handover Request (HOREQ) that requests a full rate traffic channel on GSM for speech (GSM voice). In addition, information about ciphering MS Classmark 2 and 3 (UE software capabilities) is included. Source and Target ID identify source and target cells or areas of the handover. As a rule, the 3G source is identified by Service Area Identity, for GSM cell CGI is used again. The cause of the handover request is typically "better cell," and an Old BSS to New BSS Information container carries

SCCP Addr. (E.164):
4910040001
TCAP OTID: 134218239

SCCP Addr. (E.164):
4910040002
TCAP OTID: 2768240847

TCAP CON ⇨ **MAP Return Result Last:** Local Op.= Prepare Handover, Invoke-ID=127,
Handover Number (E.164) = 4910041386
an-APDU:

▸ AccessNetworkProtocolID= ts3G-48006

– **BSSMAP HORAC**: Choosen Encrypt. Algorithm, Speech Version (GSM speech
full rate), L3 Layer Message Contents: **DTAP-RR HCOM**:

▸ Cell Description: BCC=7, NCC=3, BCCH ARFCN (high) = 0, BCCH
ARFCN (low) = 106, BCCH ARFCN = 211190367

▸ Channel Description 2: ChannelType=TCH/F+ACCHs+SACCH/F,
Timeslot number = 2

▸ HO Reference Value = 96

▸ PowerCommand and AccessType, Synchronization Indication,
Frequency List, CipherModeSetting=ciphering

Figure 4.48   Inter-3G-2G_MSC Handover/Relocation on E interface call flow 2/3.

Inter-RAT Handover Info: the start value for ciphering after changing the cell and UE radio
access capabilities (Message Example 4.1).

The answer to the handover request is sent from 2G MSC with a MAP Return Result Last
for Prepare Handover Local operation. It contains the Handover Number that will be used as
called party address by ISUP/BICC IAM (Figure 4.48).

The an-APDU contains BSSMAP Handover Request Acknowledge message with information about the chosen ciphering algorithm, chosen codec for speech information on traffic
channel (here: GSM codec), and a Layer 3 DTAP Handover Command that shall be forwarded
to the UE.

The DTAP Handover Command (HCOM) contains a detailed description of how to find
the target cell on radio interface. Base Station Color Code (BCC) and Network Color Code
(NCC) ensure that the right BTS of the right network is found. Absolute Radio Frequency
Code Number will ensure that the correct target cell out of all other cells of the same BTS and
neighbor BTSs is found.

Also a detailed description of the traffic channel and time slot number is given together with
the HO Ref and some further information.

The two following messages are BSSMAP Handover Detect and Handover Complete without any detailed information regarding the UE or the target cell. If Handover Complete is sent,
it is clear that the 3G-2G Handover was successful (Figure 4.49).

However, the TCAP dialog is not finished yet, because it seems that the UE is moving pretty
fast and so in this call flow we will see as next step a subsequent handover from 2G back to
3G.

4.6.4 Inter-2G-3G_MSC Handover/Relocation Overview

In case of the call flow example described in this section (Figure 4.50) the 3G MSC is still the
anchor MSC of the active voice call that was hand over to a GSM cell. Since the 3G MSC is

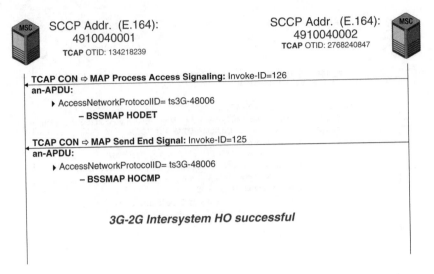

Figure 4.49 Inter-3G-2G_MSC Handover/Relocation on E interface call flow 3/3.

anchor MSC the following 2G-3G handover is a so-called subsequent handover. This is why MAP operations are different from the previous call flow.

Step 1: The handover procedure this time is triggered by frequently RSL measurement reports received by BSC via Abis interface.

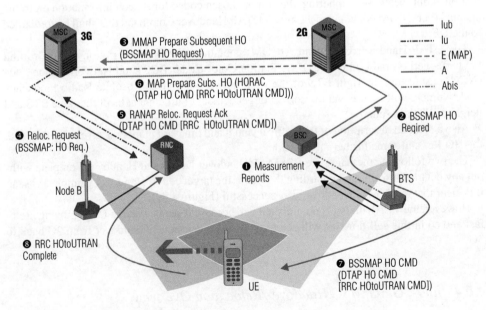

Figure 4.50 Inter-2G-3G_MSC Handover/Relocation overview.

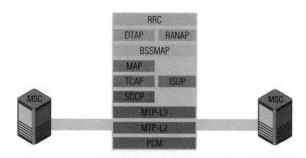

Figure 4.51   Intra-3G-2G_MSC Handover protocol stack on E interface.

Step 2: BSC decides that a handover is necessary and requires handover execution by 2G MSC.

Step 3+4: 2G MSC sends BSSMAP Handover Request to target RNC via E and IuCS interfaces.

Step 5+6+7: Target RNC assigns radio and transport resources in UTRAN and sends RANAP Relocation Request Acknowledge with embedded DTAP Handover Command that includes another message, the RRC Handover to UTRAN Command, which is forwarded together with DTAP message to UE via GSM and initiates the change back to UMTS.

Step 8: With an RRC Handover to UTRAN Complete message sent on UMTS radio and Iub interfaces, the UE confirms the successful handover to UMTS, which is the end of the procedure. It is expected that a mandatory Radio Bearer Reconfiguration procedure follows for reasons explained in message flow description.

From the protocol stack point of view, the call flow example will show that it must be expected to find DTAP or RANAP information on top of BSSMAP and embedded in DTAP or RANAP RRC Information can be transported (Figure 4.51).

4.6.5 Inter-2G-3G_MSC Subsequent Handover Messages on E Interface

The SCCP addresses and TCAP transaction IDs are still the same when 2G MSC sends its MAP Prepare Subsequent Handover request message to 3G MSC, including BSSMAP Handover Request (Figure 4.52).

The BSSMAP cause "traffic load" indicates that the GSM cell is probably not able to fulfill the QoS requirements for the call. The RANAP SourceRNC-to-TargetRNC-Transparent-Container is created by the source BSC. So this BSC needs to have a basic "knowledge" about RANAP and RRC parameters and functions. A completely new software is required to ensure this. Hence, a BSC in a GSM network interconnected with 3G RAT cannot be compared to a BSC used for plain GSM. With this impression it also becomes clear why 3GPP does not only care about UTRAN specification, but also about new GERAN standards.

The reception of SoureRNC-to-TargetRNC-Transparent-Container triggers NBAP Radio Link Setup on Iub interface. For this radio link setup procedure the same radio configuration identities and parameters will be used that can be found in RRC Handover to UTRAN Command message.

SCCP Addr. (E.164):   SCCP Addr. (E.164):
4910040001   4910040002
TCAP OTID: 134218239   **TCAP** OTID: 2768240847

TCAP CON ⇨ **MAP Prepare Subsequent HO:** Invoke-ID=124,
Target MSC Number = 4910040001,
Target RNC ID: MCC=262, MNC=99, LAC=301, RNC ID = 301
an-APDU:
- AccessNetworkProtocolID= ts3G-48006
- **BSSMAP HOREQ**: ChannelType= Full Rate TCH/Speech
  - EncryptionInfo: Encrypt. Algorithm, Encrypt. Key, Classmark Info 2
  - Cell ID (source): MCC=262, MNC=99, LAC=120, Cell Identity=133
  - Cell ID (target): PLMN ID = 262 99, LAC = 301, RNC ID = 301
  - Cause: Traffic Load, Classmark Info 3, Speech Version = GSM speech full rate
  - **RANAP SourceRNC-to-TargetRNC-TransparentContainer**:
    - **RRC Container** (MS Radio Capabilities, MS Classmark)
    - Number of Iu instances = 1, relocation type = UE-involved
    - Target Cell ID = 19756877

Figure 4.52   Inter-2G-3G_MSC Handover/Relocation on E interface call flow 1/2.

Once again the MAP Return Result Last for Prepare Subsequent Handover operation contains the DTAP Handover Command including RRC Handover to UTRAN Command with a new U-RNTI and some predefined radio configuration identities for physical and transport channels. Each RNC owns some general reserved resources (codes, identifiers) for Inter-RAT Handover procedures from different technologies like GSM. The two most outstanding of these reserved resources found in RRC Handover to UTRAN Command are:

- U-RNTI Short – This U-RNTI consists of SRNC-ID plus S-RNTI-2 that has a 10-bit length. After UE receives S-RNTI-2 it constructs a 20-bit standard S-RNTI by adding 10 "0"-bits in the most significant positions of the identifier.
- Long uplink scrambling code identified by a reduced scrambling code number – This reduced scrambling code number identifies as subset of uplink scrambling codes (value $= 0 \ldots 8191$) reserved for initial use upon handover to UTRAN.

After successful handover indicated by RRC Handover to UTRAN Complete message, SRNC triggers a NBAP Synchronized Radio Link Reconfiguration procedure and executes successive RRC Radio Bearer Reconfiguration procedure to assign a new uplink scrambling code with normal code number and – if necessary – to adapt the QoS parameters to current needs.

After successful subsequent handover procedure the TCAP dialog is finished by sending TCAP End message. The Return Result Last for Invoke-ID=125 confirms executed operation for all invokes sent after this ID (Figure 4.53).

SCCP Addr. (E.164):
4910040001
TCAP OTID: 134218239

SCCP Addr. (E.164):
4910040002
TCAP OTID: 2768240847

<u>TCAP CON ⇨ MAP Return Result Last: **Local Op. = Prepare Subsequent HO, InvokeID=124,** an-APDU:</u> ▶
▸ **AccessNetworkProtocolID= ts3G-48006**
▸ BSSMAP HORAC: **Layer 3 Info:**
    – DTAP HO Command: Cell-ID=PLMN-ID+LAC+RNC-ID, L3MessageContents:
      » RRC HandoverToUTRAN Command
      new u-RNTI: SRNC-ID + S-RNTI-2
      cipheringAlgorithm = uea0
      RAB-ID=1, UL_reducedScramblingCodeNumber=291
      DL_ChannelizationCode = 5
      uARFCN_UL, uARFCN_DL}

TCAP END ⇨ MAP Return Result Last: **InvokeID=125**

2G-3G Intersystem HO successful

Figure 4.53   Inter-2G-3G_MSC Handover/Relocation on E interface call flow 2/2.

4.7 Customized Application for Mobile Network Enhanced Logic (CAMEL)

It is foreseeable that in the future more and more services will be introduced in PLMNs that require a higher level of network intelligence. To make a telephone network intelligent basically means to install in the network databases that store and software applications that process customer-related service subscription data. Typical services are, for instance, intelligent call routing as in case of mobile number portability and flexible charging services for GPRS depending on duration of connection and/or volume of transmitted data.

In the early days of prepaid services in PLMN, network equipment manufacturers offered proprietary protocols derived from fixed network Intelligent Network Application Part (INAP) as defined in ITU-T standards. Meanwhile, the CAMEL Application Part (CAP) is running in most mobile networks, which is especially necessary to exchange intelligent services information between different network operators. Without CAMEL it would be quite impossible for a visited network to charge a roaming prepaid subscriber in real time while his or her prepaid account is administrated by the subscriber's home network only.

In addition, CAMEL is necessary to guarantee a Virtual Home Environment (VHE). If a German subscriber is roaming in a Chinese network and an announcement shall be played following the ideas of VHE, it must be ensured that the announcement is not only in German language, but also has the same information, the same typical voice, etc. The easiest and most secure way for the Chinese operator to play exactly the same announcement is to play exactly the same announcement: a voice channel is set up to the German Home PLMN of the subscriber and so the German announcement the customer hears while he or she is in

China is really the announcement from his or her home network. To ensure that this service is working, it is necessary to analyze information coming from the subscriber, e.g. IMSI that contains information about home country and home network of the subscriber, and to trigger a call setup to the announcement machine in the German network. So it is necessary to have software that excerpts service relevant data from the signaling messages exchanged between subscriber and network and a database that contains additional information like the address of the announcement machine to set up the voice channel.

Since it was difficult to adapt the previously installed proprietary INAP protocols to the new CAMEL standard, this adaptation process was divided into different steps. Each step is reflected by a CAMEL Phase. CAMEL Phase 1 included only seven messages for intelligent call routing. CAMEL Phase 2 introduced full control of all circuit-switched-related IN services and finally CAMEL Phase 3 (which is currently installed in 3G networks) includes all CS services specified for fixed networks in INAP Capability Set 1 (CS-1) plus full support of IN services for SMS and GPRS. A CAMEL Phase 4 is specified in 3GPP Release 5 standards and it will include many features of ITU-T INAP CS-2.

This section gives only a short overview about CAMEL concept and signaling. To discuss all aspects and details, it would be necessary to write another book.

4.7.1 IN/CAMEL Network Architecture

To talk about INs, it is necessary to have a different look at the network architecture. Indeed, the network elements that are known like MSC, SGSN, etc., are still there, but the IN architecture point of view will assign new names to these elements according to their IN-specific functions. The CAMEL architecture overview shows where software and stored data that are related to the intelligent services can be found (Figure 4.54).

The Service Switching Function (SSF) detects incoming IN calls and requests orders for call processing from the Service Control Function (SCF). The SSF is installed on a service switching point, which can be described in a popular way as an exchange with installed IN software. Mostly the SSP is the same physical equipment as the MSC plus installed additional SSF.

The SCF, which is another software, together with the Service Data Function (SDF), which is a database, forms the Service Control Point (SCP). In the CAMEL concept this SCP is also called CAMEL Service Entity (CSE).

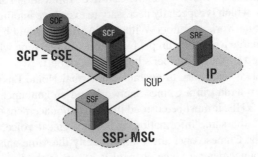

Figure 4.54 Elements of an intelligent CAMEL network.

In addition, especially for circuit-switched IN services there is a Specialized Resource Function (SRF) running on an equipment called Intelligent Peripheral (IP; this has nothing in common with the IP!). The SRF is used to detect DTMF dialing tones from voice circuits and transfer these tones into signaling (dialing) information and it is the above-mentioned "announcement machine." This is why – if this is necessary – ISUP signaling is exchanged between SSP and IP to set up and release voice channels.

4.7.2 CAMEL Basic Call State Model

To have full control of call handling it must be defined when it is time for the IN software to act. It must be also possible to follow up what happens to a call to make a clear decision whether, e.g., the customer can be immediately charged or not.

To analyze and control the call the Basic Call State Model (BCSM) was introduced. The earliest ideas about such a model have been developed by Bell Laboratories in the 1960s when they did research for AIN 0.1 standards. AIN is the North American counterpart of ITU-T INAP.

INAP specifications have adapted the AIN BCSM. 3GPP CAMEL standards are based on the INAP BCSM.

In all standards there is an originating BCSM that looks at the call from the point of view of the calling party and a terminating BCSM that represents the point of view of the called party.

In the BCSM there are Points in Call (PICs) and Detection Points (DPs). DPs have numbers while PICs have only names. DPs 1 and 8 are missed, because they have been specified for INAP CS-1, but not adapted for CAMEL Phase 3. By the way, the overall number of DPs is seen as a criterion of how sophisticated an IN is. Also in the fixed network not every IN software needs to implement all possible DPs. Many IN applications are tailored customer-specific solutions and include only a subset of INAP CS-1 DPs.

When the call enters a DP the call proceeding of the exchange (e.g. MSC) is interrupted and SSF may request SCF for orders on how to continue.

Best way to explain is an example: A mobile subscriber wants to make a call and dials a number. As long as there was no DMTAP SETUP message received by MSC the call is in state O_Null&Authorize_Origination Attempt_Collect_Info (Figure 4.55).

When SETUP is received, SSF detects the called party number. This is the entry event for DP 2: Collected_Information. Now the SSF checks if the called party number is valid. If not, the call is aborted by network, which is indicated by O_Exception PIC. If the number is valid, the call can be processed and the next DP (no. 3) Analyzed_Information is entered.

To reach DP 3 means to stop the call once again to get the routing information that can be provided by SCP.

Then the Routing&Alterting PIC is reached. The call can be answered (DP 7 is entered), misrouted (DP 4), called party is busy (DP 5), or called party does not answer (DP 6).

In addition, during both PICs, Analyze_Info and Routing&Altering, the calling party can decide to stop the call, for instance because it recognizes that the wrong number was dialed. This will then trigger the entry to DP 10 (O Abandon).

If the called party accepts the call, DP 7 is entered and following this the call becomes active (O_Active PIC).

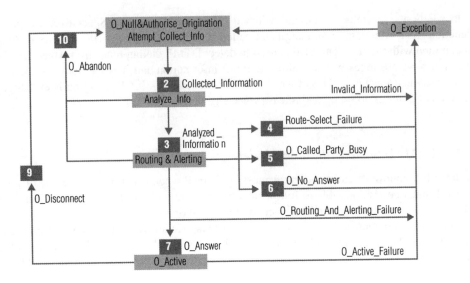

Figure 4.55   CAMEL Phase 3 originating BCSM.

4.7.3 Charging Operation Using CAMEL

As already mentioned, CAMEL is especially important for charging services. Like MAP, CAP is a user of SCCP and TCAP. Hence, all statements about SCCP addressing and TCAP signaling made in former chapters are valid for CAMEL as well. The whole communication between SSF and SCF is included in a structured TCAP dialog, but more often as in MAP, it is seen that several TCAP operations are transmitted with just one TCAP Begin or TCAP Continue message.

A typical example of CAMEL charging is running in six steps (Figure 4.56):

Step 1: A SETUP or IAM message is sent from the calling party to the SSF and triggers Initial Detection Point (DP 2).
Step 2: CAP Initial DP (IDP) operation is sent from SSF to SCF to request instructions on how to handle the call.

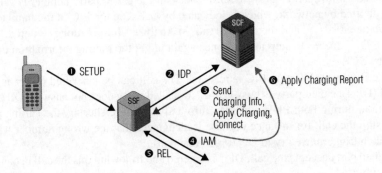

Figure 4.56   CAMEL charging operations overview.

Step 3: SCF sends Send Charging Information to indicate all parameters that will be relevant for charging as well as the advice of charging characteristic that describes which actions, e.g. tones, are sent to the charged subscriber to indicate that a specific charging event took place. An example of this is a tone that is sent if a prepaid account value reached or passed a minimum threshold. Mostly included in the same TCAP message the Apply Charging operation is sent to SSF to instruct the SSF to send charging reports to SCF if a charging event was triggered. The Apply Charging operation contains a list of all charging events, which are relevant for this call, e.g. timer values for time-dependent charging. A Connect operation is also sent by SCF to instruct the SSF how to complete the call.

Step 4: IAM or SETUP message is sent by SSF to the called party. The called party number was derived from Connect messages received from SCF.

Step 5: The call is released, which could be one of the events that trigger a charging report.

Step 6: Apply Charging Report is sent from SSF to SCF to indicate the occurrence of a charging event as defined by SCF and sent with Apply Charging message. Apply Charging Reports may also be sent during an established connection, e.g. if a charging timer in SSF expires. A typical example for a scenario like this is prepaid calling card charging, where the incoming charging reports lead to an immediate decrease of the subsriber's prepaid amount.

Note: In case that SSF should write a CDR for this call an additional Furnish Charging Information message would be sent from SCF to SFF together with Send Charging Information and Apply Charging messages.

4.7.4 CAMEL Signaling Example for GPRS Charging

There are different SSFs for GSM and GPRS defined in CAMEL standards. The difference is that gsmSSF is located in MSC while gprsSSF is located in SGSN. The gsmSCF is the same for both and installed in a central network location.

The scenario begins with an Initial DP GPRS operation embedded in a TCAP BEGIN message (Figure 4.57). The Initial DP (DP 2) is entered when SGSN receives an Activate

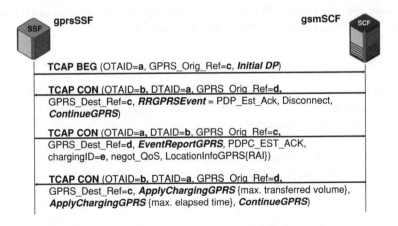

Figure 4.57 CAMEL PS call control and charging call flow 1/3.

PDP Context Request message via IuPS interface. By sending this Initial DP GPRS, the gprsSSF requests instructions from gsmSCF on how to handle the PDP Context requested by the subscriber. It should be noted that SCCP provides connectionless Class 1 transport service for TCAP dialogs. In SCCP header the two E.164 addresses of gprsSSF and gsmSCF can be found.

TCAP Origination Transaction ID and Destination Transaction ID link all TCAP messages of a single dialog together and can be used as filtering parameters in case of troubleshooting monitoring.

The Dialog Portion of this first TCAP Message contains Application Context information and a CAP GRPS Reference Number, which will link all CAP messages belonging to the requested PDP context. The CAP messages are as follows:

> TCAP **BEGIN** (Origination Transaction ID = **a**, Dialog Portion [Application Context = CAP-gprsSSF-gsmSCF, GPRS Reference Number Originating Reference = **c**], Component Portion [Invoke-ID = 1 -> Local Operation = *Initial DP GPRS*, GPRS Event Type = *PDP Context Establishment, MSISDN, IMSI*, Time_and_TimeZone, PDP Type Number = IPv4 Address, Access Point Name = *gprsservice.server.gprs*, *RoutingAreaIdentity*, *SGSN Number*, PDP Initiation Type = *MS initiated*])

The BEGIN message is answered with TCAP CONTINUE containing two more CAP operations. First, the gprsSSF is requested to inform gsmSCF if the PDP context is established and due to monitor mode settings this report will be sent as TDP-R that will trigger a new CAMEL procedure for charging on gsmSCF side. With ContinueGPRS, the gprsSSF is requested to proceed the PDP context establishment:

> TCAP **CONTINUE** (Origination Transaction ID = **b**, Destination Transaction ID = **a**, Dialog Portion [GPRS Reference Number Originating Reference = **d**, GPRS Reference Number Destination Reference = **c**], Component Portion [Invoke-ID = 35 -> Local Operation = Request Report GPRS Event -> PDP Context Establishment Acknowledge, Disconnect, Monitor mode = interrupted ("report event as TDP-R"), Invoke-ID = 36 -> Local Operation = *ContinueGPRS*])

Then gprsSSF sends an EventReportGPRS operation that indicates that PDP context was established. A ChargingID is provided to link this PDP context with a charging process on SCF side. In the Location Information GPRS IE, the Routing Area Identity (RAI) is included that consists of Mobile Country Code (MCC), Mobile Network Code (MNC), Location Area Code (LAC), and Routing Area Code (RAC).

> TCAP **CONTINUE** (Origination Transaction ID = **a**, Destination Transaction ID = **b**, Dialog Portion [GPRS Reference Number Originating Reference = **c**, GPRS Reference Number Destination Reference = **d**], Component Portion [Invoke-ID = 2 -> Local Operation = EventReportGPRS, GPRS Event Type = PDPContextEstablishmentAcknowledge, Message Type = Request, Access Point Name = *gprsservice.server.gprs*, Charging ID = **e**, Negotiated QoS {QoS Parameter},

gprsSSF   **gsmSCF**

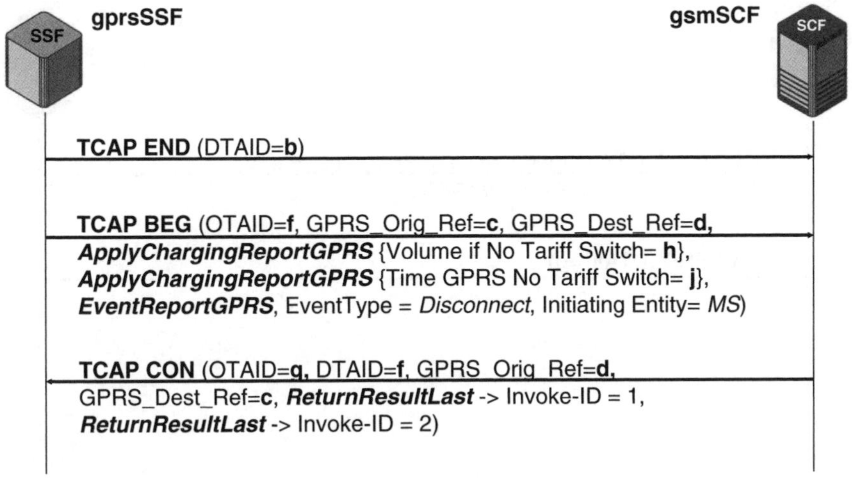

TCAP END (DTAID=**b**)

TCAP BEG (OTAID=f, GPRS_Orig_Ref=**c**, GPRS_Dest_Ref=**d**,
ApplyChargingReportGPRS {Volume if No Tariff Switch= **h**},
ApplyChargingReportGPRS {Time GPRS No Tariff Switch= **j**},
EventReportGPRS, EventType = *Disconnect*, Initiating Entity= *MS*)

TCAP CON (OTAID=**q**, DTAID=**f**, GPRS_Orig_Ref=**d**,
GPRS_Dest_Ref=**c**, *ReturnResultLast* -> Invoke-ID = 1,
ReturnResultLast -> Invoke-ID = 2)

Figure 4.58   CAMEL PS call control and charging call flow 2/3.

Location Information GPRS {RAI = **MCC, MNC, LAC, RAC**}, SGSN Number
{E.164}, Time_and_TimeZone)

With the next message, gprsSSF receives charging parameters and the order to proceed with
the PDP context processing.

TCAP **CONTINUE** (Origination Transaction ID = **b**, Destination Transaction
ID = **a**, Dialog Portion [GPRS Reference Number Originating Reference = **d**, GPRS
Reference Number Destination Reference = **c**], Component Portion [Invoke-ID =
39 -> Local Operation = ApplyChargingGPRS{max. transferred volume}, Invoke-ID
= 40 -> Local Operation = ApplyChargingGPRS{max. elapsed time}, Invoke-ID =
41 -> Local Operation = *ContinueGPRS*])

With TCAP END message the first TCAP dialog is finished (Figure 4.58).

TCAP **END** (Destination Transaction ID = **b**)

A new TCAP dialog is started with TCAP BEGIN message containing ApplyChargingRe-
ports and one EventReportGPRS that indicates that MS wants to deactivate the PDP context.

TCAP **BEGIN** (Origination Transaction ID = **f**, Dialog Portion [Application
Context = CAP-gprsSSF-gsmSCF, GPRS Reference Number Originating
Reference = **c**, GPRS Reference Number Destination Reference = **d**], Component
Portion [Invoke-ID=1 -> Local Operation = *ApplyChargingReportGPRS* {Volume
if No Tariff Switch = **h**), Invoke-ID=2 -> Local Operation =
ApplyChargingReportGPRS {Time GPRS No Tariff Switch = **j**), Invoke-ID=3 ->
Local Operation = *EventReportGPRS*, GPRS Event Type = *Disconnect*, Disconnect
Specific Information -> Initiating Entity = *MS*])

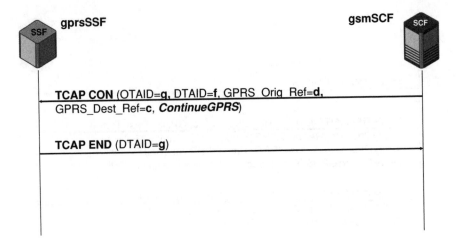

Figure 4.59 CAMEL PS call control and charging call flow 3/3.

The gsmSCF delivers two Return Result Last to gprsSSF to indicate that both the Apply Charging Report GPRS operations (compare Invoke-ID values with those in previous message!) have been received and the execution of a charging operation was successful.

TCAP **CONTINUE** (Origination Transaction ID = **g**, Destination Transaction ID = **f**, Dialog Portion [GPRS Reference Number Originating Reference = **d**, GPRS Reference Number Destination Reference = **c**], Component Portion [ReturnResultLast -> Invoke-ID = 1, ReturnResultLast -> Invoke-ID = 2])

Another TCAP CONTINUE message is sent from gsmSCF to gprsSSF. It contains a ContinueGPRS operation to order the gprsSSF to proceed with PDP context deactivation (Figure 4.59).

TCAP **CONTINUE** (Origination Transaction ID = **g**, Destination Transaction ID = **f**, Dialog Portion [GPRS Reference Number Originating Reference = **d**, GPRS Reference Number Destination Reference = **c**], Component Portion [*ContinueGPRS*])

TCAP END message finishes this second TCAP dialog.

TCAP END (Destination Transaction ID = **g**)

Glossary

1G	First Generation
2G	Second Generation
3G	Third Generation
3GPP	Third Generation Partnership Project
4G	Fourth Generation
8PSK	Eight Phase Shift Keying

A

AAA	Authorization, Authentication, and Accounting
AAL	ATM Adaptive Layer
AAL2	ATM Adaptation Layer Type 2
AAL5	ATM Adaptation Layer Type 5
ABR	Available Bit Rate Service
AC	Admission Control
ACK	Acknowledge
ACM	Address Complete Message (ISUP)
ACTS	Advanced Communications Technologies and Services
AESA	ATM End System Address
AICH	Acquisition Indicator Channel
AK	Anonymity Key
AKA	Authentication and Key Agreement
ALCAP	Access Link Control Application Part
AM	Acknowledged Mode in RLC
AMC	Adaptive Modulation and Coding
AMF	Authentication Management Field
AMPS	American Mobile Phone System
AMR	Adaptive Multi-Rate
ANM	Answer Message (SS#7)
ANSI	American National Standards Institute
AOA	Angle of Arrival Positioning
AP	Application Part

UMTS Signaling   Ralf Kreher and Torsten Rüdebusch
© 2005 Tektronix, Inc.   ISBN: 0-470-01351-6.

API	Application Programming Interface
APN	Access Point Name
AR	Authentication Reader
ARIB	Association of Radio Industries and Businesses
ASE	Application Service Element
ASN.1	Abstract Syntax Notation One
AT	Attention Command
ATD	Absolute Time Difference
ATM	Asynchronous Transfer Mode
AuC	Authentication Center
AUTN	Authentication Token
AUTS	Authentication Synchronization
AV	Authentification Vector

B

BAT	Bearer Association Transport
BCC	Base Station Color Code
BCCH	Broadcast Control Channel
BCFE	Broadcast Control Function Entity
BCH	Broadcast Channel
B-CSCF	Breakout-Call Session Control Function
BCSM	Basic Call State Model (INAP, CAMEL)
BEC	Backward Error Correction
BER	Basic Encoding Rules
	Bit Error Rate
BIB	Backward Indication Bits
BICC	Bearer Independent Call Control
B-ISDN	Broadband ISDN
BMC	Broadcast/Multicast Control
BRAN	Broadband Radio Access Network
BS	Base Station
BSC	Base Station Controller
BSIC	Base Station Indendity Code
BSN	Backward Sequence Number
BSS	Base Station Subsystem
BSSAP	Base Station Subsystem Application Part
BSSAP+	Base Station Subsystem Application Part PLUS (special set of BSSAP messages used on optional Gs interface)
BTS	Base Transceiver Station

C

CA	Certificate Authorities
CA-ICH	CPCH Channel Assignment Indicator Channel
CAMEL	Customized Application for Mobile Network Enhanced Logic

CAP	CAMEL Application Part
CATT	China Academy of Telecommunication Technology
CBR	Constant Bitrate
CBS	Cell Broadcast Service
CC	Call Control
	Country Code
CCCH	Common Control Channel
CCITT	Comité Consultatif International Téléphonique et Telecommunication
CCPCH	Common Control Physical Channel
CCS 7	Common Channel Signaling System 7
CCS#7	See CCS 7, SS#7, SS7
CCTrCH	Coded Composite Transport Channel
CD-ICH	CPCH Collision Detection Indicator Channel
CDMA	Code Division Multiple Access
CDMA2000	Third Generation Code Division Multiple Access
CDR	Call Detail Record
CF	Call Forwarding
CFN	Connection Frame Number
CG-GW	Charging Gateway
CGF	Charging Gateway Function
CGI	Cell Global Identity
CGN	Charging Gateway Node
CI	Cell Identity
CIC	Circuit Identification Code
C-ID	Cell Identifier, Charging ID (in GTP$'$)
c-ID	Cell Identifier in ASN.1 source codes
CID	Connection Identifier
CK	Cipher Key
CLP	Cell Loss Priority
CLPC	Closed Loop Power Control
CM	Communication Management
CMSREQ	Connection Management Service Request (a CS NAS message)
CN	Core Network
COMC	Communication Control
CP	Control Plane
CPC	Centralized Power Control
CPCH	Common Packet Channel
CPICH	Common Pilot Channel
CRC	Cyclic Redundancy Check
CRNC	Controlling RNC (Radio Network Controller)
C-RNTI	Cell Radio Network Temporary Identity
CS	Circuit Switched
	Convergence Sublayer (ATM)
CSCF	Call Session Control Function
	Call Server Control Function
	Call State Control Function

CS-CN	Circuit Switched Core Network
CSE	CAMEL Service Environment
CSICH	CPCH Status Indication Channel
CSPDN	Circuit Switched Public Data Network
CTCH	Common Traffic Channel
CTrCH	Common Transport Channel
Cu	Interface between TE and USIM
CW	Call Waiting
CWTS	China Wireless Telecommunication Standard Group

D

D-AMPS	Digital AMPS
DCCH	Dedicated Control Channel
DCFE	Dedicated Control Function Entity
DCH	Dedicated Channel
DECT	Digital Enhanced Cordless Telephone
DES	Data Encryption Standard
DGPS	Differential Global Positioning System
DL	Downlink
DLR	Destination Local Reference Number (SCCP)
DoS	Denial of Service
DP	Detection Point (in INAP/CAMEL BCSM)
DPC	Destination Point Code
DPCCH	Dedicated Physical Control Channel
DPCH	Dedicated Physical Channel
DPDCH	Dedicated Physical Data Channel
DRM	Digital Rights Management
DRNC	Drift Radio Network Controller
DRNS	Drift Radio Network Subsystem
D-RNTI	Drift RNC Radio Network Temporary Identifier
DRX	Discontinuous Reception
DS-CDMA	Direct Sequence CDMA
DSCH	Downlink Shared Channel
DTCH	Dedicated Traffic Channel
DTE	Data Terminal Equipment

E

E2E	End-to-End
EC	Echo Cancelling Equipment
ECF	Establish Confirm (ALCAP message)
ECT	Explicit Call Transfer
EDGE	Enhanced Data Rates for GSM Evolution
EFR	Enhanced Full Rate
E-GPRS	Enhanced GPRS

E-HSCSD	Enhanced HSCSD
EIR	Equipment Identity Register
EIRP	Equivalent Isotropic Radiated Power
E-RAN	EDGE Radio Access Network
ERQ	Establish Request (ALCAP message)
ESP	Encapsulation Security Payload
ETR	ETSI Technical Report
ETS	ETSI Telecommunication Standard
ETSI	European Telecommunication Standards Institute

F

FACH	Forward Access Channel
FBI	Feedback Information
FCCH	Frequency Correction Channel
FDD	Frequency Division Duplex
FDMA	Frequency Division Multiple Access
FEC	Forward Error Correction
FER	Frame Error Rate
FH-CDMA	Frequency Hopping CDMA
FIB	Forward Indication Bits
FP	Framing Protocol
FPLMTS	Future Public Land Mobile Telephony System
FRAMES	ACTS Future Radio Wideband Multiple Access System
FSN	Forward Sequence Number

G

Gb	GPRS Interface Between SGSN and GSM BSS
Gc	Interface Between GGSN and HLR/AuC
GERAN	GSM/EDGE Radio Access Network
Gf	Interface between SGSN and EIR
GGSN	Gateway GPRS Support Node
Gi	Interface between GGSN and External Network
GMLC	Gateway Mobile Location Center
GMM	GPRS Mobility Management
GMSC	Gateway MSC
GMSK	Gaussian Minimum Shift Keying
Gn	Interface Between Two GSNs
Gp	Interface Between Two GGSNs
GPRS	General Packet Radio Service
GPS	Global Positioning System
Gr	Interface Between SGSN and HLR/AuC
Gs	Interface Between SGSN and Serving MSC/VLR
GSM	Global System for Mobile Communication
GSM-R	GSM Railway

GSMS	GPRS SMS
gsmSCF	GSM Service Control Function
GSN	GPRS Support Node
GTD	Geometric Time Difference
GTP	GPRS Tunneling Protocol
GTP-C	GTP Control
GTP-U	GTP User

H

H.323	ITU-T protocol suite originally developed as standard for videotelephony via LAN, later enhanced as standard for Multimedia over IP/Voice over IP
HARQ	Hybrid Automatic Repeat Request
HCS	Hierarchical Cell Structure
HDR	Header
HE	Home Environment
HEC	Header Error Control
HFN	Hyper Frame Number
HLR	Home Location Register
HM-CDMA	Hybrid Modulation CDMA
HO	Handover
HON	Handover Number
HSCSD	High Speed Circuit Switched Data
HSDPA	High Speed Downlink Packet Access
HSPA	High Speed Packet Access
HSS	Home Subscriber Server
HTML	HyperText Markup Language
HTTP	HyperText Transfer Protocol
HW	Hardware
Hz	Hertz, cycles per second

I

IAM	Initial Address Message (ISUP)
ICC	Integrated Circuit Card
ICGW	Incoming Call Gateway
ICO	Intermediate Circular Orbits
I-CSCF	Interrogating-CSCF
ID	Identifier
IDEA	International Data Encryption Algorithm
IE	Information Element
IEC	International Electrotechnical Commission
IETF	Internet Engineering Task Force
IK	Integrity Key
IKE	Internet Key Exchange
IMEI	International Mobile Equipment Identity
IMS	IP Multimedia Subsystem

IMSI	International Mobile Subscriber Identity
IMT-2000	International Mobile Telecommunications at 2000 MHz
IMUN	International Mobile User Number
IN	Intelligent Network
INAP	Intelligent Network Application Part
IP	Internet Protocol
IPDL	Idle Period Downlink
IPSEC	IP Security Protocol
IPSP	IP Signaling Point
IPv4	Internet Protocol Version 4
IPv6	Internet Protocol Version 6
I/Q	In-phase/Quadrature
IS-95	Interim Standard '95, North American Version of the CDMA Standard
ISCP	Interference Signal Code Power
ISDN	Integrated Services Digital Network
ISO	International Organization for Standardization
ISP	Internet Service Provider
ISUP	ISDN User Part
ITU	International Telecommunication Union
ITUN	SS7 ISUP Tunneling
Iu	UMTS Interface Between 3G-MSC/SGSN and RNC
Iub	UMTS Interface Between RNC and Node B
Iu-CS	UTRAN Interface Between RNC and the Circuit Switched Domain of the CN
Iu-PS	UTRAN Interface Between RNC and the Packet Switched Domain of the CN
Iur	UMTS Interface Between RNCs
IWF	Interworking Function

K

KAC	Key Administration Center
kbps	Kilobits per second
kHz	Kilohertz
KPI	Key Performance Indicator
KQI	Key Quality Indicator

L

L1	Layer 1 – Radio Physical Layer
L2	Layer 2 – Radio Data Link Layer
L3	Layer 3 – Radio Network Layer
LA	Location Area
LAC	Location Area Code
LAI	Location Area Identity
LAN	Local Area Network
Lc	Interface Between GMLC and GSMSCF
LCS	Location Services

Lg	Interface Between GMLC and MSC/SGSNs
Lh	Interface Between GMLC and HSS
LI	Length Indicator
LLC	Logical Link Control
LMU	Position Measurement Unit
LOS	Line of Sight
LSSU	Link Status Signal Unit

M

M3UA	MTP Level 3 User Adaptation Layer
MAC	Medium Access Control
	Message Authentication Code
MAC-I	Message Authentication Code for Data Integrity
MAP	Mobile Application Part
MAPSEC	MAP Security Protocol
MBMS	Multimedia Broadcast and Multicast Service
Mbps	Megabits per second
MC	Multi-Carrier
MCC	Mobile Country Code
MC-CDMA	Multi-Carrier CDMA
MCE	Multiprotocol Encapsulation
Mcps	Megachips per second
MCU	Multipoint Control Unit
MD5	Message Digest #5
MDTP	Multinetwork Datagram Transmission Protocol
ME	Mobile Equipment
MEHO	Mobile Evaluated Handover
MExE	Mobile Execution Environment
MGCF	Media Gateway Control Function
MGW	Media Gateway
MHz	Megahertz
MIB	Master Information Block
MIMO	Multiple-Input Multiple-Output
MM	Mobility Management
MNC	Mobile Network Code
MNRG	MS not Reachable for GPRS
MO	Mobile Originated
MOBC	Mobility Control
MOC	Mobile Originated Call
MP3	MPEG 1 Audio Layer 3
MPEG	Moving Pictures Expert Group
MRC	Maximum Ratio Combining
MRF	Media Resource Function
MS	Mobile Station
MSC	Mobile Switching Center
MSE	MExE Service Environment

MSISDN	Mobile Subscriber ISDN Number
MSN	Mobile Subscriber Number
MSRN	Mobile Station Roaming Number
MSS	Mobile Satellite System
MSU	Message Signal Unit
MT	Mobile Termination
MTC	Mobile Terminated Call
MTP	Message Transfer Part
MTP3	Message Transfer Part Level 3
MTP3-b	Message Transfer Part Level 3 Broadband for Q.2140
MTU	Maximum Transmission Unit

N

NAS	Non-Access Stratum
NBAP	Node B Application Part
NCC	Network Color Code
NDC	National Destination Code
NE	Network Elements
NEHO	Network Evaluated Handover
NLOS	Non-Line Of Sight
NMS	Network Management Subsystem
NMT	Nordic Mobile Telephone
NNI	Network-to-Network Interface
Node B	UMTS Base Station
NRI	Network Resource Identification
NRT	Non-Real Time
NSAP	Network Service Access Point
NSS	Network Subsystem
NT	Network Termination
NW	Network

O

O&M	Operation and Maintenance
ODB	Operator Determined Barring
OHG	Operator Harmonization Group
OLPC	Open Loop Power Control
OMC	Operation & Maintenance Center
OPC	Originating Point Code
OQPSK	Offset Quadrature Phase Shift Keying
OSA	Open Service Architecture
OSI	Open System Interconnection
OSPC	Originating Signaling Point Code
OSS	Operation Subsystem
OTDOA	Observed TDOA
OVSF	Orthogonal Codes with Variable Spreading Factor

P

PC	Power Control
PCCH	Paging Control Channel
P-CCPCH	Primary Common Control Physical Channel
PCH	Paging Channel
PCM	Pulse Code Modulation
PCPCH	Physical Common Packet Channel
PCS	Personal Communication System
P-CSCF	Proxy-Call Session Control Function
PCU	Packet Control Unit
PD	Protocol Discriminator
PDC	Personal Digital Communication
PDCP	Packet Data Convergence Protocol
PDH	Plesiochronous Digital Hierarchy
PDN	Packet Data Network
PDP	Packet Data Protocol (e.g. PPP, IP, X.25)
PDR	Plesiochronous Digital Hierarchy
PDSCH	Physical Downlink Shared Channel
PDU	Packet Data Unit
PEM	Privacy Enhanced Mail
PER	Packed Encoding Rules
PHS	Personal Handyphone System
PIC	Point in call (of INAP/CAMEL BCSM)
PICH	Paging Indicator Channel
PIN	Personal Identification Number
PKI	Public Key Infrastructure
PLMN	Public Land Mobile Network
PMR	Private Mobile Radio
PN	Pseudo-Noise
PNFE	Paging and Notification Function Entity
POC	PSTN Originated Call
	Push-to-Talk Over Cellular
PPP	Point-to-Point Protocol
PRACH	Physical Random Access Channel
PRY	Physical Layer
PS	Packet Switched
PSC	Primary Synchronization Code
P-SCH	Physical Shared Channel
PS-CN	Public Switched Core Network
PSPDN	Packet Switched Public Data Network
PSS	Packet Switched Streaming Services
PSTN	Public Switched Telephone Network
PSVC	Permanent Switched Virtual Connection
PT	Payload Type
PTC	PSTN Terminated Call

P-TMSI	Packet TMSI
PUSCH	Physical Uplink Shared Channel
PVC	Permanent Virtual Connection

Q

| QoS | Quality of Service |
| QPSK | Quadrature Phase Shift Keying |

R

R	Interface Between TE and MT
R4	Release 4 of 3GPP UMTS Standard
R5	Release 5 of 3GPP UMTS Standard
R99	Release 1999 of 3GPP UMTS Standard
RA	Routing Area
RAB	Radio Access Bearer
RAC	Routing Area Code
RACE	Research in Advanced Communications in Europe
RACH	Random Access Channel
RADIUS	Remote Authentication Dial-In User Service
RAI	Routing Area Identity
RAN	Radio Access Network
RANAP	Radio Access Network Application Part
RAND	Random Number (Used for Authentication)
RAS	Registration, Admission, and Status
RAT	Radio Access Technology
RAU	Routing Area Update
RB	Radio Bearer
REL	Release Message (ISUP)
RES	Response in Authentication
RF	Radio Frequency
RFC	Request for Comments in IETF
RFE	Routing Functional Entity
RL	Radio Link
RLC	Radio Link Control
	Release Complete Message
RLCP	Radio Link Control Protocol
RLP	Radio Link Protocol
RNBP	Reference Node Based Positioning
RNC	Radio Network Controller
RNS	Radio Network Subsystem
RNSAP	Radio Network Subsystem Application Part
RNTI	Radio Network Temporary Identity
ROHC	Robust Header Compression
RR	Radio Resource

RRC	Radio Resource Control
RRM	Radio Resource Management
RSA	Public-key Security Algorithm by Rivest, Sharmir, and Adleman
RSL	Radio Signaling Link (protocol on GSM Abis interface)
RT	Radio Termination
	Real Time
RTCP	Real-Time Control Protocol
RTD	Relative Time Difference
RTP	Real-time Transport Protocol
RTT	Radio Transmission Technology
	Round Trip Time
RX	Receiver

S

SA	Security Association
	Service Area
SAAL	Signaling ATM Adaptation Layer
SAC	Service Area Code
SAI	Service Area Identifier
SAP	Service Access Point
SAR	Segmentation and Reassembly Sublayer
SB	Scheduling Block
SCCP	Signaling Connection Control Part
	SCCH Synchronization Control Channel
S-CCPCH	Secondary Common Control Physical Channel
SCE	Service Creation Environment
SCF	Service Control Function
SCH	Synchronization Channel
SCI	Subscriber Controlled Input
SCP	Service Control Point
S-CSCF	Serving-CSCF
SCTP	Stream Control Transmission Protocol (as defined in RFC 2960)
SDH	Synchronous Digital Hierarchy
SDO	Standard Developing Organization
SDU	Service Data Unit
SET	Secure Electronic Transactions
SF	Spreading Factor
SFN	System Frame Number
SGSN	Serving GPRS Support Node
SGW	Signaling Gateway
SHA-1	Secure Hash Algorithm #1
S-HTTP	Secure Hypertext Transfer Protocol
SIB	System Information Block
SIF	Service Information Field
SIM	Subscriber Identity Module

SIO	Service Information Octet
SIP	Session Initiation Protocol
SIR	Signal-to-Interference Ratio
SLIP	Serial Line Internet Protocol
SLR	Source Local Reference Number (SCCP)
SLS	Signaling Link Selection
SLTA	Signaling Link Test Acknowledge Message (MTP L3)
SLTM	Signaling Link Test Message (MTP L3)
SM	Session Management
SMCP	Short Message Control Protocol
SME	Short Message Entity
S-MIME	Secured Multipurpose Internet Mail Extension
SMLC	Serving Mobile Location Center LC
SMPP	Short Message Peer-to-Peer Protocol
SMpSDU	Support Mode for Predefined SDU sizes
SMS	Short Message Service
SMSC	Short Message Service Center
SMTP	Short Message Transport Protocol
SN	Sequence Number
	Serving Network
	Subscriber Number
SOCKS	Socket Security
SPC	Signaling Point Code
SQN	Sequence Number
SRB	Signaling Radio Bearer
SRNC	Serving RNC (Radio Network Controller)
SRNS	Serving Radio Network Subsystem
S-RNTI	SRNC Radio Network Temporary Identity
SS	Supplementary Service
SS7	CCS 7 (Common Channel Signaling System No. 7)
SS 7	See SS7, CCS 7
SSCF	Service Specific Coordination Function
SSCF-NNI	Service Specific Coordination Function – Network Node Interface
S-SCH	Secondary Synchronization Channel
SSCOP	Service Specific Connection Oriented Protocol
SSCS	Service Specific Convergence Sublayer
SSDT	Site Selection Diversity Transmission
SSF	Service Switching Function
SSL	Secure Socket Layer
SSN	Subsystem Number
SSSAR	Service Specific Segmentation and Reassembly Sublayer
STC	Signaling Transport Converter
STM	Synchronous Transfer Module
STM1	Synchronous Transport Module – Level 1
STP	Signaling Transfer Point
SUGR	Served User Generated Reference

SUT	System Under Test
SVC	Switched Virtual Connection
SW	Software
SYSINFO	System Information

T

TA	Terminal Adaptation
TACS	Total Access Communication System
TAF	Terminal Adaptation Function
TAID	Transaction Identifier in NBAP, RNSAP, RRC messages
TBS	Transport Block Set
TC	Transcoder
TCAP	Transaction Capabilities Application Part
TCP	Transmission Control Protocol
TD-CDMA	Time Division – Code Division Multiple Access
TDD	Time Division Duplex
TDMA	Time Division Multiple Access
TDOA	Time Difference of Arrival positioning
TD-SCDMA	Time Division – Synchronized Code Division Multiple Access
TE	Terminal Equipment
TEID	Tunnel Endpoint Identifier
TETRA	Terrestrial Trunked Radio Access
TFC	Transport Format Combination
TFCI	Transport Format Combination Indicator
TFI	Transport Format Indicator
TFS	Transport Format Set
TH-CDMA	Time Hopping CDMA
TI	Transaction Identifier in PS NAS messages
TIA	Telecommunications Industry Association
TIO	Transaction Identifier in CS NAS messages
TLS	Transport Layer Security
TMSI	Temporary Mobile Subscriber Identity
TN-CP	Transport Network-Control Plane
ToA	Time-of-Arrival
TPC	Transmission Power Control
TR	Technical Report (3GPP, ETSI)
TRAU	Transcoder and Rate Adaptation Unit
TRX	Transceiver
TS	Technical Specification (3GPP, ETSI)
TTA	Telecommunications Technology Association
TTC	Telecommunications Technology Committee
TTP	Traffic Termination Point
Tu	Interface between NT and RT
TX	Transmitter

U

UBR	Unspecified Bitrate Service
UDP	User Datagram Protocol
UDT	Unitdata (an SCCP message)
UE	User Equipment
UICC	UMTS Integrated Circuit Card
UL	Uplink
UM	Unacknowledged Mode in RLC
UMSC	UMTS Mobile Switching Center (the integration of the MSC and the SGSN in one physical entity (UMTS+MSC=UMSC)
UMSC-CS	UMSC Circuit Switched
UMSC-PS	UMSC Packed Switched
UMTS	Universal Mobile Telecommunications System
UNI	User-Network Interface
UP	User Plane
URA	UTRAN Registration Area
URL	Uniform Resource Locator
U-RNTI	UTRAN Radio Network Temporary Identifier
USAT	UMTS SIM Application Toolkit
USIM	UMTS Subscriber Identity Module
USSD	Unstructured Supplementary Service Data
UTRA	UMTS Terrestrial Radio Access
UTRAN	UMTS Terrestrial Radio Access Network
Uu	UMTS Air interface

V

VAS	Value Added Service Platform
VBR	Variable Bitrate Service
VCC	Virtual Channel Connection
VCI	Virtual Channel Identifier
VHE	Virtual Home Environment
VLR	Visitor Location Register
VMS	Voice Mall System
VMSC	Visited MSC
VoIP	Voice Over IP
VPI	Virtual Path Identifier
VSC	Videotext Service Center

W

WAP	Wireless Application Protocol
WARC	World Administrative Radio Conference
WCDMA	Wideband Code Division Multiple Access
WLAN	Wireless Local Area Network

WLL	Wireless Local Loop
WML	Wireless Markup Language
WTLS	Wireless Transport Layer Security
WWW	World Wide Web

X

X.25	An ITU-T Protocol for Packet Switched Networks
X.509	Internet X.509 Public Key Infrastructure
XMAC	Expected Message Authentication Code
XRES	Expected user Response
XUDT	Extended Unitdata (an SCCP message)

Bibliography

Technical Specifications

3GPP Technical Specifications; http://www.3gpp.org
The 3GPP specifications referred to in this book are from the Release 99, Release 4, Release 5, and Release 6 set of
  specifications.
European Telecommunication Standards Institute; http://www.etsi.org
Internet Engineering Taskforce Specifications; http://www.ietf.org
International Telecommunication Union; http://www.itu.int

Extract of UMTS-Related Specifications

3GPP 21.133   Security Threats and Requirements
3GPP 23.110   UMTS Access Stratum Services and Functions
3GPP 25.301   Radio Interface Protocol Architecture
3GPP 25.321   Medium Access Control (MAC) Protocol Specification
3GPP 25.322   Radio Link Control (RLC) Protocol Specification
3GPP 25.323   Packet Data Convergence Protocol (PDCP) protocol
3GPP 25.324   Radio Interface for Broadcast/Multicast Services
3GPP 25.331   Radio Resource Control (RRC) Protocol Specification
3GPP 25.401   UTRAN Overall Description
3GPP 25.410   UTRAN Iu Interface: General Aspects and Principles
3GPP 25.411   UTRAN Iu Interface Layer 1
3GPP 25.413   UTRAN Iu Interface: RANAP Signaling
3GPP 25.420   UTRAN Iur Interface: General Aspects and Principles
3GPP 25.423   UTRAN Iur Interface RNSAP Signaling
3GPP 25.430   UTRAN Iub Interface: General Aspects and Principles
3GPP 25.433   UTRAN Iub Interface NBAP Signaling
3GPP 29.061   GPRS Tunneling Protocol (GPT) across the Gn and Gp interface CCITT Rec. E.880 Field data
                 collection and evaluation on the performance of equipment, network, and services
3GPP 33.102   3G Security Architecture
3GPP 33.105   Cryptographic Algorithm Requirements
3GPP 33.120   Security Principles and Objectives
3GPP 35.201   f8 and f9 Specifications
3GPP 35.202   KASUMI Algorithm
ETSI ETR 021   Advanced Testing Methods (ATM); Tutorial on protocol conformance testing (especially
                 OSI standards and profiles) (ETR/ATM-1002)
ETSI GSM 12.04   Digital cellular telecommunication system (Phase 2); Performance data measurements

UMTS Signaling   Ralf Kreher and Torsten Rüdebusch
© 2005 Tektronix, Inc.   ISBN: 0-470-01351-6.

GSM 04.05	Protocols Layer 2
IETF M3UA	G. Sidebottom *et al.*,[QA1] "SS7 MTP3-User Adaptation Layer" (M3UA draftietf-sigtran-m3ua-02.txt) (Work in Progress), IETF, March 10, 2000
IETF SCTP	R. Stewart *et al.*,[QA1] "Simple Control Transmission Protocol," draft-ieftsigtran-sctp-v0.txt (Work in Progress), IETF, September 1999
IETF RFC 791	Internet Protocol
IETF RFC 768	User Datagram Protocol
IETF RFC 1483	Multim Protocol Encapsulation over ATM Adaptation Layer 5
IETF RFC 2225	Classical IP and ARP over ATM
IETF RFC 2460	Internet Protocol, Version 6 (IPv6) Specification
ITU-T I.361	B-ISDN ATM layer specification
ITU-T I.363.2	B-ISDN ATM Adaptation Layer Type 2
ITU-T I.363.5	B-ISDN ATM Adaptation Layer Type 5
ITU-T Q.711	Functional description of the Signaling connection control part
ITU-T Q.712	Definition and function of Signaling connection control part messages
ITU-T Q.713	Signaling connection control part formats and codes
ITU-T Q.714	Signaling connection control part procedures
ITU-T Q.715	Signaling connection control part user guide
ITU-T Q.716	Signaling Connection Control Part (SCCP) performance
ITU-T Q.2100	B-ISDN Signaling ATM Adaptation Layer (SAAL) – overview description
ITU-T Q.2110	B-ISDN ATM Adaptation Layer – Service Specific Connection Oriented Protocol (SSCOP).
ITU-T Q.2130	B-ISDN ATM Adaptation Layer – Service Specific Coordination Function for Support of Signaling at the User Network Interface (SSCF at UNI)
ITU-T Q.2140	B-ISDN ATM Adaptation Layer – Service Specific Coordination Function for Signaling at the Network Node Interface (SSCF AT NNI).
ITU-T Q.2150.1	B-ISDN ATM Adaptation Layer-Signaling Transport Converter for the MTP3b
ITU-T Q.2150.2	AAL Type 2 Signaling Transport Converter on SSCOP (Draft)
ITU-T Q.2210	Message transfer part level 3 functions and messages using the services of ITU-T Recommendation Q.2140
ITU-T Q.2630.1	AAL Type 2 Signaling Protocol (Capability Set 1)
RFC 791	Internet Protocol Specification

ETSI SAGE Task Force for 3GPP Authentication Function Algorithms, General Report on the Design, Specification and Evaluation of the MILENAGE Algorithm Set: An Example Algorithm Set for the 3GPP Authentication and Key Generation Functions, 2000-11-22 VERSION 1.0

Literature

Bekkers, *Mobile Telecommunications Standards: UMTS, GSM, Tetra, & Ermes*, Artech House, United Kingdom, 2001.

Bostelmann, *UMTS Signaling and Protocol Analysis/UTRAN and User Equipment*, Artech House, United Kingdom, 2004.

Business Interactive, *UMTS Essentials*, Wiley, United Kingdom, 2003.

Castro, *The UMTS Network and Radio Access Technology*, Wiley, United Kingdom, 2001.

Dohmen and Olaussen, *UMTS Authentication and Key Agreement*, Agder University College, Norway, 2001.

Hillebrand, *GSM and UMTS*, Wiley, United Kingdom, 2002.

Holma and Toskala, *WCDMA for UMTS*, Wiley, United Kingdom, 2001.

Kaaranen, Naghian, Laitinen, Ahtiainen, and Niemi, *UMTS Networks*, Wiley, United Kingdom, 2001.

Kappler, *Course UMTS Networks*, XV UMTS Evolution, WS03/04, TKN TU, Berlin, 2003.

Laiho, Wacker, and Novosad, *Radio Network Planning and Optimisation for UMTS*, Wiley, United Kingdom, 2002.

Langnes, Aamodt, Friiso, Koien, and Eilertsen, *Security in UMTS – Integrity*, Telenor R&D, Norway, 2001.

Lempiaine and Manninen, *Radio Interface System Planning for GSM/GPRS/UMTS*, Kluwer Academic Publishers, Germany, 2001.

Lescuyer, *UMTS. Origins, Architecture and the Standard*, Springer Verlag, Germany, 2003.

Tsuji and Tokita, Proposal of MISTY1 as a Block Cipher of Cipher Suites in TLS, *48th IETF*, 2000.

Walke, Seidenberg, and Althoff, *UMTS – The Fundamentals*, Wiley, United Kingdom, 2003.

Walker, *On the Security of 3GPP Networks*, Eurocrypt, 2000.

Wisely, Eardley, and Burness, "*IP for 3G*", Wiley, United Kingdom, 2002

Other Web Sources

3G and 4G Wireless Resources; http://3g4g.co.uk

UMTS World; http://www.umtsworld.com

mpirical, Telecoms Training, Telecoms Courses UK; http://www.mpirical.com

SearchMobileComputing.com, Source for news and tips on wireless and mobile computing; http:// searchmobilecomputing.com

Whatis.com, Computer technical encyclopedia, dictionary and glossary; http://whatis.com/

Index

UMTS Signaling   Ralf Kreher and Torsten Rüdebusch
© 2005 Tektronix, Inc.   ISBN: 0-470-01351-6.